D0576898

The Wisdom of the Body

ALSO BY SHERWIN B. NULAND

The Origins of Anesthesia
Doctors: The Biography of Medicine
Medicine: The Art of Healing
The Face of Mercy
How We Die

THE WISDOM OF THE BODY

Sherwin B. Nuland

Chatto & Windus
London

First published 1997

1 3 5 7 9 10 8 6 4 2

Excerpted text on pages 339-345 and 300-325, respectively, of this work was originally published in *Discover* magazine as two articles, "The Sacred Disease" (February 1991) and "Beast in the Belly" (February 1995). Excerpted text on pages 221-248 of this work was originally published in *The New Yorker* as 'Transplanting a Heart" (February 19,1990). Grateful acknowledgment is made to Appleton & Lange for permission to reprint material by Sherwin Nuland from *Physical Diagnosis*, edited by Siegfried Kra (New York: The Medical Examination Publishing Co., a division of Elsevier Science Publications Co., 1987), copyright © 1987. Reprinted by permission of Appleton & Lange.

First published in the United Kingdom in 1997 by
Chatto & Windus Ltd,
Random House, 20 Vauxhall Bridge Road, London SW1V 2SA

Random House Australia (Pty) Limited
20 Alfred Street, Milsons Point, Sydney,
New South Wales 2061, Australia

Random House New Zealand Limited
18 Poland Road, Glenfield,
Auckland 10, New Zealand

Random House South Africa (Pty) Limited
Endulini, 5A Jubilee Road, Parktown 2193, South Africa

Random House UK Limited Reg. No. 954009

A CIP catalogue record for this book is available from the British Library

Papers used by Random House UK Limited are natural,
recyclable products made from wood grown in sustainable forests.
The manufacturing processes conform to the environmental
regulations of the country of origin.

ISBN 0-7011-6672-X

Printed and bound in Great Britain
by Mackays of Chatham PLC

TO YOU, MY CHILDREN—

TORIA

DREW

WILL

MOLLY

Whether one at a time or all together, you show me worlds I never knew

Men go forth to wonder at the heights of mountains, the huge waves of the sea, the broad flow of the rivers, the vast compass of the ocean, the courses of the stars; and they pass by themselves without wondering.

St. Augustine, *Confessions*,
Book X, chapter 8

CONTENTS

ACKNOWLEDGMENTS

No author writes alone. Whatever his degree of solitude, the silent influence of seemingly long-forgotten words is always at his elbow, and so are the men and women who spoke them; the reading and the talk of a lifetime converge on his pages. Filtered through the individuality of a writer's mind, the distant echoes of experience become ever more insistent until they make themselves known, and find form at his fingertips though they may never rise into full, unveiled consciousness.

And so it has been with the writing of this book. I have allowed the submerged to reveal itself and it has come forth, telling me what portion of it must be expressed—often without doing the telling in any overtly recognizable pattern. And this too is one of those miracles of the human spirit of which these chapters treat: that we somehow know and act with the shrouded knowledge of countless past moments thought and unthought, conscious and unconscious; that something within us comprehends the muffled messages brought upward out of that pool of hidden memory from which the present is formed, never having lost its power to *in*form as well. In such amazements is the brain of humankind distinct from that of any of our fellow animals. In such amazements is every human mind a singularity, distinct from any other that has ever been or will ever be. In such amazements are to be found the elements from which the writing of a book is shaped.

Still, though each of the long-unthought thoughts may become as perceptible as a clearly heard phrase, they are never in themselves sufficient to accomplish a writer's entire design. What is required in addition is yet another process—a different variety of retrieval—which must take place in the very forefront of consciousness. That process is the planned, purposeful search for overt information, and it calls for such elements as deliberate preparation, research, and review. There is much in this book that demanded study and analysis; that demanded, in fact, new learning on my part, or at least new understandings. And there is also much that demanded the conscious exchange of ideas with others, and the seeking of advice. Like all authors, I have sought out the best advisers I could find, and in this I have been advantaged by the freely given gifts of immensely talented colleagues. Great swaths of this book are the outcome of conversation, not infrequently over prolonged meals of a wide variety of digestibility. But what was always deliciously digestible was the nourishing content of these successive communions by which my understanding was gradually broadened, not only about the wholeness of the human body but as well of the tiniest constituents of which it is composed and the manner in which the entirety becomes the functioning fabric of the human organism.

Though my professional life has been spent in constant enlargement of my acquaintance with that organism, there is no replacement for the discerning eyes of others, especially those who have become expert on one or another specific aspect of it. I am honored to thank the friends at Yale and elsewhere who have so carefully reviewed the sections of this book dealing with areas of their expertise, and done their best to keep me from error. It is hardly necessary to add that if I am found anywhere to have strayed from the straight and narrow of factual accuracy, it is only because I have eluded their meticulous screening and wandered off on my own. The names of the following men and women will be far from unfamiliar to readers who work in their respective fields: Sidney Altman, Lawrence Cohen, Edmund Crelin, Alan DeCherney, Thomas Duffy, Rosemarie Fisher, Gilbert Glaser, Bernard Lytton, Margretta Seashore, and William Stewart. I cannot emphasize too strongly that my request of every expert was to verify the facts of biology and medicine insofar as there is general agreement about the current state of knowledge. Wherever interpretations appear, or speculation, or what some might even call day-dreaming—it is mine alone.

I have not hesitated to contact other knowledgeable colleagues when help of a highly specific nature was needed, as in verifying some obscure datum, tracking down a bibliographical reference, rendering an opinion when some piece of evidence was in dispute, or soliciting a fresh point of view. At such times of need, I have at least once and sometimes much more than once turned to Toby Appel, Sharon Baca, Saul Benison, Jerome Bylebyl, Joseph Fruton, Paul Fry, Rafaella Elaine Grimaldi, Gail Harris, Majlen Helenius, John Hollander, Elena Rose Kagan, Katherine Landau, Robert J. Levine, Regina Kenny Marone, Peter McPhedran, Emanuel Papper, James Ponet, Gordon Shepherd, Stephen Waxman, and my consistently dependable source of new insights, Ferenc Gyorgyey. Not to be forgotten is Wendolyn Hill, to whom I am indebted for focusing the full force of her medical artistic skill on my illustrations.

As the major themes of this book were taking shape, I twice presented them at seminars of my fellow Fellows of the Whitney Humanities Center at Yale. A more formal presentation occurred as the 1995 John P. McGovern Award Lecture of the American Osler Society, a group of academic physicians committed to medical education and the humanities, taking place at the society's annual meeting in Pittsburgh. I am grateful to the membership of both these groups for the lively discussions that ensued on each occasion, the echoes of which will be discernible to the participants in the text that follows.

Several of my friends and neighbors, none of them a doctor or scientist, have graciously read the entire manuscript from the viewpoint of the general reader. The comments of each of them have sharpened my thinking and helped me to clarify issues. They are Judith Cuthbertson, Alexander Sommers, and Sarah Tyler.

To reckon him who taught me this Art equally dear to me as my parents, to share my substance with him, and relieve his necessities if required. These are the first words of the Hippocratic Oath, following immediately after the invocation of the gods Apollo, Aesculapius, Hygeia, and Panacea. No matter the dedication and skill of professors and colleagues, every doctor knows that his best teachers are his patients. It is through them that he comes to appreciate the fullness of the body's splendor, and the dangers of its frailties; it is through them that so much of his understanding of the human spirit arises. Thousands of men and women have entrusted their health and often their lives to my hands. This now seems the proper place, and at last the proper time, to thank them for showing me the way toward my philosophy, and theirs.

And as always, there are those who have held my hand throughout, worried with me, and broadened my perspective of what a book like this could be, while also broadening my perspective of myself. Of three in particular, I will now sing.

I discovered the wisdom and warmth of Jay Katz less than half a dozen years ago, and his rigorous intellectual honesty as well. He is among those few in whose presence one feels diminished should so much as a single thought reach consciousness that lacks the purity of analytical focus characterizing his own relentless search for equality of justice for all people. From Jay I have learned that a sure path toward such justice is to be found in the uncompromising scrutiny of motivations, mine and those of others. It is a habit not easily acquired, nor is the degree of humility that must perforce accompany it. Most would agree that teaching humility to a surgeon is a job akin to cleaning the Augean stables: it requires diverting entire streams of self-regard. But I have tried to benefit from the example of Jay Katz. *His contributions to this book are immeasurable.*

It is only partially in jest that I refer to Robert Massey as my guru—in many ways, I want to be just like him when I grow up. He has read every word of the manuscript of this volume, as he did of *How We Die.* The margins of each are filled with notes and suggestions that arise from his distinctive combination of erudition with the empathetic perceptions of an experienced clinical physician and teacher. It doesn't hurt either, that he is a wonderful writer. Among the most rewarding aspects of the making of both books have been those many times when Bob Massey and I, as part of our long-established practice of coming together to puzzle over the world we are trying to fathom, incidentally exchange thoughts on what I have written. *His contributions to this book are immeasurable.*

Although we have spent many hours discussing this book as it evolved, Vittorio Ferrero, my brother in all but the biological sense, did not read any of it until publication, but he has been reading its author for about a quarter of a century. He is my guide to the story and humanity of us all. Together we have studied myth, memory, and the realities of our time. I have learned insight from Vittorio's insight, loyalty from his loyalty, and the truths to be found in the lessons of history, be it the history of classical literature and thought, or the history of me. *His contributions to this book are immeasurable.*

Whatever else a book may be, it is ultimately an offering to its potential readers, and it must be presented to them in a form that is accessible and attractive. A vast symbolic distance often separates the head of an author from the eyes of a reader, and those who seek to bridge that gulf or at least to narrow it bear a weighty responsibility. Theirs is an undertaking at once intellectual, artistic, and nurturing, while at the same time sensitive to the zephyrs and gales that blow so capriciously on the Rialto. It can only succeed in the hands of men and women who revere books and their making. Of those there are greater and lesser, which is why an author published under the imprimatur and imprint of Alfred A. Knopf must never forget his good fortune.

My experience at Knopf has been an endless source of pleasure to me. Sonny Mehta has intuited from the start just what it is that my writing strives to be. He has understood it, believed in it, guided it, and been its strongest advocate. It is not possible for me to write badly when Sonny is confident that I will write well.

The seventeenth-century poet George Herbert has written that "Reason lies between the spur and the bridle." True of life, Herbert's aphorism is equally true of writing. A good literary rider needs an effective spur on each boot, and my two are the sharpest, in every good sense of that word. Lynn Chu and Glen Hartley have twice propelled my efforts forth from the starting line at a velocity of enthusiasm that surprised even me. But I have needed a bridle too, and I am securely endowed with one that fits perfectly in the literary steed's mouth, ensuring not only comfort and direction but also the confidence that comes when both horse and horseman are provided with precisely the right degree of restraint—and that is my editor, Dan Frank.

As for reason . . . the best of sweet reason has been bestowed on me: my wife, Sarah Peterson. We have now been running this race of life side by side for many years, and during all that time no word of mine has ever remained on the page—and no deed of mine has ever been done—that has not been true to the standard she sets for us both. She brings reason to my existence. Reason indeed! That Peterson woman is my reason for being.

S.B.N.

INTRODUCTION

Centuries ago, when little was known of science, the mystery of the body's internal machinery enthralled ordinary people and tantalized the educated. It seemed a miracle, this bustling edifice of thought and action—beyond the capacity of mere mortals to comprehend, and yet providing here and there a hint that the inscrutable might somehow be understood if only properly directed efforts were made. In time, the right direction was indeed found and the efforts were rewarded, yet the tantalizing and the mystery not only did not lessen; they actually grew. The more became known, the more miraculous seemed the intricacies of the whole and the more urgent the drive to expand our knowledge of organs and tissues.

As knowledge grew of cells and the turbulence of chemistry within them, it became evident that the seeming chaos of our tissues has about it an overarching purpose—in this sense, not a theological or philosophic purpose, but one based on the simple biological principle of survival. If an organism is to survive, every activity within it must in some way be part of the effort. Moreover, it is imperative that there be total coordination if the outcome is to be the singular momentum that is ongoing life.

The integration of all parts of this effort has a seeming wisdom about it, by which the multiplicity of processes is somehow guided into a harmonious whole. The essence of success is the dynamism that allows each cell to respond instantaneously to even the most minor threat to its integrity and therefore to the integrity of the entire organism. There can be no chemical complacency. A high degree of radical readiness—to the point of instability, in fact—is required to allow the immediate change that corrects a tendency toward imbalance. Every disturbance sets off the calling forth of compensatory mechanisms that neutralize it. Our steadiness is a dynamic equilibrium.

To coordinate all of the instabilities in all of the cells requires that the far-flung parts of an organism be in constant communication with one another, over long distances as well as locally. In the scale of animals that includes humankind, this is accomplished by messages sent via nerves, in the form of electrical energy we call impulses; via the bloodstream, in the form of the chemicals we call hormones; and—to nearby groups of cells—via the specialized substances we call local signaling molecules. As each of these methods of communication was discovered, researchers became increasingly impressed with how well integrated is the entire array of apparently disparate activities. They came to recognize the inherent wisdom of the body. Not only did they recognize it; the term also began to appear in their writings.

The first scientist to use the metaphor of wisdom to characterize the body's seemingly intuitive integration of its diverse faculties was Ernest Starling, one of the two codiscoverers of hormones. Delivering the prestigious Harveian Oration of the Royal College of Physicians in 1923, he spoke of the regulation of bodily processes, their adaptability, and the contribution of hormones toward integrating them into a single unified system. For his epigraph, Starling chose a verse from the Book of Job (38:36): "Who hath put wisdom in the inward parts? or who hath given understanding to the heart?"

He entitled his oration "The Wisdom of the Body."

There is a certain irony in Professor Starling's epigraph. In the Hebrew original, the word translated as "heart" is *sechvi*, a term so distinctive that this is the only place in the entire Bible where it occurs; and the term's precise definition has been the focus of much learned discussion. According to some rabbinical authorities, it is equivalent to "mind," perhaps from the belief, shared with Aristotle, that the heart is the seat of the mind. The wonder of the human body is not only the wisdom of its physiology but also the breadth of its mind.

Nine years after Starling's oration, the great explicator of the body's automatic control over vegetative functions (such as digestion, circulation, and temperature control) Walter B. Cannon, of Harvard University, used the same title for a popular book he wrote. He chose to follow Starling's lead as a way of recognizing the contribution made by the English scientist to the understanding of the body's regulatory mechanisms. In the introduction to his book, Cannon quoted the French physiologist Charles Richet, who in 1900 had stated that "instability is the necessary condition for the stability of the organism."

A stable system is not a system that never changes. It is a system that constantly and instantly adjusts and readjusts in order to maintain such a state of being that all necessary functions are permitted to operate at maximal efficiency. Stability demands change to compensate for changing circumstances. Ultimately, then, stability depends on instability.

"The Wisdom of the Body" was to appear one more time, as the title of the fourth of a series of twelve Gifford Lectures presented by Sir Charles Sherrington at the University of Edinburgh in 1937–1938, several years after he won a Nobel Prize for his studies of the ways in which the nervous system coordinates bodily functions. He began the fourth lecture by suggesting to his audience that they see the human body as he did after a long career of studying it: "Wonder is the mood in which I would ask to approach it for the moment." The lectures have been collected into a single volume which encapsulates the quest on which Sherrington based his life's work; it is called *Man on His Nature*.

The book you are holding is written by yet another man awestruck with the amazement that is us. I offer it not in imitation of the great scientists who have preceded me, but in homage to their vision.

For thirty-five years, my hands have been deep within the body of humanity. In a sense, my head has been there, too—my *sechvi*, in fact. I expect never to recover from the thrill of discovery that accompanied my first look into the living body of an anesthetized fellow human being. *Revelation* is not too grand a word for what I experienced at that instant. The multicolored, multitextured fabric of tangible, pulsating reality that is our innermost sanctum represents, to me, nature's most exquisite artistry. To practice that other art, the art of maintaining and restoring the integrity of its diverse faculties, is the privilege that became the guiding force of my life.

Not alone the structure of us, but also the infinite variety of processes by which we maintain that singular constancy and unity of moment-to-moment life, has inspired me to write this book—I want everyone to know what I have come to know.

I write as a physician, albeit one who is the subject of an enchantment. Though this book treats of scientific specifics, I am not a scientist. I am a clinician, and my natural interest lies in people. I have written this book as I would embark on a journey seeking the basis upon which our species has developed the qualities that make us human beings.

Any clinical physician's personal observations will be uniquely his own. Clinicians have long realized that each of us will view just a little differently from any other of us those of our brothers and sisters who entrust themselves to our care. In science, emotional distance, objectivity, and reproducibility count for everything. In clinical medicine, the observer—the clinician—uses himself and his distinctive experience of life as the lens through which everything is to be seen, the mirror in which all is reflected. What would be a weakness in the laboratory is a strength at the bedside.

In certain respects, the clinical physician does have advantages over the basic scientist, and one of them is the perspective from which he evaluates new information. The handicap of less intimate knowledge of every one of a dense cluster of trees seems more than offset by the possession of an overview of the entire forest's relationship to the countryside of which it is a part. Another seeming handicap that I believe to be more apparent than real is the necessity that the clinician not delay before acting. He has not the basic scientist's luxurious leisure to restudy and retrench. Sometimes, in the face of the patient's need, he must act even when incompletely ready. At any given moment, he is required to have digested masses of often unrelated information and to have constructed a worldview through which to diagnose and treat those of his fellows who have come to him to be healed. Though his viewpoint on any issue of human biology may be as yet insufficiently formed and in the process of evolving, he must nevertheless function and be accountable on the basis of its status in his thinking at the instant of decision-making at the bedside. When the clinician is wrong, there is no remedy and no way to retrace his steps along the life of the patient he has harmed.

During a long career, I have become accustomed, as have all clinicians, to the necessity of choosing a course of action in spite of unavoidably incomplete information, an as-yet-insufficient state of the supporting science, my own acknowledged subjectivity, and a constantly shifting landscape of medical culture. The urgencies of the bedside demand it. It is of this that the art of medicine consists.

I have required of myself that such a course of action be consistent with the most current scientific thinking, the facts of the individual case as I understand them, findings verifiable by other observers, and the examined experience of a lifetime devoted to studying the human condition. Above all else, it must make sense.

My perception of the human body and its mind has been formed in precisely this manner. Accustomed to acting when certainty cannot be possible, the clinician cannot shrink from committing himself to a conception of our species's biology, though he may still await further substantiation. The book you are beginning is about my image of the workings of your body and mine.

And so this book was written with every experience of my life woven into it. As Michel de Montaigne said in the preface to his *Essays, "Je suis moi-même la matière de mon livre"*—"I am myself the substance of my book."

It is my thesis that we are greater than the sum of our biological parts. Not only that, but we have it in us to be still better than we are—and the choice lies to a significant degree in our own hands, as I have tried to demonstrate by telling the stories of people whom I have been privileged to know during my surgical career.

No matter how similar our parts to those of other animals, there are to be found within them some characteristics that make us uniquely human. Whether as overtly evident as the power of the brain that has been given us or as taken for granted as certain adaptations within our skeletal system, we are possessed of elements that make us different from any other living thing that has ever existed on this earth.

If this is true of our bodies, how much more must it be true of the qualities of emotion and thought that we alone of all the animal kingdom possess? Although focused primarily within the coordinating tissues of the brain, powers such as these originate in the contributions of every segment of us. Like everything else about human function, the myriad fragments that coalesce to form human feeling arise from the workings of the physical structure that has been given to us. The uniquely human mind is a property of the organic characteristics of the uniquely human body, just as all function is the accompaniment of the physical structure from which it arises. There is no duality of mind and body—all is one. Were it possible for René Descartes to reappear at this hour, today's neuroscience would quickly convince him of the unity. In that one, which is the summation of our humanness, I include the quality that I find most remarkable, the quality which above all else makes us distinctively what we are. By this, I mean a biologically inherent particularity that is both miracle and mystery, the particularity I choose to call the human spirit.

Notwithstanding the tragedies that humankind has visited on itself individually and collectively, and the havoc we have wreaked on our planet, we have become endowed nevertheless with a transcendent quality that expands generation upon generation, overcoming even our tendency toward self-destruction. That quality, which I call spirit, has permeated our civilization and created the moral and esthetic nutriment by which we are sustained. It is a nutriment, I believe, largely of our own making. Without it, we would wither and lose our way in the wilderness.

As I define it, the human spirit is a quality of human life, the result of living, nature-driven forces of discovery and creativeness; the human spirit is a quality that *Homo sapiens* by trial and error gradually found within itself over the course of millennia and bequeathed to each succeeding generation, fashioning it and refashioning it—strengthened ever anew—from the organic structure into which our species evolved so many thousands of years ago. It lives while we live; it dies when we die. Whatever else of a man may remain to join the consciousness of eternity, this magnificence I call the human spirit does not exist a moment beyond the moment of death. It is neither soul nor shade—it is the essence of human life.

Five years ago, I embarked on a study of the various ways by which we take leave of life. What is it, I set forth to ascertain and explain, that actually happens to our bodies when we are in the process of dying, and how do we respond to the waning of the vital force? I was interested not only in those who die but also in those who remain behind—their loved ones and their medical caregivers. I learned a great deal in the process of searching and writing, but I was later to learn even more from the thousands of people throughout the world who wrote and spoke to me during the years following the publication of *How We Die*. I was taught much more than I had ever anticipated about the human spirit. Now even more than before, I have become convinced that it often plays a far greater role than commonly imagined in those events of our lives and our deaths that at first glance may appear to involve only the strictly physiological, or even anatomic, aspects of our existence.

Some might question that a skeptical physician should have the temerity to undertake such a mission, and of those I would ask only that they entertain the possibility that a belief in the spirit of which I speak requires neither religious faith nor a reliance on the supernatural. In my judgment, the human spirit is the result of the adaptive biological mechanisms that protect our species, sustain us, and serve to perpetuate the existence of humanity. It is as inseparable from the body as the mind is inseparable from the brain—it is, quite simply, a quality that originates from our genetically determined structure and function. Some see God in it, and some see only biology. And there are others who will very clearly see both.

This is a very personal book. Two of the researchers of earlier centuries whom I have most admired, John Hunter and Claude Bernard, made a point in their investigations of never reading the publications of those of their contemporaries and recent predecessors whose field of study matched their own. Their aim was to remain unfettered by the influence of other thinkers. In writing that has a speculative aspect, I have always followed their course of action, reading the opinions of others only after I have formed my own. Like Hunter and Bernard, I prefer to examine the facts directly, and not after they have been filtered through the minds of others. Montaigne felt the same:

> *There is more ado to interpret interpretations than to interpret things, and more books upon books than upon all other subjects. We do nothing but comment upon one another.*

My philosophy is my own, and I well recognize that it may reveal more about its founder than its founder knows. Nietzsche reminds us that every philosophy is "the self-confession of its originator, a kind of unintentional, unconscious *memoires*."

A word about style: Throughout, I have stayed away from what I consider the obnoxious awkwardness of such trendy constructions as are represented by "he or she" or "he/she," and the variety of similar burdens that have of late been imposed on language in an ill-advised attempt at gender parity. Being male, I have chosen "he," and it must be understood by it that I include both sexes in my meaning. I do this not because of insensitivity, but simply in order to avoid the choppiness inherent in doing otherwise. As an unwavering believer in the equality of men and women, I would urge my female colleagues to write of "she." Such a sensible convention is long past due.

And finally: I have had immense enjoyment in the writing of this book, and have deepened my understanding of things previously less than clear to me. It is my hope—my fervent wish—that everyone who undertakes this journey with me will find the same joy, and the same inspiration, that I try to reflect in the way I have told the story that is us.

Sherwin B. Nuland
New Haven, January 1997

AUTHOR'S NOTE

The names of several physicians and patients have been altered to preserve confidentiality. They appear as Drs. Jorge Mendez, Kevin Foley, and Charlie Harris, and the Tailors, the Cretellas, and Mr. Trouble. Also somewhat altered are a few details in the story of Max Tailor, but only sufficiently to make the identities of family members less recognizable to readers who may at any time have known them.

THE WISDOM OF THE BODY

1 | THE WILL TO LIVE

"A *little lower than the angels, and crowned with glory and honor"*: In such sonorous cadences have orators and singers of psalms ever extolled the wonder that is a human being. Since we mortals began recording contemplations on the singularity of our own selves, we have crafted phrases of endless awe to laud the miracle that is us. "I am fearfully and wonderfully made," continued the Psalmist, never thinking to doubt the excellence of God's creative artistry, the ultimate perfection of which, most surely, is to be found in humankind. "Marvellous are thy works," he proclaims to his Creator. The most marvelous, of course, are ourselves.

What a paradox it is, then, that we so often take the marvels of our bodies for granted and express shock when an occasional imperfection makes itself known. For we are, of necessity, miracles with flaws. There is fascination in the flaws and in nature's manner of dealing with them, just as there is fascination in the truly marvelous aspects of how we are made. We do well to have knowledge of both the marvels and the mistakes—we profit by learning what we can of our kind, and of ourselves.

Here follows a clinical history. It tells of an event that would never have occurred but for one of the many flaws that sometimes lie in wait to upset the entire delicately balanced equilibrium that is harmonious health. It is a story about the inherent compensatory mechanisms that have been incorporated into our bodies by the very nature that has left us just a bit flawed—mechanisms that help us to counteract the flaws and even to overcome them; it is a story about some of the ways in which the art and science of modern medicine can intervene to deal with the results of nature's imperfections, and to make use of nature's compensatory powers; it is a story of an outcome that is nothing less than a triumph of the human body, and of the human spirit that is its issue.

Diseases have a wide variety of ways by which they first make their presence known. Sometimes they lurk unknowable within us for weeks or months—even years—daily extending their malign domain until some vague puff of minimal perception brings subtle warning, perhaps only an intimation, that something may not be quite right. The initial evidence of such a sickness may be so insubstantial that its victim is unaware that any change has taken place, requiring the eyes or touch of others in order to notice it. But other sicknesses are strangers to subtlety. They burst forth all at once, taking patient and family so completely by surprise that an overwhelming sensation of imminent catastrophe is commonly their accompaniment. All too often, that sensation conveys an accurate assessment of reality; not infrequently, patients so stricken die before proper measures can be taken.

It is with this latter, quite extreme, presentation of uncontrolled instant chaos that the story of Margaret Hansen's ordeal begins.

Mrs. Hansen was a cheerful optimist, a devout believer in God, and a contented woman. She had been blessed with five healthy children and a sturdy, handsome husband who adored her. Except for the near-universal condition of never having quite enough financial security, she reveled in the life that had been granted her. Although trained as a schoolteacher, Marge had stopped working fifteen years before the events of this story took place, when Mary, her first child, was born. At forty-two, Marge Hansen was proud to be thought of as a good old-fashioned American housewife and mom.

The Hansen clinical history started off early on the oppressively humid Sunday afternoon of a typical August dog day. Almost since dawn, the unrelenting fierceness of the hazy sun had been thickening the still, steamy air that saturates every stifling community along the Connecticut rim of the vast, tepid bathtub that is Long Island Sound in late summer. Simply put, it was a disgustingly hot and muggy day. Marge had played tennis that morning—unwisely but well—at a cabana club about ten miles from her home in New Haven. The game had left her as breathless and without energy as the day itself. It was only at the insistence of her friend Ann that she reluctantly forced her enervated body to start swimming laps in the club pool. Both women had agreed to swim regularly as part of their summer exercise program. Marge had been in the water only a short time when the first chapter of her illness suddenly began. This is how she described it to me:

> *I had done several laps. I was midway in the pool when I had the strangest sensation. I've never been shot, but that's what it felt like—like an explosion inside of me, like . . . like . . . [I watched her face intently as she hesitated, visibly searching for the precise way to make clear what had happened within her. She found it, then burst out with a single loud exclamation] BOOM! It was here [she was pointing to an area just below her left rib cage]. I tried to stand up in the water, but I couldn't. Each time I straightened up, I felt like I was about to faint. I felt suddenly very weak. After that first blast, I don't remember the pain as much as I do the feeling that I was going down.*
>
> *Ann helped me out of the water and onto a lounge chair. Once I got there, I felt better. I stayed in the chair for a bit and continued to improve. After a while, I got up and went to the picnic area. Once again, I had the sensation come over me that I was going to pass out. I hunched over and then sat myself under a tree. Everyone around me seemed alarmed at what I looked like, more so than I was. They were yelling things like "Call nine-one-one—do something!" Someone yelled, "She looks like she's having a heart*

attack." There was all this hubbub, and I just wanted to get out of there. So I asked Jack to take me home.

Jack Hansen has not forgotten "how lousy and pale" his wife looked when he got to her. Rather than risk the twenty-minute ride home, he took her to the nearby house of a friend. Once there, she began to feel somewhat better, but each time she got up, the faintness returned. It was several hours before she was well enough to return to New Haven.

But then, after we got home, the pain came back, and it was different. Every time I tried to move, I had a hit of pain. It shot up my back and into my left shoulder. I thought to myself, I am *having a heart attack, because I remembered hearing of people with pain in their shoulder and arm. Soon I was doubling over with it.*

Jack took me to the emergency room at the Hospital of St. Raphael, and I tried to explain to the doctors what had happened. But I couldn't lie down on the gurney. Every time I tried, the pain became agonizing. Standing up for the X rays made me faint, but lying down made the pain become awful. I didn't know what to do.

They couldn't find anything, and they decided my problem was muscle spasms in the back. They did give me a shot that was pretty powerful, and a prescription for Percocet.

Marge was discharged from the emergency room and went home to a good night's sleep, provided by the strong dose of Demerol she had been given. She felt somewhat better the next day, but still had enough pain that she tried the Percocet, which only made her nauseated. After two days of this, she decided that she had better see a doctor:

I hadn't seen a doctor in years, except my obstetrician, Dr. O'Connell. So I sat down at the kitchen table that morning with my telephone book in front of me and I called I don't know how many doctors to see if I could get an appointment—and I couldn't. No one would see me unless I was willing to have a thorough physical exam and new-patient workup beforehand, and there was a four-

> *to six-week waiting period. What do you do? I was very upset. I remember sitting and crying, right there at the kitchen table. I knew something was wrong, but there was nothing I could do about it except go back to the emergency room, where they had already told me my problem was only muscle spasm.*

The faintness had just about disappeared by that time, and the pain began gradually to lessen over the course of the next several days. Because the discomfort was apparent only when Marge stood up straight, she spent most of her time in a hunched-over position, and slept upright, with a thick pillow behind her back. When she finally became completely comfortable, she came to the conclusion that the emergency room doctors must have made the correct diagnosis after all. Nonetheless, she never lost the vague suspicion that something was not quite right inside of her. One day four weeks later, when Jack suggested a game of tennis, she got as far as lacing up her sneakers when a sense of foreboding made her think better of it. Somehow, her body was telling her that all was not well.

Later that same morning, October 1, 1980, Marge was pushing open a stubborn window when the sharp pain suddenly returned to her abdomen and left shoulder. For the rest of the day, she felt not quite herself, although the pain abated somewhat. So much was she not herself, in fact, that she uncharacteristically elected to absent herself from the first PTA meeting of the school year, scheduled for that evening.

Shortly after supper, Marge went upstairs to help her youngest child, five-year-old Tommy, with his bath. While washing him, she began to feel as though the energy was rapidly draining out of her. By the time the boy's bath was completed, a profound sense of weakness had so overcome her that she felt "like I was floating." Somehow, she managed to stand up and make her unsteady way toward her bedroom. Once there, she staggered to the edge of the bed and collapsed onto it in a sitting position, not letting herself lie down because she sensed that being supine would worsen the pain, which by then was increasing.

> *This time, I knew it wasn't a heart attack. My mother had had a stroke, and I thought this must be the same thing because my mind was fuzzy and there was this terrible feeling of weakness and of losing ground.*

Marge was able to summon just enough strength to call out to Jack, who rushed upstairs and found his wife pallid and so faint that she couldn't do more than respond feebly to his alarmed questions.

Because the Hansens live close to the center of New Haven, it was only a few minutes until the emergency crew responded to Jack's 911 call. When they arrived, they found a patient still responsive enough to know she was upset when they cut her favorite green jersey off of her, but nevertheless unable to answer their questions coherently. She refused to lie down, insisting on maintaining an upright, somewhat crouched position, because every time she attempted to flatten herself out, the pain became more severe. Finally, the paramedics gave in and simply picked her up and sat her in a chair. In this way, she was carried down the stairs and gingerly slid into the ambulance, chair and all. The paramedics were worried about the wisdom of letting her travel that way because, as Jack would later put it so succinctly, "When they took her blood pressure, there was hardly any."

On arrival at the emergency room of New Haven's Hospital of St. Raphael, the "hardly any" was found to be fifty, less than half what it should have been. Marge's skin had little more color than the sheet covering the gurney upon which she was quickly placed, and her pulse rate was 130, almost twice normal. She was quite obviously in shock. When it was observed that the lower part of her abdomen was noticeably bloated, the emergency room doctors were certain that she was bleeding internally at a rapid rate, and would have to be taken to the operating room as quickly as possible. Because she was by then too far gone to understand the wording of the surgical permission form, it was thrust in front of Jack and he scrawled his name on it as fast as he could. He had been repeating over and over again (because he was asked over and over again), "No, she's not pregnant. As far as I know, her periods have been normal." He could tell that the doctors did not trust the accuracy of his information, because they had already contacted Dr. Jorge Mendez, a staff obstetrician/gynecologist, as soon as they received the call from the ambulance that a pallid forty-two-year-old previously healthy woman was being sped to the hospital in a state of deep shock. No one seemed to question that Marge Hansen was bleeding from a ruptured tubal pregnancy.

To confirm that impression, Marge was hastily put up in stirrups and a needle was passed into her pelvis through the back wall of her vagina, a procedure called culdocentesis. When the attached syringe quickly filled with fresh blood from within the pelvic cavity, the diagnosis appeared beyond doubt. By this time, Mendez, who had been having dinner at his country club five miles away, had arrived at the hospital.

In the intensity and speed of the events progressing so relentlessly as their patient lapsed more deeply into shock and became by the minute less responsive, no one seemed to care about, or even to know or remember, the event of five weeks earlier. As for Marge, her flickering bit of consciousness was fixed on the immediate present, and the decreasing probability that there would be a future.

> *As I arrived in the emergency room I had a sense that I was in real trouble—that I was losing ground. I knew I had to pray very hard and very fast, and I did. I almost feel stupid mentioning that I saw the bright light that people talk about, because it may have only been the regular spotlight that hangs over every gurney. But there* was *a point when everything went very, very bright. I remember blinding bright light—I remember feeling like I was floating. But I knew it wasn't the right time for me. I felt that I might easily have floated away, and I didn't want to. I was praying to God to give me more time, telling Him that I loved my life and I felt I had a lot. I didn't want to leave.*

Those were Marge's last conscious thoughts. She was rushed upstairs to the operating room, where the scrubbed, gowned, and gloved Mendez awaited her, his surgical team in perfect readiness to cut open her abdomen within seconds of the completion of the induction of anesthesia and the laying on of sterile drapes. Because his patient's condition was so precarious, Mendez had taken the wise precaution of calling in the chairman of St. Raphael's gynecology department, Dr. Kevin Foley, who even then was speeding in from his home in Madison, some twenty miles up the Connecticut shoreline.

Events in the emergency room had unfolded at such a breakneck pace that less than half an hour elapsed between the ambulance's siren-shrieking arrival there and the dead-run sprint to the OR. There hadn't been sufficient time for blood to be properly cross-matched for transfusion. At first, large-bore intravenous lines had been placed in both

Marge's forearms, and salt solution poured into her as fast as her veins could carry it. But salt solution is only a substitute for the real thing, and when bleeding is very rapid it can be a poor one. Because Marge's blood pressure continued to drop, transfusions of O-negative type had been started on the way to the OR elevator by a nurse galumphing awkwardly alongside the rushing gurney while hanging the plastic bags onto the intravenous setup that swayed precariously from side to side with each revolution of the wheels. O-negative, the so-called universal donor blood, even when not specific for the recipient, carries a likelihood of seriously adverse reaction that is small enough to be acceptable, especially when the alternative to its use is almost-certain death, and soon. Emergency room doctors use this stratagem only rarely, and they knew they were taking a chance. But to have done anything less in Marge's case would have left it virtually certain that their patient would continue her swift descent into irreversible shock—*bleed out* is the term commonly used—before a perfect match could be made. While the universal-donor transfusions were pouring into Marge, blood bank technicians were scrambling in their basement lab, trying to complete accurate cross-matchings as quickly as possible.

By the time Marge's gurney was wheeled into the OR, the blood pressure obtained by the anesthesiologist was zero and the pulse was barely perceptible: "It's thready, damn it, and too weak to be sure of." Within minutes, he could no longer feel it. Mendez didn't wait for him to complete the very hurried induction of anesthesia. As soon as his by-then-moribund patient was lifted onto the table, he began to cleanse her increasingly distended belly with iodine and then draped it. Even before the breathing tube had been thrust forcefully down into Marge's inert windpipe, the scalpel was in the surgeon's hand—he couldn't hesitate a moment longer, and she seemed too far gone, in any event, to feel anything. Not even sure his patient was fully anesthetized, he pressed the scalpel firmly through the skin just below her navel and ran it in a straight line downward toward the pubis. She was in such profound shock that the small vessels of the underlying fatty layer barely oozed as the blade passed through them.

The muscle and fibrous layers covering both sides of the abdominal wall meet in the midline of the belly and fuse together to form a strong vertical band of tissue that stretches from the tip of the breastbone all the way down to the pubis. It was into this thick band (called the linea alba, the white line) that Mendez cut next, in order to open directly into the capacious organ-crammed container that is the abdominal cavity. Immediately on putting his scalpel through it, he reflexly jerked his head and shoulders back, recoiling from the surprise of being hit by a spurt of blood that shot out under pressure from within the blood-choked belly. Because of the distension, he had expected plenty of hemorrhage, but not this—there had been so much bleeding that the abdomen could no longer contain the tightly compressed volume of its contents.

Startled by his discovery, Mendez hastily completed his navel-to-pubis incision and then tried frantically to expose the pelvic organs by plunging the tubing of his suction apparatus down into the overflowing red torrent that now appeared before him. Even with the just-arrived Foley's help, he couldn't aspirate enough blood to enable him to see clearly. The operating table had been tilted head-down to help counteract the shock by taking advantage of gravity to keep blood from pooling in the legs. This also helped maintain blood flow to the brain, and now the surgeons wanted more steepness than they already had. The table's mechanism, however, only allowed for the feet to be elevated to an angle of thirty degrees higher than the head. It was barely enough, but the resultant contribution of gravity and the addition of a second suction apparatus enabled the team to ascertain that the blood was not coming up out of the pelvis, as had been thought. It was pouring down from some unknown and inaccessible site in the upper abdomen, thus ruling out any possibility of an ectopic pregnancy. With that sudden realization, Foley felt himself out of options and desperately needing whatever help he could get. Almost shouting, he instructed the senior nurse to order an emergency page for a general surgeon—any general surgeon who might answer—to get into the operating room as quickly as possible. It was just then that the anesthesiologist announced that the patient was bleeding faster than he and his assistant could pump transfusions into her. So deep was her shock, he reported, that she was requiring hardly any anesthetic. Resigned to the inevitable, he uttered the last words Foley heard before the page operator began blaring out her urgent summons: "We're losing her, Kevin."

While all this turmoil was hurtling toward its climax, I was arriving at the hospital to begin the unhurried visitations of my regular evening rounds. I pulled into a parking space, casually walked some fifty yards to the building, and strolled leisurely through the emergency room entrance, having spent a minute or two exchanging small talk with the security guard stationed there. And then, all at once—

In the forty years since I first began training as a surgeon, I have never heard a more startling page message (or a more startled page operator's voice) than the one that just then blared out from the overhead speakers, at the very instant I stepped into the hospital doorway: "Any general surgeon! Any general surgeon!" The page operator was more shouting than announcing it. Her frenzied, high-pitched tone gave the insistently repeated message the clarion sound of a firehouse bell: "Go immediately to the operating room—immediately—any general surgeon!"

As alarming and jarring as the page was, it was at the same time strangely exhilarating. At once supplicating and commanding, part outcry for help and part call to arms, the clamorous message spoke to me like some suddenly recalled ancestral imperative. Almost instanteously, I made the only decision I could. I paused only long enough for confirmation that I wasn't hallucinating the whole thing, or perhaps misinterpreting it. But as the call blared out yet another time, its reality was proven by the sight of visitors and staff personnel gaping at one another in what appeared to be disbelief or confusion. Feeling as though I had been set off by some self-starting high-speed internal dynamo over which I had no authority, I raced off toward the nearest staircase and bounded up the steps. I was forty-nine years old at the time and still capable of taking them three at once, but I'm sure I must have done even better than that, because I arrived at the second-floor OR suite in what seemed an instant or two.

A waiting nurse shouted me in the right direction and ran ahead of me up the short length of corridor toward, toward . . . I could only guess what. Obviously contaminating the sterile sanctity of the whole scrubbed scene with my germ-laden street clothes, I stepped across the threshold of Operating Room 6. The sight that greeted me in that hectic arena was even more startling than the page's message had foretold.

Kevin Foley was positioned at the right side of the operating table, his back to me and his feet planted firmly about ten inches apart, as though to stabilize his rangy, square-shouldered frame against the fearful challenge of what lay before him. The sheen of wet, sticky blood glistened on the sides of his white OR shoes, and even from behind, I could see that the arms of his surgical gown were soaked with it, as well. From the opposite side of the table, Foley was being assisted (if that is the right word) by two overwrought gynecology residents who prattled like a couple of frenzied adolescents, near hysteria and yet competing with each other to impress the chief with their sangfroid under fire. Behind them, Mendez had stepped away from the action, and was pacing rapidly back and forth, his hands clasped together under the sterile towel that covered them. In addition to the two scrub nurses passing instruments, several other young women were scurrying about, trying to keep the area near Foley free of the blood-soaked sponges he was pulling out of the depths of the abdomen and flinging one after another onto the floor. Near his feet, a harried aide knelt on her knees before a two-quart suction bottle, whose freshly hemorrhaged contents she was anxiously attempting to empty into a large bowl set down beside her, as the frothy liquid spilled onto her unsteady hands. A pair of anesthesiologists stood ramrod-straight at the head of the table, both of them furiously hand-pumping transfusions into their failing patient. Their worried eyes stared down over blue strips of gauze mesh that covered jaws clenched with the dreadful tension pervading the room. The pandemonium lacked only a latter-day Hogarth to record it for posterity.

In his lightly accented English, Mendez was in the midst of hallooing some suggestion to Foley when he noticed me in the doorway, and called out my name. As Foley turned to speak to me, I saw that the entire front of his gown was soaked in blood, and more was spilling over the sides of the widely spread abdominal incision, flowing onto the already-saturated surgical drapes.

Only a surgeon could be expected to understand, I think, why it was that in some paradoxical and perverse way I found this scene of near-catastrophe oddly reassuring, comforting, in fact. Until that moment, I had had no way of knowing what to expect—while bounding up the stairs, a dimly perceived but very real apprehension had entered my mind. I might encounter a situation that was beyond me, something that I might even make worse, something that would cause me to regret for the rest of my life that I had answered the page's insistent call instead of simply turning my back on its urgency and slinking off to my car before anyone noticed I was there. Whatever my ambivalent ruminations, though, there had never been the slightest possibility that I would do such a thing—

walking away from that kind of cry for help would have violated every precept taught me by my life and my training, and every bit of moral sense I had. Also, like so many surgeons, I bear an obsessive preoccupation with accepting responsibility, amounting really to a compulsively neurotic sense of duty. And yet, in some jarringly unforeseen and unwelcome way, the incongruous thought had become more overt in the few seconds it took me to get to the OR: Am I about to botch something up? Will I, in one quick stroke of ineptness and fate, bring my career crashing down around my feet, and with it my sense of what I am? Am I on my way to destroy an unknown patient and myself at the same time?

This was the shadowed setting of trepidation against which I surveyed that OR scene of chaos and found it reassuring. Abdominal bleeding is something I know about. There is not a major blood vessel in the belly I don't have an intimate acquaintance with—sometimes I feel as though the entire warm, moist vastness of the abdominal cavity belongs to me. It has always welcomed me as a helpful friend, and I have never responded to its wide-open embrace with anything less than a calling up of every quantum of cerebral and technical competence I possess, in order that I might justify its unconditional trust. We have had a rewarding relationship, the belly and I.

And so, seeing that it was my old familiar friend that was in trouble, my abrupt outburst of apprehension was stilled, and I was ready to go to work.

Foley quickly filled me in as the residents continued to suck away blood into a fresh collecting bottle. The usually confident surgeon was unsettled by his inability to find the source of bleeding. His sentences were coming in short bursts, as though from a man breathless from chasing something, or being chased. There is no worse moment for a surgeon than the very tiny span of time during which he begins to realize that his patient is bleeding to death and he can do nothing about it.

I said a few words meant to be reassuring, then sprinted to the locker room to change into a scrub suit. Trusting to the probity of three cleanup men who were arguing football as they idly mopped, I left my clothes in a heap on the moist floor in front of my locker. There seemed no time even to spin the three numbers of the combination. When I got back to Room 6, I bypassed the scrub room and simply stepped into the gown a nurse was already holding open for me. The first scrub nurse pulled a pair of gloves onto my unwashed hands and snapped them over my wrists as I took Foley's place at the table. Residents from the general surgical service had already been called to take over from the two beleaguered young men standing opposite me.

The incision was being held open by the curved blades of a large steel spreading instrument called a self-retaining retractor. Forty-five minutes into the surgery, eight units of transfusion and a great deal of fluid had already been pumped in, and the pulse and blood pressure were still unobtainable. The eyes of the senior anesthesiologist looked grim as he told me there were no numbers to report, and then he shook his head slowly from side to side in somber expression of his certainty of the futility of further efforts.

With nothing, really, to lose, I called for heavy scissors and cut away the blood-sodden drapes that hid the upper abdomen from view. In doing this, I ignored the fact that most of the skin had not been cleansed by germicidal prepping solution. What I did next was a gross violation of yet another principle of aseptic surgery, but I couldn't afford to lose so much as the few precious seconds it would have taken to slosh some iodine onto the field. The antibiotics begun in the emergency room would have to be relied upon to make up for my breaches of sterile orthodoxy.

The young first scrub nurse slapped a scalpel into my outstretched palm, and I swept it from the lowermost tip of the breastbone down to the upper edge of the gynecologists' incision, in one movement laying open the unprepped skin and its thin layer of underlying fat and exposing the clean white fibers of the midline, as Mendez had done when he had earlier cut into the lower part of the abdomen. Using firmer pressure, I repeated the vertical sweep through the linea alba, joining with Mendez's incision so that the abdomen was now open from top to bottom, from sternum to pubis. A second self-retaining retractor in the upper portion of the extended opening gave me plenty of room to look for the source of the hemorrhage, or at least I hoped so.

With the entire abdomen now lying wide open, and its cut edges spread far apart by the retractors, the blood could be more efficiently sucked away, although it was still not possible to see the organs well enough to ascertain the origin of the bleeding. In that instant, it crossed my mind that the liver might be the source; such a heavy flow of blood could be caused either by an unrecognized injury or by a specific type of liver tumor that is occasionally found in women of childbearing age. I groped wrist-deep into the opaqueness of the scarlet lake, then guided my submerged left hand blindly upward and to the right until I held between thumb and index finger the thick curtain of tissue called the porta hepatis (given that name by the ancients because it is the passageway through which the liver's main arteries and vein enter and the bile duct leaves). I squeezed off the entire major blood supply to the liver, but this so-called Pringle maneuver had no effect on the hemorrhage. Obviously, the liver was not the guilty party.

Something had to be done to slow the flow of blood at least enough to increase visibility—one of the first principles every beginning surgeon learns is that exposure is the key to safety and success. I had tried to provide it by doubling the size of the incision, but thus far to no effect. It was time to shut off the entire abdominal blood supply directly at its source. Not only was this necessary for visibility; the patient had by then reached the point where a few more minutes of uncontrolled blood loss would have resulted in cardiac arrest and death. The anesthesiologists had already decided, they would tell me later, that I had failed.

Every cavity of the body receives its blood supply directly or indirectly from the aorta, a huge conduit into which the most powerful chamber of the heart, the left ventricle, pumps its contents at an average rate of seventy-two times per minute. Each stroke delivers some 70 cubic centimeters, or 2.3 ounces, into the aorta. The vessel descends like a pulsing hose down the back of the chest, giving off branches to the upper body and the head, before passing through an opening in the back of the thick transverse sheet of muscle and fiber that is the diaphragm, in order to enter the abdomen. Through its entire length, the aorta lies directly on the rounded bodies of the spinal column's segments. I now reached up with my left hand to the point immediately below the diaphragm and pressed the rubbery hose shut by flattening it firmly against the thick bone behind. With that step, the hemorrhage stopped almost completely. Although there was still some moderately brisk ooze from a place a few inches below my obstructing hand, it now became possible to see that the residual bleeding was coming from a relatively localized area in the upper-left quadrant, near the entrance to the spleen, that purple-red fist-sized organ that lies far out laterally under the rib cage,

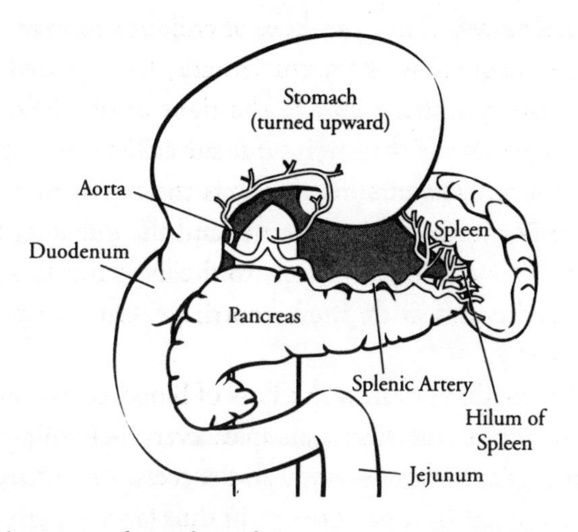

Anatomic relationships of structures in the upper abdomen. The stomach has been rotated upward, to expose the pancreas lying behind it.

tucked beneath the back of the diaphragm, just beyond the tail end of the pancreas.

The pancreas is about six inches long, and its inverted bowl-shaped head and elongated body make it look like a salmon-colored snake lying transversely across the back of the upper abdomen, its tail pointing to the left, toward the center, or hilum, of the concave inner surface of the spleen. The spleen's main blood supply, the splenic artery, leaves its source in the aorta at a right angle and travels leftward with its accompanying vein along the upper surface of the pancreas until it reaches the tail. The distance from the tail of the pancreas to the hilum is approximately two inches. To complete this remaining part of the journey to their destination, the two vessels enter a fold of tissue called the splenic pedicle, which carries them from the pancreatic tail directly into the hilum. Except for this final short distance, it is common for the lower third of the circumference of both artery and vein to be imbedded in the upper edge of the pancreas. In essence, then, the tail of the pancreas and the hilum of the spleen are connected to one another by the splenic pedicle containing the vessels.

By compressing the aorta, I had deprived the entire abdomen and lower body of their blood supply, except for what relatively small flow still came in via some collateral vessels from the chest, bypassing the main channel and entering from the sides. With such a marked decrease in inflow, it was possible to see that the residual ooze was coming from somewhere near the pancreatic tail and hilum of the spleen. So, although the massive hemorrhage had been stopped, the pool of blood brought in by the collateral vessels still kept welling up so obstinately that it was not yet possible to pinpoint the exact site of trouble. This localized bleeding could be stopped by packing the area with large sponges, and this I proceeded to do.

When blood is accumulating quickly from some area in the vicinity of the spleen, it will be found in the vast majority of cases to originate in a tear of that fragile organ. This now became the presumed source of the problem.

Perched though my patient was on the very brink of death, it was nevertheless clear that maintaining pressure on the aorta would buy time for the anesthesiologists to catch up with most of her blood loss. I kept my fingers pressed tightly against the great vessel while four more pints of transfusion were pumped in, raising the blood pressure gradually to 110 and stabilizing the situation, at least for the moment. The delay bought me about ten minutes to think at relative leisure, and to prepare the rest of the team for what I intended to do next. Until that point, all conversation had consisted of short comments by various members of the group and the several commands I had issued to the general surgery residents who had come in to assist.

I asked about the patient's output of urine. A catheter had been put in Marge's bladder in the emergency room, but I was informed by one of the nurses that its attached drainage bag was empty. The absence of urine meant that there was so little circulating blood left in her body that not enough was reaching the kidneys to allow them to excrete waste products, an ominous sign if its cause is not soon corrected.

While waiting for the transfusions to be completed, I described my proposed plan of action to the two assistants and assigned each of them his role. Everything had to go off perfectly, or we would lose our patient in spite of having come so far, and both of them knew it. The second assistant was standing alongside and up-table of me, on my left. I put my right hand on top of his and guided it carefully to the aorta so that he could replace me in applying the pressure. This allowed me ten free fingers with which to work. Presuming the spleen to be ruptured near its hilum, I planned, with the first assistant's help, to dissect and separate the organ from the bed of tissue in which it lay attached to the back wall of the abdomen and to bring it up into the wound. Having done that, I would tie off the pedicle, containing its artery and vein in the short space between hilum and tail of pancreas, and then remove the organ. It would be necessary to carry this off with considerable speed, because it had to be done through the pool of blood that would reaccumulate as soon as I removed the packing that had been pressed down into the area in which I would be dissecting. Without taking that packing out, there was no way to expose the site from where the blood had been coming. All of these maneuvers, done very rapidly and in perfect sequence, would take precise coordination of everyone's efforts.

I told the scrub nurses exactly which instruments I would need, and in what order. Having made sure the suction lines were free of clot, I waited for the senior anesthesiologist to let me know that our patient was stable enough for us to make some very fast moves during which she might again lose blood rapidly, especially if the spleen was further injured during its hurried removal. He announced a gradual drop in pulse rate from 150 to 100. I glanced over the screen at him to see whether I had permission to go ahead. He nodded his assent and grunted a few words of encouragement, in the category of what is usually yelled to encourage the home team pitcher when the bases are loaded with none out—the OR equivalent of "Chuck easy, baby—you got 'em on the run." In sequence, I looked directly into the eyes of each scrubbed member of the team to be sure they were ready. Then, trying to sound as casual as possible, and just loudly enough for everyone in the room to hear me, I said, "OK, let's do it."

In the flash of an eye, I slid my left hand behind and above the spleen, drew it forward and downward, away from the diaphragm, and told the first assistant to make a short incision into its thin fibrous underneath attachments by using the long scissors that he was holding poised and ready in his palm. As soon as he had done that, I slipped the index finger of my right hand down into the slit he had made and quickly passed it upward and then downward with exactly enough force to tear the layer of anchoring tissue three-quarters of the way around the circumference of the organ. Done blindly and fast, the maneuver is not very elegant, but it works well when a surgeon is in a hurry. I was flying by the dead reckoning learned in a hundred meticulous, painstaking dissections I had carried out in this anatomic region over the years.

I knew exactly where to put my fingers, even without a clear view of the tissue planes that I was opening so hastily. In a few seconds, all of the binding attachments had been cut or torn away and I had the spleen freed and held upward in my hand so that only the pedicle, containing its artery and vein, connected it to the pancreas. With the packs removed, the entire area was so inundated with rapid ooze that I still couldn't identify the exact source of the continued bleeding. But by holding the spleen up, I had a better view of the vessels. Suddenly, it became apparent that the bleeding was not out of the spleen itself but from a spot in its feeding artery about four inches to the right of where it entered the organ as it lay partially buried in the tissue of the pancreatic tail. I had the second assistant momentarily lighten the pressure on the aorta, and a forceful spurt of blood shot out at me through a hole in the vessel wall measuring about an eighth of an inch. With both assistants concentrating their suckers on that spot, I could see clearly enough to recognize something I had only read about but had never before encountered. In fact, I have never encountered it again in the years since that night.

The vessel wall around the hole in the artery was frayed, dilated, and obviously pathological. The nature of the disease that had almost killed Marge Hansen became immediately apparent, although no member of the team now staring at it in disbelief could have guessed the diagnosis until that instant. Our patient had an aneurysm of the splenic artery—a weakened area where the vessel wall was ballooned out to form a saclike bubble. The bubble had burst, so that each heartbeat had been pumping a thick stream of blood out into the abdominal cavity. By compressing the aorta, we had slowed the loss of blood sufficiently to stabilize things, but there were enough collateral vessels feeding into the artery that the heavy ooze had persisted.

It was a simple matter to throw a silk stitch into the damaged part of the artery and have the first assistant tie it down so tightly that all bleeding stopped instantly. When the other resident was told to take his hand off the aorta, the field remained dry. The sigh of relief that filled the room may have been only figurative, but it seemed audible nevertheless.

Within seconds, the atmosphere in that place had been transformed. The tension dispersed, bodies relaxed, and a few small jokes were made. I remember one of the residents saying something about the discretion shown by laundries that wash surgeons' underpants. Mendez and Foley became particularly voluble—almost to the point of silliness—and it was easy to see why. The rest of us still had work to do, but soon after the situation was under control, the two gynecologists took their leave, calling out congratulations as they went out the door. I asked Foley to stop in the waiting room to tell the patient's family that things were now stable.

OR 6 was filled with the euphoria of having saved a life. The breath-holding rush of suspense was over, and there were smiles and more wise-cracks, even from the anesthesia team. We settled down to the fastidious business of continuing to dissect around the tail of the pancreas until it was so free that it could be held up with the spleen and its pedicle. To accomplish this, my team and I resumed our accustomed role of minutely meticulous technicians, working slowly and in precise coordination until the tail and the contained diseased part of the artery were hanging so freely that they seemed to beg us to cut them loose from the patient. Using several silk threads, I tied off the artery and vein between the aorta and the aneurysm, then cut cleanly and safely across the pancreatic tissue where the body of the pancreas narrowed into the tail. The specimen I handed off to the nurse was a single block of tissue consisting, from right to left, of the tail of the pancreas with the stitched aneurysm partially buried in it, the splenic pedicle, and the spleen.

Lest the protein-eating enzymes made by the pancreas spill out into the abdominal cavity from the organ's open cut end, it was necessary to stitch it closed, which I did with a row of fine silk sutures. As a final step to prevent the corrosive juices from digesting the surrounding tissues, I stitched a nearby wad of fat down onto the closed pancreatic stump. These maneuvers took time, but by then we were in no hurry and I was determined that there should be no technical complications of the surgery. Obviously, there remained significant danger of infection because so many rules of asepsis had been broken, but I hoped to minimize that risk by leaving all tissues so cleanly dissected that there would be no postoperative accumulation of blood, tissue juices, or bits of damaged protoplasm.

With that in mind, I made a tiny incision in the skin of the left flank and threaded a long plastic tube of half-inch diameter through it, later to be attached to a low-suction aspirator that would carry off any detritus or fluid that might accumulate during the postoperative days. After all sponges had been removed from the abdomen, the two self-retaining retractors were taken out and the linea alba was closed with stout polypropylene stitches. To minimize the possibility of infection in the incision itself, I laid several soft latex drains into the thin layer of fat underlying my patient's skin, then completed the closure by placing a row of fine plastic sutures in her epidermis.

Only then did I ask my patient's name—up to that moment, I had not wanted to know it. Her anonymity had cloaked her from me, kept her a creature of organs and tissues, and protected us both with an emotional distancing that made somewhat easier the perilous journey we had just taken together. Not knowing who she was allowed my mind to remain free of intrusive personal thoughts. Better that I was unaware of five-year-old Tommy and the other kids, and of Jack; better that no one had told me that Jack and Marge's four brothers and their assembled families were pacing back and forth from waiting room to hospital chapel, and trying to bring some comfort to her weeping eighty-year-old father. Had I not remained ignorant of that distressed family, I might have been less able to maintain my surgical dispassion about the clinical catastrophe occurring in my patient's belly and have thought too much about the personal catastrophe occurring in her life, and theirs. That is not a chance I would have wanted to take. There are reasons both medical and emotional that surgeons drape an incisional site so closely that nothing else human can be seen, as well as hide their patient's sleeping face behind a cloth screen. Those reasons go far beyond the prevention of infection, and the most critical of them is to maintain detachment from the intimidating reality of what they are doing to a man or woman made very much like themselves. The greater the danger, the greater the need for distance. Never is that more true than when we are fighting for a life.

Marge had received fourteen units of blood, two of plasma, and two packs of platelets, the cells that aid in clotting. Almost all of the blood had been transfused during the first hour. In addition, she had been given a huge volume of intravenous fluid, and she was now beginning to produce urine again. As the dressing was being put into place, the anesthesiologist announced that the blood pressure was up to 140.

I stepped back from the operating table and almost tripped over the rectangular metal footstool on which Foley had stood while peering over my shoulder at the action. Several such stools were scattered around the periphery of the table, because a small army of observers had been constantly mounting and dismounting them, alternating with one another to share a view of the drama taking place in the belly of this unknown woman. Some of them, like the extraordinarily efficient blood bank technicians, had good reason to spend a few hurried minutes there, but there were others who seemed to have wandered in and out for the sheer excitement of it. During the total of more than two hours of surgery, word had filtered out among the hospital's evening staff that something dramatic and rare was going on in the OR, and a parade of residents and technicians of various sorts had donned scrub suits and entered Room 6 to see for themselves. Respectful of OR protocol, they had for the most part been so unobtrusive that none of us on the scrubbed team, intent as we were on the task before us, had been aware of their presence. By the time Marge's dressing was being applied, they had all dispersed and returned to their own responsibilities, leaving a debris of flat metal observation posts scattered around the room.

With a series of short kicks, I pushed Foley's abandoned footstool toward the wall and followed close behind each forward slide. When it reached its destination, I turned around and sat heavily down on it, using the wall to guide my back as I eased myself down in a slow slide. Almost languidly, I stripped off my blood-streaked gloves and my cap and mask, then just sat there staring down at the clotting scarlet stain that covered the entire front of my gown. I had begun to feel just a bit tired as the closure stitches were going in, but the full force of my exhaustion now hit me all at once. My mind was exhilarated, even euphoric, at what I had done—I wanted to relive the ineffable excitement of the hours now ended, as though it had taken place in a dream of adventure from which I had just awakened. Something within me wanted to sing and shout, to dance carefree and make love, to acclaim my triumph to the heavens and the ages—a woman's life had been saved, and I would always remember the wonder of this night. But I was so drained of energy that my soaring thoughts seemed unconnected to my enervated, motionless body. I sat there sapped and immobile, my back propped wearily against the blue-green OR wall.

Fatigue's hold on me could not be permitted to continue for long; there were details that needed attending to. With a look of gentle concern on her pretty young face, the first scrub nurse, barely five feet tall in her thick-soled operating shoes, stepped solicitously in front of my footstool and thrust her tiny hand downward at me. As I reached up to take it in my own, it flashed through my mind that this delicate childlike structure was the same sturdy, dependable hand that for more than two hours had been slapping the correct instruments into my palm with the unhesitating precision of an expert, often without my having to ask for them. What had been accomplished in that operating room would have been impossible without the skill of those fingers. It seemed fitting that this talented little hand's last action in the drama was to help me to my feet.

I thanked every member of the team that had performed so well, and I slapped the back of my first assistant, an enthusiastic surgical resident who had rushed in to replace his gynecological counterpart as soon as he got word that I had taken over. We briefly discussed the orders he would write to cover the first postoperative day, and I carefully shed my surgical gown, which was by then becoming stiff with Marge Hansen's coagulated blood. Not wanting her family to see my scrub suit, which was in much the same condition, I went out to the locker room and changed into a fresh one. I found my street clothes still in a heap on the floor, exactly as I had left them.

It is said that a person on trial can tell a jury's verdict simply by looking at its foreman's face as he and his eleven peers file back into the courtroom. If that is true, how much more must it be so for an anxious family awaiting the arrival of the surgeon at the end of a difficult operation. Except for Jack Hansen (who looked like a Viking), every one of the dozen or so people who leaped up at my entrance into the waiting room could very well have stepped out of one of those television commercials promising discount phone calls to Ireland. Not only were Jack and Mary and Marge's brothers and father there, but so were her sisters-in-law and some of their older children. I must have had a broad grin on my face, because the atmosphere in that small space became noisy and joyful as soon as they saw me.

My first words were simple and directly to the point: "She's O.K., she's fine. You'll have her home in about ten days, good as new." This was hardly the moment to share my worries about the possible complications that might result if there was leakage from the cut end of the pancreas; if there was a delayed complication caused by the massive volume of transfused blood, some of which had not been cross-matched; or, of most concern, complications resulting from my violations of sterile technique. I just wanted to assure my patient's family that she had survived her ordeal and was on her way to the recovery room. That accomplished, I flopped myself down on a lounge chair and proceeded, with the family standing crowded around me, to tell them what had happened and what I had done. I finished with a few words that, even as I said them, sounded trite. But they conveyed the full force of my awe at what I had witnessed in that operating room.

What I said was, "Her will to live was the thing that saved her." I omitted the rest of the sentence, although it was rolling through my mind: ". . . and our team's determination not to lose this battle." Mrs. Hansen had not been allowed to die because I was able to call on a force within myself and the others that arose from reserves that are even now beyond my power to understand. Since the members of our species first became defined by our distinct genetic characteristics, we have called upon those reserves in as-yet-incomprehensible ways in order to struggle against the looming threats that are everywhere the constant accompaniments to our existence on this planet. Marge Hansen was saved by that of the human spirit which was in her and in those OR personnel who refused to allow her life to end.

An hour later, Marge was fully responsive, although not yet sufficiently alert to appreciate what she had been through. It would not be until after I had completed the following morning's OR schedule that she and I would have our first talk. Shortly before 2:00 a.m., I left her bedside in the surgical intensive care unit and drove slowly home, along the broad tree-lined avenue that leads out of New Haven and continues into my suburban town. During the brief ride of no more than fifteen minutes, I found myself in a kind of reverie, reliving not only the medical events but everything that surrounded them, including the coincidence that I had found myself at the Hospital of St. Raphael at the precise instant when I was needed.

I have always cared for the great majority of my patients—more than 90 percent of them—at Yale–New Haven Hospital. Sometimes weeks would go by without my doing a single operation at St. Raphael's. Not only that, I am almost never in either of New Haven's two hospitals after 7:30 p.m. unless an emergency keeps me there or brings me back in from home; barring unexpected circumstances, my late rounds are invariably completed by 7:00. On the evening of Marge Hansen's arrival in the emergency room, guests had come to dinner, an unusual event in mid-week. At 8:30, realizing the lateness of the hour, I excused myself and drove into New Haven to see my patients. Because only two of them were at St. Raphael's, both with very straightforward medical problems, I went there first so that I could later feel free to wander the wards and X-ray department of Yale–New Haven completely at leisure. And so it was an extraordinary coincidence that took me across the threshold of the emergency room at exactly the moment when the page began calling.

And there is more. Of all the general surgeons who might have been in the hospital at that hour (several of whom perhaps were), why me? Why, specifically, was the one person who was there to hear the call and heed it also the one general surgeon in New Haven who had for more than ten years nurtured a special interest in the surgery of the spleen and had, because of that interest, acquired more experience with that organ than anyone else in the medical community? In 1969 and for the next three years, it had fallen to me to perform virtually all of the splenectomies for adult patients referred to the Yale radiotherapy and oncology units for treatment of Hodgkin's disease. I had published seven clinical articles on the subject in scientific journals, and given many lectures to medical groups. The coincidence not only of being there but of being there, like the Civil War general Nathan Bedford Forrest, "fustest with the mostest" tantalized my thoughts and added a bewildering sense of an element I have always refused to truck with, the element that some are pleased to call predestination.

With all of this running through my mind, I was hardly aware of reaching my own street and bringing the car to a stop in the driveway of my house. But as I turned the key in the lock of my front door and thought about reconstructing the night's events for my wife, Sarah, reverie's glow gave way to an expanding immensity of triumphant feeling that could not long be contained within me. It grew stronger as I mounted the two flights of stairs to the bedroom where Sarah lay sleeping. I stood at her bedside, gazing down at her as though she, too, was part of this night's wonderment. Apparently she sensed my presence—she opened her eyes and looked up at me. Later, she would tell me what she saw.

> *When you're charged up about something, you've always sent off some kind of almost electrical impulse—and you were elated that night, physically wired and charged by this experience you'd just been through. You radiated this sense of energy and accomplishment. And you had, in fact, accomplished an amazing thing. Something transmitted the charge to me, and that must have been what awakened me. There you were, your face beaming as though someone had put a spotlight on it, and you were smiling. You stood there with your hands on your hips the way you do at times like that, with your fingers curled almost into a loose fist. I remember exactly what you said: "You won't believe what's just happened."*
>
> *It's not often—in fact, it's been very rare in our life together—that you talk in terms of snatching someone from the jaws of death, but there was a profound sense in this that you really had saved someone's life.*

I sat down on Sarah's side of the bed and began to tell her, in the minutest detail, the saga of that night. I told her about the emergency page, about my jumbled thoughts as I raced to the OR, about the operation itself, and then about meeting the Hansen family. When I was all through, it was impossible for either of us to sleep, and so I continued to talk. From time to time, I stopped to answer a question Sarah had, or to respond to an observation she made, but mostly I just went on and on, because she kept wanting to hear more. As if I was narrating a story, I told her about blood vessels and how aneurysms form; about the spleen and pancreas and their functions; about the ways in which the heart and circulation try to compensate for major hemorrhage; about the clotting of blood and the healing of wounds; and finally, I told her, for the umpteenth time since we first met, that I believe there is a palpable but yet-unexplored factor in human biology that accounts for a patient's will to live and a doctor's ability to save her. In the following chapters, I will complete the narration of everything I said that night, and more.

2 | THE CONSTANT SEA WITHIN MAINTAINS THE CONSTANCY WITHIN

The salvation of Margaret Hansen exemplifies as well as anything I have ever witnessed the ways in which our bodies rally to face the ordeals brought on by life-threatening events.

Scientists use the catchall words *responsiveness* and *adaptability* to refer to the myriad methods used by biological structures in reacting to changes within and without themselves. The myriad methods have but a single purpose: to enhance the probability of survival. No less pessimistic an observer of human nature than Arthur Schopenhauer went so far as to state his certainty that the only God discernible in nature is the will to live. The naturalistic pragmatism of many of today's dispassionate scientific researchers persuades them that the primary and perhaps the only driving force of every living thing is the protection of its DNA for transmission to the next generation.

While it is difficult to fault this rigorously practical viewpoint on its own terms, even the most unimaginative of researchers must surely appreciate that there is a great deal more to life than mere biochemistry. The glory of nature is the extent to which many animals and even plants have evolved beyond that most basic of tasks and developed ever more complex abilities. In considering humankind, *abilities* is inadequate to convey a meaning perhaps better expressed by such words as *powers* or *talents*, or even the *wonders* and *marvels* of the Psalmist.

The result of primitive life's eons-long developmental process has been the begetting of organisms for which the elements of survival and reproduction were only the starting point toward generating hosts of specialized capacities, including at last the capacity to feel and savor conscious pleasure.

Not pleasure alone but also the ability to anticipate it and reflect on it are distinctive qualities of our species. So far have we gone beyond our basic self-protective and propagating drives that we have in the process become endowed with such wonders and marvels as abstract reasoning; the personal and communal wherewithal to build and preserve highly organized civilizations; a level of altruism so far removed from simple instinctual self-preservation that it sometimes runs counter to it, and by doing so becomes an exemplary societal ideal—to sacrifice for others; and perhaps above all, a sense of beauty and even spirituality. Without question, these are all qualities magnificently transcendent to the simple requirements of DNA transmission. So transcendent are they, in fact, that they seem a reward for having persevered—an extra we have granted ourselves, or been granted, as a gift for making proper use of the extraordinary brain with which nature has favored us.

But perhaps even our search for beauty, it might be argued, can in a sense be traced back to the primitive drive to preserve DNA. Pleasure is the eventual trophy for obeying our instincts of survival; pleasure makes life valuable and worth sustaining. The varied forms in which it has been discovered and created by humanity cannot be fully considered except in the context of biologic adaptability and an understanding of the fabric of the human body. And yet, in the full range and power of esthetic thought, there is a quality unexplainable on the mere basis of sustaining life.

I choose to say *fabric* advisedly here. When the greatest of all anatomists, the Belgian Andreas Vesalius, published his didactic masterpiece *De Humani Corporis Fabrica* (*On the Fabric of the Human Body*) in 1543, he used the word in the kinetic sense in which it was understood during the Renaissance, to mean not only the edifice but also the workings that take place within it. The human spirit, I believe, is something we have created from the fabric of our human body.

The threat to the fabric of Margaret Hansen's body was a weakness in the wall of her splenic artery. Such a defect is called an aneurysm, from the Greek *aneurysma*, "a widening."

There is no structure in the body that can be described as simple. Even an artery, whose only responsibility would seem to be the carrying of freshly oxygenated and newly nourished blood to the tissues, is a thing of quite impressive complexity—impressive, but hardly intimidating; complex, but hardly incomprehensible. Separated into its three concentric layers, the wall of an artery is seen to be formed by nature to carry out functions more valuable than might be expected were its only duty to be a passive conduit.

Some of those functions are in the service of responsiveness. They come into play as part of the body's constant readjustments intended to support living processes by maintaining internal equilibrium—a state of well-ordered steadiness which biologists like to call homeostasis, from two Greek words meaning "to keep things the same." The essence of homeostasis and its importance for the organism is to be found in the words of the physiologist Walter B. Cannon, who introduced the term in the 1920s: "As a rule, whenever conditions are such as to affect the organism harmfully, factors appear within the organism itself that protect it or restore its disturbed balance." The maintenance of the body's homeostasis requires the integrated coordination of every tissue within it, but it is the vascular system (the arteries and veins) that, along with the immune, endocrine (hormonal), and nervous systems, links them all together in the common enterprise of sustaining life.

For tissue to function in anything resembling normal fashion, the environment around each cell must remain stable. It is the responsibility of the vascular system to maintain stability by ensuring a constant supply of the molecules required by the tissues, and by dependably carrying off the waste products that result from life processes. By such means are the surroundings maintained in which each cell can perform optimally.

Until not very long ago, the study of a cell's surroundings was thought to be the most fruitful approach to developing a thorough understanding of life processes. Not only is its environment crucial to a cell's existence; it is the means by which it maintains contact with the world outside the organism. Part of the reason for emphasis on the environment of a cell was the practical issue that it was far more accessible than its inside. But the past four to five decades have witnessed the development of new analytical techniques and instruments that allow direct investigation of intracellular processes and structures, including the study of the large molecules that are involved in so much of the cell's work. A new scientific discipline has been born.

To designate this baby branchlet of science as it was first making its appearance, the term *molecular biology* was introduced in 1938. But its acceptance, at first gradual, only became rapid with certain improvements in the technology of X rays and the electron microscope, enabling the molecular approach to become indispensable in such areas of research as biochemistry, genetics, and structural chemistry.

It had long been known that small differences in proteins are the reasons for the differences between various species. These differences obviously have a chemical basis, and it became the role of the molecular biologist to elucidate it. It is largely through the study of these proteins, the so-called master molecules of life, that so much of cellular function has come into the realm of scientific knowledge. The cell, far more than its neighborhood, is now the focal point of biological research. But none of this could have happened had it not been preceded by almost a century of study of the surroundings in which cells exist.

In the mid-nineteenth century, the French physiologist Claude Bernard coined the term *milieu intérieur*, or "internal environment," to describe the state of affairs that permits life to remain constant and independent of whatever is happening in the external surroundings in which an animal finds itself. Bernard pointed out that every multicellular animal really has two environments, the *milieu extérieur*, "external environment," which is the air or water around it, and the *milieu intérieur*, formed by the fluids bathing all the cells within the body. "A complex organism," he wrote, "should be looked upon as an assemblage of simple organisms which are the natural elements that live in the liquid *milieu intérieur.*"

This liquid's chemical resemblance to seawater has been the source of many a late-night lucubration about the continuity between today's complex animals and their single-celled marine ancestors of about 3½ billion years ago. To this day, it appears, cells still need something of their original primordial surroundings in order to sustain life. By keeping its internal environment stable, the animal makes itself independent of its external surroundings, thus allowing all function within its cells to proceed normally regardless of what is happening outside its skin. Our cells live not in the air through which we walk, but in the fluid in which they are bathed. The protoplasm within us is separated from the external environment by a leathery covering of skin, whose outer layer is composed of dead tissue. Inside, we are wet. It is our internal wetness that gives us life. Except where it touches its neighbor, every cell (other than those of the brain and spinal cord) lies surrounded by the nurturing fluid of the *milieu intérieur.*

The *milieu intérieur,* then, consists of the circulating blood plasma and the nutrient liquid that bathes our cells. It enables a kind of contact, albeit buffered and indirect, between what is outside of us and what is inside, while at the same time providing protection for the cells by maintaining the stability that surrounds them. In order for the *milieu intérieur* to stay in this state of dynamic, self-righting equilibrium, it is necessary for the body's various organs and tissues to be able to respond to the variations to which we are always being exposed by moment-to-moment events both inside and outside of us. The process, therefore, is a two-way street of mutual dependence. The various cells, tissues, and organs serve to maintain the stability of the *milieu intérieur,* which in turn is the reason cells can continue to function. It is clear that life is maintained by a huge number of constantly compensating alterations by every tissue and organ, all with the purpose of perpetuating the stability of the *milieu intérieur,* which in turn perpetuates the stability of the cells. The seventeenth-century English poet Abraham Cowley may have been presciently describing the world within us when he wrote:

The world's a sea of changes, and to be
Constant, in nature, were inconstancy.

"It is the constancy of the *milieu intérieur,*" wrote Bernard, "which is the condition of free and independent life. All the mechanisms of life, no matter how varied they are, have only one object, to keep the conditions of life constant in the internal environment." Had Schopenhauer been a biologist, he might have added that the God he discerned in nature must certainly be doing much of His work in the *milieu intérieur.*

Bernard's was a brilliant leap of understanding, all the more convincing for being confirmed in thousands of ways during succeeding generations of scientific discovery. It is homeostasis, the dependability and steadiness of the internal environment, that keeps us alive. The structure of the arteries and veins and the manner in which they function are based on their critical role in maintaining the equilibrium of the *milieu intérieur.* They are the main channels of the transportation system that delivers the goods and carries off the detritus of our inner selves.

The full name of the group of structures that includes heart and all of our veins and arteries is the cardiovascular system, but the present discussion deals only with the vessels. Oddly enough, *artery* derives from a Greek word meaning "air duct," based on the ancient belief that these vessels contain not blood but an airlike essence called pneuma, drawn into the lungs with each breath and from there transported to the left side of the heart and then out into the arteries. To explain away the obvious fact that arteries bleed when cut, the Greeks wishfully postulated a series of connections between them and the blood-filled veins. In this fanciful formulation, blood was said to rush into an artery as soon as it was injured. Not until the experiments of the great physician Galen in the second century A.D. was the misconception corrected, but by that time *arteria* was firmly entrenched in the language of medicine.

As previously noted, an artery, far from being a simple inert tube, is a dynamic three-layered flexible hose that has the capability of widening or narrowing its diameter and capacity anywhere along its length to accommodate the fluctuating needs of the tissue it supplies, as well as those of the entire body. Although its inner lining consists of a single circumferential layer of flattened cells, the middle of its three coats is packed with elastic and muscle fibers. Because the muscle fibers contract without our conscious control, they are called involuntary; because of a certain evenness in their microscopic appearance, they are called smooth. This distinguishes them from the muscles we use at will, which are called voluntary and are striated with fine microscopic lines.

Wherever smooth muscle is found in the body, whether in the arteries, the gut, or elsewhere, it functions without our deliberate effort and usually without our knowledge. In other words, its action is autonomous, independent of our conscious will. For this reason, the part of the nervous system that controls it is called *autonomic*. The autonomic nervous system carries messages to glands, smooth muscle, and the muscle of the heart—that is, it controls those processes like glandular secretion, peristalsis, and heart rate which must proceed automatically in order for our bodies to maintain healthy functioning.

Unlike its voluntary counterpart, smooth muscle is usually in a state of partial contraction, which physiologists call *tone*. As required by the ever-changing needs of the body, the autonomic nervous system can signal the contraction to lessen or to become more complete, and it may even produce sustained spasm of a tubular structure. It is sustained spasm of smooth muscle in the wall of their feeding arteries that makes our hands cold; it is sustained spasm of smooth muscle in the wall of the gut that gives us cramps.

Clearly, the middle coat of the arterial wall is the dynamic layer. Surrounding that flexible muscular and elastic encirclement is a thin tunic of rather wispy fibrous tissue that forms the outer coat, providing a degree of protection to the vessel while at the same time anchoring it somewhat to similar fibrous material in nearby structures.

The largest artery in the body is the aorta, a vessel at least as big around as a man's thumb. The aorta rises upward out of the powerful pumping chamber of the heart (the left ventricle) and makes a U-turn downward and to the left, toward the abdomen. Along the curve of this arch it gives off large branches to the head and arms. Continuing down and lying on the vertebral column of the chest, it pierces the back of the diaphragm and enters the abdominal cavity. It was precisely at this point that I cut off the circulation to the lower half of Marge Hansen's body, by pressing the aorta flat against the vertebrae with my hand just beneath the diaphragm.

At the level of the navel, the aorta divides into an inverted Y, each of whose limbs passes diagonally down through the pelvis and then into one or the other leg and eventually to the toes. Along the entire length of the aorta and its two pelvic divisions, arteries are constantly leading away from the main channel like so many roads coming off a major highway, each heading toward the organ or area of tissue dependent on it for blood supply.

Or an equally apt analogy: Like a limb coming off the trunk of a tall tree, each artery arising from the aorta gives off branches of its own, dividing and subdividing again and again into ever smaller vessels. It is for this reason that the whole system is referred to as the arterial tree.

The penultimate and smallest arterial passageway is a microscopic vessel called the arteriole. By contraction or dilatation of the smooth muscle in its wall, an arteriole controls the amount of blood entering the capillaries (the final conduit) just beyond it. It might be said, in fact, that an arteriole is the gatekeeper for the bed of capillaries it feeds. So efficient is the gatekeeping mechanism that only some 10 percent of the body's many miles of capillaries contain blood at any given moment. The whole system allows for transfers of significant amounts of blood from one tissue to another—from one part of the body, in fact, to another—so that the constantly changing needs of the various organs can be accommodated.

The structure of a capillary wall is similar to the inner lining of an artery in that it consists, except in the brain and spinal cord, of not much more than a single rather porous layer of flattened cells. Between and through these cells, the molecules of oxygen, carbon dioxide, and nutrients, as well as the waste products of cellular function, pass back and forth. They go into and out of the fluid around the cells, allowing it and the cells to be bathed, nourished, and cleansed at the same time. The equilibrating two-way transferences serve to maintain the stability of the *milieu intérieur*, and therefore support homeostasis.

The capillaries are the critical vessels for which all the rest of the mechanism of heart and arteries exist. The function of the entire cardiovascular system is only to bring the contents of the blood to the capillaries in order to maintain the homeostasis that keeps cells functioning normally. These minuscule vessels are 1/4000 of an inch (6 microns) in diameter, and they are everywhere. They exist in a fine meshlike framework, so intimately positioned between cells and groups of cells that they are like the streets and alleyways within the multitude of cellular neighborhoods that are inside every organ.

In the fact that blood enters individual groups of capillaries only as it is required for local cellular function lies one of the true amazements of animal biology. By an extraordinary system of messages transmitted by hormones and nerve impulses, the various structures of the body make known their moment-to-moment needs. By constant communication, they are able instantaneously to inform one another about exactly what major or minor adjustments must be made, exactly where, and exactly how to fulfill the need. One form of such communication is illustrated by a mechanism involved in maintaining normal blood pressure when hemorrhage suddenly occurs, as in the case of Marge Hansen.

Nestled within the wall of the aorta just above its takeoff from the heart lies a group of tiny structures called aortic bodies. These little sensors are known as receptors because they receive and are sensitive to messages about changes in the quantity and quality of blood coursing through that huge vessel. Those of the aortic bodies affected by pressure are called baroreceptors (as in barometer); those affected by changes in concentration of oxygen and carbon dioxide are called chemoreceptors. Additional baroreceptors are also found in the arteries that ascend from the aorta to supply the head and neck, called the carotids. (The word *carotid*, incidentally, is derived from the Greek *karotikos*, or "stupefying," which is the effect resulting from squeezing tightly on those arteries. For obvious reasons, *garrote* has the same origin.)

When blood flow is normal, the baroreceptors are subjected to a constant and predictable degree of stretch, caused by the pressure of the regularly pulsating blood within. In response to the stretch, a steady series of impulses is sent via the autonomic nerve fibers to the part of the brain called the medulla, one of whose functions is to help regulate the cardiovascular system. The medulla shares this role, as will be explained more fully in Chapter 4, with another part of the brain called the hypothalamus.

When hemorrhage results in significant loss of blood, the pressure on the aortic wall drops, putting the baroreceptors under less tension. Consequently, fewer impulses are sent to the brain. The brain centers, primarily in the hypothalamus, respond by firing warning signals over the wide-ranging system of autonomic nerve fibers to the heart and all blood vessels of the body, in an attempt to maintain the normal level of pressure.

These signals accomplish their mission in several ways. The heart responds to them by speeding up its rate and strengthening the force of its beat, thereby increasing the total volume of blood it drives into the aorta and carotids; the arteries respond by contraction, especially of the arterioles, thereby decreasing access to the capillaries—and by this means lessening the total capillary space that must be filled; the veins (which have some smooth muscle in their walls) also constrict, and by doing so they achieve two objectives: Constriction decreases the total capacity of the venous bed so much that a smaller volume of blood is required to fill it, and constriction also squeezes blood back to the heart more effectively so that it can rapidly start out once more on its cycle to the tissues. Using all of these strategies, the usefulness of the lessened quantity of blood is maximized. The body acts wisely by compensating in ways that permit the pressure to be maintained. This is a perfect example of what biologists mean when they speak of responsiveness.

There is a price in all of this. When arterioles constrict, the capillaries and therefore the tissues they supply must temporarily make do with less blood. With the exception of three organs, actually, a period of decreased flow is easily tolerated. These are the heart, the brain, and, to a lesser extent, the lungs. All three need a constant high level of freshly oxygenated blood in order to stay alive. Here, too, there is wisdom in the body. The arterioles feeding the three vital structures do not participate in the generalized contraction. They remain wide open, sparing brain, cardiac, and lung cells and allowing them to benefit from the temporary sacrifice being made by the rest of the tissues.

The kidney is, of course, one of the sacrificial organs whose blood flow is much decreased by constriction of its arterioles, and this is the reason Marge Hansen stopped producing urine until her pressure rose as her bleeding came under control. Not only the arterioles but the peripheral arteries also constrict when the blood volume is low. This is why a person in shock from blood loss is cold to the touch, a condition made even more extreme by rapid transfusion of refrigerated bank blood, which is often being pumped in too fast to allow for proper warming.

Because of decreased total blood volume and spasm of arteries, Marge's pulse was impossible to feel during the period of her bleeding. In response to the surveillance of the baroreceptors, her heart rate rose to 170 during the time her pressure was unobtainable; had her cardiac output been monitored, it would have been found to be quite high, the result of increased rate and increased force of the heartbeat. By this means, the blood is recycled more rapidly and a smaller volume is made to do more work.

The mechanisms initiated by baroreceptors are not the body's only way of rallying against sudden blood loss. There are other useful armaments in nature's store, and every one of them is used to fight off the life-threatening forces of hemorrhage. The *milieu intérieur* itself becomes involved in the process. When capillary blood flow is decreased, the pressure relationship changes between the tiny vessel's inside and the fluid in the interstices between the cells, the part of the *milieu intérieur* called the interstitial fluid. The balance then comes to favor the interstitial fluid. To restore the balance, fluid shifts into the capillaries, causing a rise in the liquid contents of the circulatory system. Although this process dilutes the blood, it does increase its total volume. The possible magnitude of such a shift can be appreciated by considering that the cells of a person weighing 150 pounds are soaking in 22 pounds of interstitial fluid. A shift into the capillaries of only a small percentage of this can have a significant effect on the amount of blood in the circulatory system. Even when diluted, a pint of blood weighs nearly a pound.

Thus far, the responses described have dealt in general with biophysical phenomena such as electrical impulses and pressure relationships, in which chemical actions have only a secondary part. But in other ways, chemistry does play a major role. Among its methods is the body's production and use of hormones and signaling molecules to help maintain homeostasis.

The term *hormone* was introduced in 1902 by two researchers at University College, London, to categorize the action of what they called "some chemical substance" they had discovered in the small bowel. Because the substance, produced by certain cells of the bowel wall, acted on the pancreas to make it secrete digestive enzymes, they named it secretin. The two, William Bayliss and his brother-in-law Ernest Starling, decided to call secretin a hormone, from the Greek verb *hormaein*, "to excite," or "to set in motion." Since then, the word has been applied to any chemical produced within certain tissues or organs that is then carried by the bloodstream or other body fluids to quite different tissues or organs in order to regulate the action of certain types of cells at the distant site. Hormones "excite" those cells to carry out specific duties—they "set in motion" the activities of those cells. A well-known example is the hormone made by the thyroid gland, which regulates the body's rate of metabolism; another is insulin, produced in specialized cells buried in the pancreas and carried all over the body to control the utilization of sugars and fats. There are dozens of hormones, and new ones are constantly being discovered. They provide a mechanism by which information is integrated and transmitted from one part of the body to another.

Hormones are manufactured by a variety of glands scattered all over the body, as well as by specialized cells in the digestive tract, liver, pancreas, kidneys, ovaries, testicles, and the placenta of pregnant women. The glands are the pituitary (often called the master gland, because it exerts a regulatory influence on the others), thyroid, parathyroid, adrenal, pineal, and thymus. This entire group of glands and cells is called the endocrine system, from two Greek words meaning "internal" and "separate," a designation originally intended to indicate that they secrete internally (into blood vessels rather than the cavities of hollow organs such as the gut) and form a separate method of control over the body's functioning.

As it turned out, the initial belief that hormones function separately was not borne out by subsequent research. We now know that they, the nervous system, and certain local signaling molecules work as a coordinated whole to influence one another in determining the internal workings of the body.

As an example of coordination between nervous system and hormones, consider the small structure in the lower part of the brain called the hypothalamus. The hypothalamus exerts a major influence on the pituitary gland, to which it is attached on its lower surface. The integration between the endocrine function of the pituitary and certain nerve tissue in the hypothalamus is illustrated by the role played by a hormone called vasopressin in the maintenance of blood pressure.

Vasopressin is a word easily recognized as referring to something that "presses," or constricts, blood vessels. As noted earlier, the messages conveyed along nerve fibers from baroreceptors travel to the medulla and the hypothalamus. When blood pressure drops, the hypothalamus responds, in the manner described on page 36, to the decreased number of impulses reaching it. But it also signals the back, or posterior, part of the pituitary to release vasopressin into the bloodstream, which carries it to the arterioles, causing them to constrict.

The structure of the pituitary itself is an example of the way in which the nervous and endocrine systems are so intertwined with each other. Unlike its forward, or anterior portion, the posterior part of the pituitary is not really an endocrine gland—it consists of fibers and the endings of nerve cells located in the hypothalamus, which is immediately above. And so the hormone vasopressin is really being secreted not by an endocrine gland but by the hypothalamus, a part of the brain.

Another name for vasopressin, incidentally, is antidiuretic hormone, because it also acts to transfer water from the secreting ducts of the kidney back into the blood, raising the blood volume and decreasing the amount of urine produced. This, too, was a factor in Marge Hansen's lack of output. In all of these ways, nature conspires to maintain homeostasis.

Had Marge Hansen been transfused with enough blood to bring her blood pressure up but yet an insufficient amount to restore her normal quantity of red cells, still another hormone would have come into play. The formation of red blood cells is called erythropoiesis (from two Greek words meaning "I produce red"), and it occurs in the bone marrow. Among the factors that determine the rate of erythropoiesis, the amount of oxygen in the blood is most crucial. Since almost all of the blood's oxygen is carried in the huge iron-containing protein molecule called hemoglobin within the red cells, hemorrhage results in less oxygen because there are fewer red cells. This decrease stimulates certain cells in the kidney to produce a hormone called—naturally—erythropoietin, whose function is to enter the bloodstream and signal the bone marrow to make more red cells.

With the exception of erythropoiesis, the entire panoply of compensatory mechanisms automatically springs into action as soon as the blood pressure drops. The panoply's purpose is to assure that sufficient nutrients and oxygen are supplied to protect the brain from dying and the heart from stopping. Unless the compensatory mechanisms are overwhelmed, as they are when blood loss is simply too fast or too massive, the system functions predictably and well—the body's defenses do exactly what they are supposed to. Marge Hansen emerged from her anesthesia without any damage to either her heart or her lungs. Her kidneys resumed normal function as soon as the blood volume rose sufficiently. The mind of this good-humored woman was as alert and quick as ever, and the sparkle in her blue Celtic eyes was undiminished. In all, there was not a shred of evidence during her recovery period or afterward of exactly how close she had come to death.

Reference was made above to local signaling molecules. These are chemicals secreted by cells in some given tissue to bring about a particular change required in the nearby area, either to counteract one or another stimulus or to enhance an event in progress (local growth of some specific tissue is such an event). Cells produce local signaling molecules in response to changes they detect in their immediate environment. A class of molecules called prostaglandins is an example of these chemicals, several of which act locally to help constrict or dilate tiny blood vessels in response to needs of nearby tissue for more or less oxygen or nutrients. It can thus be seen that a considerable array of mechanisms exists, interacting with one another to maintain homeostasis. Whatever else it may be, life is one big balancing act.

The instant responsiveness of Marge's cardiovascular system provides insight into the myriad other self-regulating mechanisms built into our bodies. Although hemorrhage is an emergency requiring prompt action, there are many other ways in which surveillance and correction are provided for tissues. Everywhere, cells produce chemical by-products of their functioning. When those by-products enter the *milieu intérieur*, they have the effect of changing the characteristics of the circulating blood in such a way as to induce responses tending to maintain equilibrium. Decreases in the levels of oxygen or increases in carbon dioxide or acidity in the interstitial fluid around voluntary muscle cells, for example, trigger immediate action to alter the amount of blood reaching the area and to bring more oxygen and carry off more carbon dioxide and acid. The amounts of circulating chemical substances such as adrenaline or various pressure-regulating hormones fluctuate from moment to moment as demanded by tissue needs, accomplishing similar ends. Our cells live by transmitting messages about their ever-changing requirements, and those messages are carried via fluid, blood, and nerve impulses to other cells that know how to respond. The change itself triggers the very mechanism that permits the subsequent return toward homeostasis. The total of the entire concatenation of constant messages and fine-tuning answers is the sum of biological life.

One more thing needs to be explained. In describing her pain, Marge Hansen was very clear in remembering how it shot up into her left shoulder, and her inability to lie down because the supine position made her discomfort increase. So much a problem did this become that she had to be transported to the emergency room upright in the ambulance. Why should a person with an abdominal injury feel pain in her shoulder, and why should it be aggravated by lying down? The answer is to be found in a peculiarity of the nervous system. The nerves carrying sensation from the diaphragm originate in the same segments of the spinal cord as the nerves carrying sensation from the tip of the shoulder. When Marge lay down, the free blood in her abdominal cavity washed up against the diaphragm and irritated it, sending a cascade of pain messages to the spinal cord, whose pain-carrying tracts transmitted it up to the brain, where it reached the level of consciousness. Because the messages from diaphragm and shoulder are carried side by side in adjacent nerve fibers, the brain's cortex misinterprets the source and thinks it is the latter that is hurting instead of or in addition to the former. This phenomenon is called "referred pain," and it is one of the first things a medical student must learn when he begins to study diagnosis.

The bit of splenic artery that came so close to doing Marge Hansen in was sent off to the pathology lab, along with her spleen and that small bit of pancreas where her aneurysm lay. The blown-out portion of the artery was stained, glued on a slide, and put under a microscope, where it was seen to exhibit what the pathologist called "the degenerative changes noted in a selected population of post-gestational females."

Marge's disease had proven to be a woman-hater. Worse than that, the misogynist preyed specifically on mothers and even more specifically on those mothers who valued the act of conception so highly that they undertook it again and again. This particular form of aneurysm is most commonly seen in women who fall into that much-blessed category called "grand multip" by the obstetrics guild—women who have had five or more full-term pregnancies, which is precisely the point at which Marge stood, or rather lay, on the emergency room gurney as that memorable evening of surgery began.

The term used to describe the weakness that caused Marge's splenic artery blowout is *fibrodysplasia*. Ordinarily, a dysplasia is an abnormality of development, but pathologists use the term to refer to any alteration in the organization of cells, including abnormal size or shape. Whether from the effect of pregnancy hormones or due to increased splenic blood flow as the fetus enlarges, the middle layer of the splenic artery of some women becomes thinned out. The thinning is due to an unexplained decrease in muscle cells and an equally obscure fragmentation of elastic fibers. It may, in fact, be abetted by pregnancy's natural increase in two hormones called elastin and relaxin, whose real purpose is to soften and therefore relax muscles and ligaments so as to accommodate the increasing size of the pregnancy and the urgent requirements of delivery.

This is an example of how our bodies sometimes betray us. There are numerous ways in which hormones intended to respond to some specific circumstance have undesirable side effects related directly to the very characteristic that makes them so useful in responding. Elastin and relaxin do their jobs by thinning fibrous and elastic tissue in the joints of pregnant women, most importantly in the pelvis. Such thinning and weakening is necessary if the pelvis is to accommodate the enlarging uterus and later make way for delivery. But as a consequence, the hormones also occasionally thin the walls of blood vessels, an action to which the splenic artery is particularly susceptible.

With repeated pregnancies, the vessel wall progressively weakens, and it finally begins to splay and bulge at some particular point, like a weakness in an old tire. In time, a thin-walled aneurysm is formed, looking like nothing so much as a protruding bubble that may be as small as a fraction of an inch in diameter, positioned anywhere on the circumference of the artery. With blood pounding through it at a pressure of 120 millimeters of mercury, only luck prevents the thinned outpouching from bursting inside of every woman who harbors it.

Marge's luck ran out on that Sunday afternoon in August when she felt the "explosion inside of me, like . . . like . . . BOOM!" She survived that first event because of the body's tendency to wall off disease, plus the fortunate anatomic location of the splenic artery. Along with the pancreas, it lies in a relatively small enclosed space in the back part of the abdomen— called the lesser sac. The lesser sac is shaped somewhat like a flattened purse whose front wall is the back of the stomach. When the aneurysm ruptures, blood quickly fills the purse. Because at first it has difficulty forcing its way out, the hemorrhage is stopped by the buildup of pressure within the lesser sac. A large clot forms, which by its very volume further compresses the torn vessel.

The pressure within Marge's lesser sac was the cause of the abdominal pain she experienced during the five to six weeks until her final collapse. Slow bleeding often continues in spite of the clot, until the expanding tension of the entrapped blood becomes so high that the lesser sac can no longer contain it. At this point, the blood bursts out into the general abdominal cavity and the patient quickly dies if nothing is done. Clinical physicians call this sequence "the double rupture phenomenon"; it occurs in some 25 percent of those few people whose splenic artery aneurysm actually ruptures.

The condition is so rare, in fact, that the most reliable figures I have been able to obtain from routine X-ray and autopsy studies indicate that less than 1 percent of the population has so much as an aneurysm, and fewer than one in five of even this small number have had so much as a single symptom, like pain. Rupture occurs in less than 2 percent of the original 1 percent; this means that at most two of every ten thousand of us are likely to have a splenic artery aneurysm rupture. The ratio of female to male is four to one, with almost half of the women being grand multips. If there ever was a candidate waiting for the onset of a rare catastrophe striking out of the blue, it was Marge Hansen. And if there ever was a woman capable of responding to nature's assault with the entire battery of mechanisms provided by that selfsame nature to maintain homeostasis in the face of its own attempts to disrupt it, it was Marge Hansen.

3 | OF NYMPHS, LYMPH, AND COURAGE IN THE FACE OF CANCER

The veins are not the only conduits carrying fluid back to the heart. As early as the golden age of Greece, the great Hippocrates had already identified an extensive system of channels draining a milky-white material from the intestine, which he correctly thought to be digested fatty material. It would be more than two millennia before the pathway into the bloodstream of these tiny vessels was properly traced, and even more centuries for their entire complexity to be unraveled from the webwork of which they are a relatively small part. The webwork would come to be called the lymph system.

In time, it was understood that the lymph, or lymphatic, system was composed of far more than just the channels draining the gut of digested fat. The purpose of the vast majority of its many miles of interconnected vessels, in fact, is to pick up excess fluid from the spaces between cells and return it to the circulation. The fluid is clear and transparent, and it is for this reason that it came to be called lymph.

"Your eyes, my own sweet darling, are limpid pools of turquoise blue." How much bad love poetry has borne the burden of some equivalent of those lines? The word *limpid* is traceable to the Latin *lympha*, a word tortuously derived from the Greek idea of nymphs, who were goddesses of springs or lakes—places of clear water. That which is limpid is clear, free from turbidity or guile. Because the physicians of antiquity noted that a watery fluid filled these little channels running in the same geographic pattern as veins, they called it lymph. It consists essentially of interstitial fluid on its way back to the bloodstream.

In the previous chapter, I referred to the transferences back and forth across the capillary wall that allow fluid and nutrients to exit into the interstitial area while waste products and fluid go back into the capillary. Of the several factors that determine the relative rate of transfer in and out of the capillary, the most important is the simple one of the pressure difference between the inside and outside of the tiny vessel—in other words, the pressure gradient between the inside of the capillary and the interstitial fluid through which it passes. At the end of the capillary closest to the arteriole (the so-called arterial end of the capillary), pressure is higher than it is nearer the point at which the capillary empties into the tiniest vein, or venule. This is for the simple reason that the blood pressure transmitted from the arteriole gradually diminishes as it traverses the bed of capillaries. The result of this is that fluid tends to be pushed out of the capillary at its arterial end, where the pressure is higher, and reenter it a bit downstream at its venous end, where the pressure is lower. The sum total is to keep the balance steady. Unlike fluid, however, the passage of chemical substances is determined by their relative concentration on either side of the capillary wall. They tend to pass from the side with more to the side with less in order to equalize the relative number of molecules.

Depending on circumstances, it frequently happens that less fluid returns to the capillary than left it, leaving a slight excess in the interstitial fluid. To compensate for this situation is one of the two jobs of the lymph system, which travels along with the veins. The lymph system begins in tiny structures that look like capillaries. These gradually coalesce into larger vessels that eventually empty into a major vein (the subclavian), entering it in the lower neck shortly before its contents return to the heart.

The lymph system, therefore, functions to drain the interstitial area of excess fluid and also to pick up proteins that may have leaked out of capillaries.

Each cell in the wall of the tiniest lymph channel overlaps the edge of its neighbor but is not attached to it. Owing to this arrangement, it can be pushed inward like a valve if the outside pressure is higher than the inside. When the opposite situation prevails, the valvelike mechanism shuts, and thus nothing can leave the lymph channel. This makes these channels particularly efficient for taking in large structures like protein molecules. Protein molecules that have been forced out of the capillaries are too large to get back into them. Owing to the greater permeability of the tiny lymph channels, such molecules can enter them and become part of the lymph. The pickup of fluid and protein is facilitated by the low pressure within the lymph channels, which is significantly less than that in the capillaries.

Once in the channels, the fluid and its contents is called lymph. Specialized lymph capillaries called lacteals exist in the intestinal lining, and they form one of the routes by which digested nutrients are picked up so that they can enter the bloodstream. It is this material that Hippocrates observed.

The larger lymphatic channels have three-layered walls similar to those of veins. Depending on their point of origin in the long journey toward their final destination in the subclavian (so named because it lies just beneath the clavicle) vein, they travel inside our bodies, up our arms and legs or down our necks. Pushed along by the movements of muscles and organs, the lymph thus makes its way from the most distant part of the periphery toward the center to rendezvous with the blood in the subclavian vein. Along their path, the channels periodically come upon groups of spongy, nut-shaped way stations called lymph nodes, into which they temporarily empty their contents.

The typical lymph node is less than a half inch in diameter and consists of a fibrous capsule surrounding a spongelike network of chambers through which the lymph percolates as though in a filter until it finally drains out of the node into a channel on the opposite side from its entry point, to continue on its voyage back into the bloodstream. A lot goes on in that little structure, about which more later. The lot has to do with the lymphatic system's other job: to prevent and fight disease.

In much the same way that tiny lymph channels are able to take in relatively large molecules like proteins, other fragments and structures may also enter them with the lymph, such as bacteria, viruses, bits of tissue debris, and tumor cells. As part of its involvement in fighting off disease, the lymphatic circulation deals with these as best it can in a usually successful attempt to destroy them. But at other times, the system not only fails but actually adds to the problem by allowing harmful structures like microbes or cancer cells to use it as a convenient highway along which to travel from one organ to another.

The role of the lymphatic system in cases of cancer is one of those examples of the sometimes ambiguous relationship that our bodies have with forces that would attempt to destroy them. Under unusual circumstances, responsiveness fails. When this happens, the very structures and functions provided by nature for our survival sometimes become conspirators in our downfall. The splenic artery aneurysm is, of course, a case in point.

Were a single characteristic to be chosen that most distinctively differentiates malignancies from any other tissues or growths, it is their ability to spread to sites not directly in contact with them. Such a spread is called a metastasis (from the Greek, meaning "a move from one place to another"), and it is responsible for most of the mortality caused by cancer. Almost always, it is the circulatory or the lymphatic system that is the channel transporting a metastasis from its place of origin to the eventual location where it attempts to find a new home.

Cancers arise because DNA in the cell of some particular tissue is altered by a chemical or physical influence called a carcinogen. This leads to the production of certain genetic products that cause uncontrolled multiplication of the involved cell. The process can be stopped at any of several early developmental points, including the very earliest one of correcting the alteration in the DNA, by a process that will be described in chapter 5. In addition, immune and other mechanisms (such as the activity of genes that suppress the multiplication of tumor cells) exist within us that, under ordinary conditions, prevent the disordered growth that is called malignancy from establishing itself.

But even when it loses its battle to prevent a cancer from forming in the first place, the body does not give up. At first, it does all it can to keep the growth contained. Although many malignant cells do enter the vascular or lymphatic capillaries after being shed from a tumor, only a very few ever reach another organ. Even fewer can so effectively overcome the organ's resistance that they succeed in implanting themselves well enough to achieve independent growth as a metastasis.

Several factors determine how many of the cells find their way into vascular or lymph channels, but the most important is probably the speed at which the original, or primary, tumor is growing. Those that grow fast shed more cells. Other significant determinants are the size of the tumor and the length of time it has existed. The vast majority of cells entering the blood vessels (probably more than 90 percent) are trapped and destroyed in the capillary beds of the liver and lung. A number of lethal forces act to destroy them, ranging from something as simply mechanical as the trauma of being beaten up by the swirling circulation to a factor as complex as the system of immunity activated by the very existence of the primary tumor. The immune system uses protein molecules called antibodies to destroy cancer cells in much the same way it attacks a bacterium or a virus, and it also activates and mobilizes a battery of cells within lymph nodes (called T cells) that have the ability to kill cancer cells with which they come into contact. (For a more complete discussion of immunity, see chapter 10.)

The immune system's role in our bodies is to distinguish between what belongs in us and what does not. Another way to put this is to say that it knows the difference between "self" and "nonself." It organizes in an attempt to eliminate the foreignness it recognizes as nonself. Because many (but not all) malignancies contain elements that are nonself, one of the major thrusts of recent cancer research has been to attempt to increase the immune system's ability to carry out surveillance and destruction.

The part played by lymph channels and nodes in malignancy is typified by their role in the metastasis of breast cancer. As the tumor grows, its cells infiltrate lymphatic channels at the capillary and larger levels. Some of the cells are picked up by the stream of lymph and carried along with the flow until they reach the nearest node or group of nodes. For most people with breast cancer, these are in the axilla, or armpit. The axillary nodes act much like a filter for the tumor particles coming to them from the breast. Not only that, but the presence of tumor stimulates the production in the node of immune cells that attempt to kill their cancerous antagonists.

The tumor causes a similar production of immune cells by distant uninvolved nodes and also in the spleen. In this way, the local lymph nodes have several strategies by which they can fight the cascade of malignancy reaching them. Unfortunately, none of these strategies is always completely effective, so a significant number of tumor cells sometimes pass through the node unharmed, then flow into the lymph channel that exits it. They enter the bloodstream with the flowing lymph, and from there they are carried through the heart and to structures such as liver, lungs, and bone.

To recapitulate: The great majority of cancer cells are killed in the bloodstream, and especially in its capillaries, by the combination of factors noted above. But if there is enough seeding of the circulation, some clumps and individual cells do get through and then settle safely within the capillary bed of some distant organ. So powerful are the body's defenses that only some one-tenth of 1 percent of circulating tumor cells survive the attacks of mechanical destruction and host defense. And even having negotiated that perilous journey, there is still no certainty that the few hardy survivors will actually implant in the new home they have finally reached. Like all planted seeds, their ultimate success depends on the conditions of soil and climate. If the new organ is hospitable, capillaries grow in to bring nutrients to the struggling implant, and it will begin to grow and multiply.

When a woman with breast cancer is found to have what are called "positive nodes" in her axilla, it is because some of the malignant cells reaching the nodes have found a home there and have begun to multiply. The tiny lymph vessels draining the breast enter larger channels that transport their contents to nodes in the armpit. Although there is some drainage to other groups of nodes, such as those under the breastbone and clavicle, the major portion of the flow is toward the armpit. When the natural defenses cannot kill off the malignant cells that have entered the lymphatic channels, some of them implant in one or several armpit nodes.

Many years ago, it was thought that breast and other cancers spread by growing directly from the site of origin toward the nodes. From this hypothesis it logically seemed to follow that surgery to remove all surrounding tissue, including the node-bearing area, should result in a high proportion of cures; and after such an operation (called radical mastectomy) was introduced in 1883 by William Halsted and others, the cure rates did go up dramatically, but they never reached the expected high percentage. It took many years to recognize that the presence of positive nodes in the axilla could be interpreted to mean no more than that the cancer cells had left the breast and were already elsewhere in the body. In no way did it indicate that the malignancy had not gone beyond the axilla, in addition to implanting there. Because of connections between blood and lymph vessels and also because breast cancer can invade vascular capillaries in addition to lymph channels, the finding of positive nodes in the axilla did not a priori mean the disease had gone no further. It came to be accepted, in fact, that positive nodes were a harbinger of the appearance of the disease in distant tissues.

Once this was recognized, the aim of surgery changed. It was now performed to excise only the original tumor, because it is the source of the shedding of cells. At the same time, the surgeon might also remove enough of the axillary nodes for study under the microscope to determine (albeit in a rather general way) how effectively the cancer cells might be producing metastases elsewhere. Chemotherapy would then be the next step, to help sterilize the blood and lymph streams and to interfere with the growth of any metastases that were forming or to kill them outright. Based on this approach, the principle of excising only the tumor and a rim of surrounding tissue (a procedure that was given the name "lumpectomy") came into being for cancers that were still relatively small. In order to be certain there were no other hidden areas of malignant cells in the remaining breast tissue, it would be treated with X-ray therapy.

When large numbers of women were treated on the basis of this new understanding, it was discovered that it worked very well, providing the tumor did not exceed a size of approximately two inches in diameter. Women whose malignancies were larger than this had better long-term survival rates when they underwent removal of all breast tissue. With this knowledge, it became possible in the 1970s virtually to "tailor" various combinations of surgery, chemotherapy, and X-ray treatment to individual patients. Determining factors in the choice of treatment were tumor size, presence of positive nodes, microscopic appearance of the malignant tissue, and certain hormonal studies that were being developed at that time.

Women with breast cancer have benefited in several major ways from the new insights into the biology of their disease. A strong emphasis on early diagnosis has made it possible to subject far fewer women to mastectomy; the cure rate for breast malignancies has shown a steady improvement; many of those women who have not been cured are living significantly longer and in much better health than ever before.

A dozen years ago, I took care of a woman who lived through the entire scenario that illustrates the role of lymphatic channels and nodes in the evolution of a cancer of the breast. But even beyond that, the story of her disease epitomizes a sequence of events that in its own way is characteristic of cancer's relationship to the body it is trying to destroy.

Even at our very first meeting, it was clear that Sharon Fisher was one of those gentle, soft-spoken women whose easy smile finds its source in a graciousness that reveals the genuine pleasure of believing in basic human goodness. As I got to know her better, I began to suspect that the goodness she saw in others was really her own, reflected back to her from the mirror of their eyes. The source of her optimism was a certain purity of expectation. Others tend to live up to such expectations, and sometimes I really do believe that events have a way of doing exactly the same thing, although not always.

A surgical examining room is a small enclosed space. Empty of its sparse clinical furniture, it contains not much more than cabinets, a wide shelf, and a sink. The focal eminence of the stark little chamber is the narrow padded table on which all of the significant physical and emotional transactions take place.

Suspended over the table like an unblinking cold-eyed Cyclops is the movable surgical lamp, ready with the flick of a plastic switch to throw the intrusive scrutiny of its narrow circle of all-seeing light into places usually hidden by modest propriety. To the probing eyes and fingers of a physician, everything is permitted. But without its duo of dramatis personae, the atmosphere in that place is all sterility and clinical readiness.

Everything changes when a doctor is there with a patient—especially when they are meeting for the first time, and most especially when the patient is a woman with a lump in her breast. From her viewpoint, the real eminence is the authoritative white-coated, stethoscope-emblazoned figure who, even today, is usually a man. Though the worried and self-conscious patient may find difficulty in meeting the physician's gaze directly, she is intensely aware of every movement made by him and finds significance in each uttered phrase—beginning as early as the initial words of greeting. His physical size seems to expand to fulfill some anxious anticipation.

For the doctor, the reverse is true, if he is any good at what he does. It is his patient who fills the room, seeming to occupy all of it with the perceived magnitude of her need and the challenge of her worry. A great deal is unclear in those early moments. The air is heavy with the awareness that decisions are to be made before that place is quitted, decisions that will affect every succeeding meeting between these two people and might vastly change the future for one of them. The doctor, sensible to the effect his professionalized symbolism is having on the new patient, does what he can to appear unthreatening. He tries to make his presence figuratively smaller, in order not to influence unduly the substance of what is transpiring between himself and the woman who has placed her care and perhaps her life in his hands. Those first moments demand of him that he have mastered a subtle substance of the art, as opposed to the science, of medicine. Some achieve that mastery more completely than others. Some, in fact, are wholly ignorant of it.

Sharon Fisher sat perched on the edge of the examining table, waiting for me. She had been given one of those disposable gowns that no one knows exactly how to put on, and now she sat there dressed in it as awkwardly as most. The date was May 2, 1985.

She was worried, she would later tell me, but she felt safe. Hadn't her referring physician told her that his only reason for sending her to a surgeon had been that the thickening in her left breast simply needed to be checked out in order to confirm his own virtual certainty that it was of no consequence? And anyway, she had good reason to be her usual optimistic self—she was in the sixth month of a serene, uncomplicated pregnancy that promised to give her and her husband, Curtis, another child as bright and beautiful as three-year-old Jessica. As soon as this routine consultation was over, she told herself, she and Curt could drive back to their home and resume the peaceful dailiness of their life. Sharon hadn't wanted to come to my office at all that day. "It's almost like I wanted to forget about it. If I didn't see a surgeon and I didn't have a biopsy, why, then, it would just go away—it wouldn't be there."

The events that had brought Sharon to my office were hardly unusual. It was, in fact, a sequence that almost invariably resulted in my being able to reassure a young woman that the new lumpiness in her breast, while a little out of the ordinary, was not sufficiently different from garden-variety pregnancy changes that it warranted further study. Usually, not even a biopsy is needed. On leaving her exercise class about ten days earlier, Sharon had noticed a soreness in the upper portion of her left breast. During her routine obstetric check a week later, she mentioned it to the midwife, who suggested a surgical consultation. When the obstetrician concurred, a call was put in to my office.

I have a long-standing policy that any patient with a breast problem must be seen during my very next office hours, regardless of how busy the schedule may be. There are few events in a woman's life that are as terrifying as the discovery of a lump. No one should have to wait a moment beyond what is absolutely necessary, either for the comfort of a benign diagnosis or for that other kind of comfort that comes with knowing that the process of diagnosis and treatment has been started.

> *The midwife and the doctor didn't think, especially with my being pregnant, that it could be cancer, but they didn't want me to take any chances. Yes, I was nervous and I was frightened, but, being pregnant, I didn't think there was any chance it would be malignant. I was young, and I associated cancer, especially breast cancer, with older people. And also, no one in my family has ever had a malignancy. So even with the nervousness, I expected that everything would be all right.*

And I did too—but only during the first ten minutes. As I took Sharon's history, I was reassured by hearing that she had been diagnosed between her pregnancies with the breast inflammation called cystic mastitis, and that it had caused enough pain to make her obstetrician put her on medication until it cleared. I was also reassured by the same considerations calming Sharon—her pregnancy, her age of only thirty-five, and her benign family history. It is unusual for a painful breast to harbor a cancer. And yet, with all of this, I also bore in mind a bit of advice I had given medical students and residents in a chapter I was just at that time preparing for a textbook on physical diagnosis:

> *For a woman having her breasts examined, there is no possible way to feel calm. Whether she expresses it overtly or somehow man-*

ages the outward appearance of composure, nothing less than a physician's reassurance that all is well can put a halt to the frightened wanderings of her imagination. She has read too much, and perhaps experienced too much, to approach this part of the physical examination with anything resembling a serene mind. Whether or not she is aware of the dreadful statistic that 1 of 10 American women will at some time in their lives be found to have breast cancer, she will certainly know that the disease is very common and that its course can be devastating. At the very least, she will fear surgery, disfigurement, and the discomforts of therapy. Her anxieties may reach much further, to terrifying fantasies that involve the loss of self-esteem, love, and life itself.

It should not surprise the physician that even the most routine of breast examinations will evoke such mental meanderings. Whether it is the yearly checkup, the pre-employment physical, or the scrutiny of a breast surgeon, there is no such thing, from the patient's point of view, as a fear-free ride. This is one of the several reasons there must also be no such thing as a cursory "going over" of the breasts. There is too much at stake ever to relegate this part of the evaluation of any patient to the realm of the superficial or hurried.

Even were a physician to neglect the advice about superficiality in examining Sharon, there was no escaping a certain fullness visible in the upper portion of her left breast, almost amounting to a protrusion. That area, encompassing some three by two inches in its diameter, was tender to the mild pressure of my examining fingers, and more nodular than the remainder of the mammary tissue. It was, in fact, more than firm—it was thick and hard.

But there was another finding, of even more obvious concern. Like the underlying breast tissue, the skin of Sharon's nipple was also hardened and thick. Its upper portion was wrinkled into irregular folds quite different from the fine lacework of softness on the opposite nipple, and the folds extended a bit beyond the edge of the pink-brown areola. The skin in that area looked and felt like orange peel, except that it had the slight reddish blush of inflammation. This is a characteristic type of swelling, and an ominous one. Its usual cause is blockage by cancer cells of the tiny lymph channels draining that portion of the breast. Of this kind of thing, I have written in my textbook chapter, "Sometimes the tumor that wreaks all of this havoc can be felt as a large mass within the tissues, but often the degree of thickness and edema [local swelling] is too great to allow anything but a generalized induration [thickening] to be appreciated."

It is in situations like this that a softened version of the long-practiced clinical poker face comes in handy. Chatting all the while, I finished the examination of Sharon's breasts and armpits and carefully felt her pregnant abdomen and the lymph nodes in her neck to be sure that there was no obvious evidence of tumor growth in those regions. For the moment at least, there was no indication that my patient's problem, as worrisome as it was proving to be, had gone beyond the confines of her breast.

The simplest and most direct next step was a needle biopsy. This is done by injecting a local anesthetic into the skin and underlying fat, inserting a wide-bore needle directly into the mass, and pulling upward on the barrel of the attached syringe. In this way, bits of tissue and fluid can be sucked up, transferred to a glass slide, and later examined under the microscope by a pathologist.

I told Sharon what I planned to do. It was at this point that her confidence in benignity was shaken: "Once the biopsy was done, I still hoped everything would be okay, but I just had the feeling it wasn't. I knew that something wasn't quite right."

Sharon was correct, even to the extent of placing *quite* before *right*. The pathologist read her slides as falling into the category called Class 4: "Probably malignant." But it is Class 5 in which *probably* is replaced by *definitely*. More decisive proof was needed. No surgeon and no patient would ever knowingly begin hazardous cancer treatment without assurance of the infallibility of their diagnosis.

Especially in view of the pregnancy, I decided to do a surgical biopsy. Two days later, Curt brought Sharon to the outpatient operating suite and I removed a wedge-shaped piece of the mass under local anesthesia. The grittiness transmitted to my fingertips as the scalpel cut through the tissue removed all doubt about its malignant nature. The pathologist froze the sample and reported his findings immediately. Not only did the tumor have an aggressive appearance but cancer cells could be seen within the breast's lymph channels, on the way to their next home, in the axilla.

I went out to the recovery room to speak to my patient, who was already dressed. She and Curt sat on two uncomfortable wooden chairs, waiting for me alongside her gurney. She had expected the bad news, and even though I couched it in more hopeful tones than perhaps was justified, its effect on her was devastating.

> *It was terrible to hear it. It's that awful sinking feeling inside. My first thoughts were of my children. Because I was pregnant, I was worried about the baby. I figured that you wouldn't want me to wait—that I'd need to have something done right away. I was at the point in my pregnancy that I didn't want to give up this baby. Where would I go from here? And my experience had been that everyone who had cancer died—that was the orientation I was brought up with. But I don't think I ever believed I would die. I just felt I would get through it. There would be a way. It was awful to hear I'd need a mastectomy, but my focus was always on survival. It was then, when I found out it was malignant, that the crying began, but not before that.*

I told the Fishers that the tumor seemed too large to permit anything less than mastectomy. But I tried to be optimistic about the probable status of the axilla. When repeatedly examining Sharon, I had never been able to feel any hard or enlarged nodes, and that obviously was a good sign. Nevertheless, in some 25 percent of cases, cancerous nodes cannot be felt in spite of careful searching, and I found myself telling that to Sharon and Curt lest they feel devastated if a tumor was afterward discovered there. It didn't make sense to me at this point to bludgeon them with the information that microscopic cancer had been seen in the lymph channels, which meant that in Sharon's situation the probability was far higher than one in four. I wanted my patient to be in as good a frame of mind as possible when she was wheeled into the OR the next day, and it seemed justified to seek temporary refuge in a statistic I was sure of. It seemed justified, that is, for both of us.

Sharon was admitted to Yale–New Haven Hospital early the following morning, and the operation was done a few hours later. "It was nerve-racking that morning. I waited in my room, crying and worried, not only about the cancer but about the baby inside of me."

The breast and all armpit nodes were removed, leaving the muscles of the chest wall intact. The procedure has traditionally been called a modified radical mastectomy, but is nowadays more commonly referred to as a total mastectomy with axillary dissection. When the pathologist examined the tissue, he found that two of the thirty-seven nodes in the surgical specimen contained clumps of cancer that had been carried to them through the lymph channels. Among pregnant women with breast metastasis in their axillary lymph nodes, only 47 percent are alive five years after surgery, and a significant number of those have evidence of cancer that will eventually claim their lives.

I had been a practicing surgeon for more than twenty years on that day when I had to tell Sharon her nodes were positive. In all that time, and since, I have still not found a way to be at ease with such things. I suspect the vast majority of my colleagues feel much the same. We go into medicine with the expectation that we will be helping people, and although we are well aware that the forces of disease will sometimes frustrate us, we nevertheless always carry inside ourselves the belief that our will and our powerful array of skills will somehow carry the day. When faced with a medical problem that with each new diagnostic or therapeutic measure reveals itself to be increasingly further beyond our capabilities, we lose a bit of faith in our faith and find ourselves wondering about the nature of the next as-yet-undiscovered piece of bad news. Somewhere in the middle of this evolving debacle, we must share what we know with the man or woman who has come to us seeking the help we feel increasingly impotent to give. And so when we impart bad news, it is with a mixture of sorrow for the patient and sorrow for ourselves—and frustration, too—that we have not been able to do better. No matter the number of times I have had to tell a patient or a family that things are not as we had hoped, I always feel that I have somehow let them down.

Bad enough at any time to be the bearer of such news, but worse, much worse still, to be giving it to a thirty-five-year-old woman in the sixth month of what had till then been a joyful pregnancy. There are reasons a doctor speaks gently at moments like this—when he is gentle with his patient, he is gentle with himself. It is strangely reassuring to look directly into another's eyes and share the burden. It helps, sometimes, to hold that person's hand, as if more than mere words are passing between two people who are, after all, companions of a sort, on a difficult journey together.

It was very scary to be told the nodes were positive, but there weren't that many, and I still kept thinking to myself that there was still a way to get through this, that people survived it

> *and—well, the fact is that I didn't have time to be really sick—I didn't have time to die. I just decided—you know, I have to get through it.*

What neither Sharon nor I knew at that point was that we had not yet heard the worst of it. When a woman has breast cancer, the tumor is routinely tested for its ability to respond to the female hormones estrogen and progesterone. In a general way, long-term outlook can be correlated with the positivity or negativity of such testing. When Sharon's tumor was studied, there was no evidence of responsiveness to either of the two hormones, another grave finding to add to what was already known. I tried to reassure myself with the knowledge that the negative results might be spurious, since the test can be affected by the high level of hormones circulating during pregnancy.

When the entire breast was studied by the pathologists, the extensive size of the tumor was confirmed. The main mass of it was three inches in diameter, but there was microscopic cancer invading the milk ducts well beyond that. In all parts of the tumor, cancer cells could be seen growing into lymph vessels.

There wasn't much reason for optimism. The large size of the cancer, its growth into lymph vessels, the spread to nodes, the implications of hormone negativity—all were bad signs. The great majority of women with such findings will have an early recurrence, usually within two years. The five-year survival figures of this degree of advanced disease were in the range of 25 percent, much worse than I had thought when I knew only about the positive nodes. In my letter to Sharon's obstetrician, I wrote, "I'm afraid the long-term prognosis is not good, considering the widespread nature of the tumor in the breast itself and its clinical manifestations. Perhaps I will be wrong, which would be a wonderful thing."

But wrong was not what I expected to be. I discussed the need for chemotherapy with Sharon and Curt, then arranged for consultation with an oncologist. He was reluctant to begin treatment until after the baby was delivered, out of a very real fear of injuring it. Even at six months, the baby might be at significant risk, he felt, and he elected to wait until the birth had taken place.

To the determined Fishers, it seemed like a good time for a vacation. They packed up Jessica and went off for a week at a rented cottage in Orleans on Cape Cod. When they returned at the end of June, I was pleased to see that the mastectomy wound was healing well and my patient was in good spirits, ready for the next phase of her treatment. As he had been throughout the ordeal of diagnosis and surgery, Curt remained strong and stoic, like the New England Yankee he is.

> *He was always right there with me through the whole thing. He's not real open about feelings, and I'm sure it was devastating to him too, but he was never anything but calm and dependable. He always said we'd just have to take this one step at a time.*

On August 22, Sharon delivered a healthy eight-pound twelve-ounce baby girl, who was named Victoria Reagan Fisher. The name particularly pleased me because Victoria is my eldest child's name. Like my Victoria, the baby would be called Tori or Toria. Tori was delivered by C-section, as three-year-old Jessica had been. A tubal ligation was done at the same time. During the postdelivery period, Sharon underwent the tests that had been delayed until the baby was safely out of the womb—chest X ray, bone scan, and a mammogram of the right breast, none of which revealed any evidence of new or residual disease. Two weeks later, chemotherapy was begun.

Sharon's oncologist chose to treat her with three drugs: cyclophosphamide, methotrexate, and fluorouracil. This combination had become standardized about ten years earlier when large-scale testing indicated that a six-month course of all three given together resulted in an optimal mix of cancer control and a relatively low incidence of intolerable side effects. Cyclophosphamide is one of a class of compounds called alkylating agents, which kill cancer cells by altering the information coded in their DNA. Because this is a kind of damage difficult for the cell to repair, cyclophosphamide is usually a very effective drug.

Methotrexate and fluorouracil are members of a class called antimetabolites, because they have a chemical structure similar to certain molecules called metabolites, which are intermediaries along the sequence of biochemical steps that leads to the creation of new DNA. This allows these drugs to masquerade, in a sense, as the real thing. By substituting for or competing with one of the needed metabolites, methotrexate and fluorouracil interfere with synthesis of DNA and prevent cancer cells from reproducing.

The chemotherapy was nasty—that's the best word for it. Actually, I didn't have as awful a time as a lot of people do. The oncologist told me I'd have to be prepared to lose my hair and be really sick. I remember saying to him that I wasn't going to go bald. My hair did thin out quite a bit, but I never lost all of it.

I was very nauseated all the time. I'd come home, crawl into bed, and sleep until the worst was over, and then I just got up and went about my usual things. There were times I was very tired, but the nausea wasn't any worse than the morning sickness I had with the pregnancies. I never actually vomited until the very last treatment—I think I must have been giving myself permission to be sick because this was it—it was over.

Sharon completed her chemotherapy in mid-February of 1986. She and her oncologist and I settled in for the long haul. He and I alternated seeing her several times a year to be certain there was neither recurrence nor new tumor, and every visit increased her determination to keep her focus always on survival. She had made that decision as soon as she learned her diagnosis, and neither she nor Curt had any intention of wavering from it.

I didn't begin to have any real sense of a good long-term outlook until Sharon passed the two-year mark without evidence of trouble. Following that anniversary, I began for the first time to believe that the unexpected "wonderful thing" might actually happen. But a year later, my small store of new optimism was threatened when Sharon appeared in my office with a tender nodular area in precisely the same position in her remaining breast as her cancer had been in the other breast. The skin over it appeared perfectly normal, and it had a more rubbery and less hard feel than its mirror image had presented. Were it not for her history, I would have felt certain it was a small focus of mastitis. I suggested that I see it again in three weeks, a few days after Sharon had finished her menstrual period.

On the next appointment, nothing had changed. Again, there was not a thing about the lumpiness that seemed worrisome—except that it had appeared in the breast of a woman who three years earlier had been found to have an extensive cancer. I decided to wait another two weeks.

When Sharon came back, the area of concern was exactly as it had been when I first saw it. I began to see my strategy of close observation as more procrastination than wise, watchful waiting, and I decided to recommend biopsy.

Almost three years to the day after Sharon and Curt had been given the bad news of malignancy, they returned to the same outpatient operating room as before, and I once again excised a sample of breast tissue. This time, the lump was small enough to enable me to remove it completely. Within thirty minutes, the tissue had been frozen and studied under the microscope. It contained no cancer—it was just an area of cystic mastitis, the perfectly benign inflammation Sharon had experienced in the past.

Twelve years have passed since the spring afternoon when Sharon and Curt Fisher sat with me in the consulting room of my office, discussing the possibilities and options that lay before us. As direct as I had been on that day, I was trying to project a confidence about treatment that was somewhat beyond what I really felt. I was careful to address every question put to me with absolute candor, but when the answers conveyed less than optimism, I gentled my words with the only certainty I have ever learned in a long clinical career in medicine: A disease presents itself in a unique form in each person it attacks, and it pursues a unique course. A good or a bad prognosis is a statement of probability, not of fact. Every individual responds to treatment differently than every other. Statistics and a physician's experience are useful only as general guidelines. Whether the numbers appear favorable or foul, the outcome of your or my sickness is ultimately decided by unknowable factors within us. They will probably remain unknowable for decades or perhaps even centuries to come, if not forever. The objectifications of biomedical science lead us ever closer to the real sources of healing, but long distances are still to be traveled.

Sharon had understood these things intuitively, and my explanation of them only confirmed what she already knew. But even at that early stage, she had begun to think about her own role in her recovery and survival. Recently, I asked her whether she had called on anything else, specifically those alternative methods of healing that have become so popular in recent years.

> *The most important thing is that you follow the medical protocol. But also, I think we find inner strength. You hear people say, "I don't know what I'd ever do if I found out I had cancer, or if some*

other awful thing happened to me," but I think that the strength is there—you just have to call on it when the time comes.

What I mean is that if you don't keep yourself positive, there's a tendency to give up, and these other, nonmedical things can help you that way. I'm talking about things like meditation and imaging and the philosophies of the gurus who preach cure by thinking good thoughts, and even the use of crystals, if that's the kind of thing you have confidence in.

I was having chemo at the same time as the mother of one of Jessica's friends. She had been diagnosed with Hodgkin's. Now, that's a disease people are supposed to do well with, most of the time. But she had simply made up her mind that she wasn't going to make it—and she didn't. She was doing well, but she needed to go into the hospital because she was feeling sick from the chemo. She was scheduled to come home, and when I called to see about going up to visit her, I was told she'd had a heart attack and died. Right from the outset, I think, she made up her mind that it was cancer and she wasn't going to survive it. So she spent a lot of time in bed with the covers pulled up over her head. I always felt you're a lot better off if you just keep on going, day by day. I've known plenty of people who've had cancer or some other awful thing happen to them. They seem to do very well as long as they keep positive, even if it doesn't actually give them any longevity.

I'm not overly religious, but I do think there's something—a Higher Being—and we all need a belief to hang on to. I was brought up a Baptist, but I got away from it, although I kept reading the Bible and tried to live my life by its teachings. We go to a Congregational church these days, but we had no affiliation during the time I was having the cancer treatment.

Giving up is a form of suicide. I don't believe we have the right to do that. There are others who need us, whether children or extended family. Maybe each of us has a purpose on this earth. Maybe we don't have a right to give up. It's a form of selfishness to decide you can't go on. I had days during the chemo when I thought, This is too awful; I won't go back for more treatment. But then I'd realize, I have to do this to survive—I need to live. Part of what got me through it was not only that I was responsible

> *for other people but also that I had another being growing in my body. I couldn't give up, not only for me and those who were already here, but for her.*

Sharon was speaking about that of the human spirit which sustained her. But it would be a mistake to rely on it as necessarily a factor in recovery from sickness. There are those who don't give up and yet die anyway—just as there are those who do give up and yet live in spite of their "wrong attitude." We tend to remember those whose outcome reflected their state of mind and to forget the others. Sharon tells the story of the young woman who paid the price of pessimism, and she puts great store by such things. She is certainly right about optimism's effect on the quality of life during whatever time remains, and she may also be right about its not infrequent effect on longevity. My clinical observation leaves me with conflicting beliefs. For every Sharon Fisher, I have known a dead optimist.

4 | SYMPATHY AND THE NERVOUS SYSTEM

At the end of the first chapter of this book, I left myself sitting on Sarah's side of our bed, relating what I called "the saga of that night." The choice of the word *saga* to describe the circumstances of Marge Hansen's rescue was deliberate. It can be traced to an Indo-European root meaning "a remark made with divine vision." The etymology of *divine*, in turn, goes all the way back to *dei*, meaning "the bright sun," or "the clear sky." To those who spoke the ur-language more than five thousand years ago, a saga was a set of remarks or a story inspired by the brightness of the sun and skies, an aspect of nature. In the evolution that produced the word *divine*, what was in those early days understood to be *natural* has been transformed until it took on the aura of the *super*natural. In this single word, *saga*, there may be a lesson for all of us who seek a unity that will encompass our differing visions of the origins of the human spirit. To some, it is God-given; to others, it is a product of nature.

I see Marge's as a story that comes from nature. What made her survival possible was the result of a series of biological mechanisms within her and within the members of her surgical team that burst forth with startling precision when called upon. These are natural phenomena. Some, like those due to adrenaline and endorphins, were so directly biochemical that they could have been measured had anyone had the inclination to take blood samples and analyze them. Others, given our present incomplete state of scientific knowledge, were not yet susceptible to the kinds of studies that can be done in laboratories. There was a time not so long ago when the human responses we finally learned to attribute to hormones and enzymes were thought to be due to the personalized moment-to-moment attention of God or the gods. Most of us are inclined to look at these phenomena differently nowadays, and many look differently on things of the spirit, too. The fact that they are proving to be explainable on the basis of science will not make them less divine.

As late as the middle of the nineteenth century, many authorities, and virtually all plain people too, believed that living things were possessed of an unknowable form of energy that made them vastly different from structures not endowed with life. Because this hypothetical energy was commonly called "the vital force," the scientists and other learned people who trusted in its existence were known as "vitalists." The concept of vitalism stood independent of religious conviction—there was nothing about it that necessarily required a supernatural or theological explanation. Although some vitalists, starting as far back as Aristotle in the fourth century B.C., identified the life force in humans with the psyche (which he believed, incidentally, to originate in the heart), many others thought that it had no connection to mind or soul. Obviously, though, there were those of a less scientific bent who embraced vitalism because of their certainty that the unknown and unknowable factor was God-given.

Many of the secular vitalists were persuaded that the origin of the life force was not explainable by the usual principles of physics or chemistry, but some few of the staunchest proponents of the theory maintained, on the other hand, that it would eventually be shown to be regulated by as-yet-undiscovered natural laws. They reasoned that if this proved to be the case, it might then become possible to carry out laboratory studies of this unique form of energy.

The belief in vitalism, at least among scientists, had lost favor by the latter half of the nineteenth century as biochemical and physiological elucidations of the characteristics of living things increasingly emerged from the burgeoning numbers of research laboratories making their appearance in Europe and, to a lesser extent, in North America. The so-called mechanists, those who, in contrast to the vitalists, sought physicochemical explanations of life processes, had won the battle and convinced all but a few diehards of their correctness.

And yet the general notion of vitalism, attenuated though it may be, lives on in the minds of any and all who refuse to believe that there is not some form of still-unexplained energy that brings more to the phenomenon of life than can be accounted for by a series of chemical reactions. From the strictly philosophic point of view, in fact, the general proposition of vitalism, no matter the by-now-huge mass of experimental evidence against it, is, by the very nature of the nonphysical insubstantiality that some of its adherents claimed for it, as ultimately irrefutable as is the existence of God.

The company of those who have not yet given up on some form of vitalism extends across the entire ideological spectrum, from some who are of unquestioning religious faith all the way to the most determined of unbelievers. On that distant end are to be found some individuals of such a highly developed quality of skepticism that they even question their own skeptical convictions. They leave open the possibility that living things *do* possess some nonmaterial quality not found in inanimate objects. They are the true skeptics, those who are prepared to accept that anything is possible, no matter the sheer volume of observations arrayed in opposition.

Ranged against the principle of vitalism, however, is a quantity of evidence so overwhelming that only a rare twentieth-century scientist has ever questioned it. To contemporary science, the physicochemical, or mechanistic, view of life has become axiomatic. Whether in single-celled organisms or in us, the characteristics of the condition we call life can be shown to result from quite verifiable natural processes, many of which have already been demonstrated by the techniques of the laboratory. Even stubborn latter-day vitalists cannot question that at the very least the mechanistic bases of life become even more evident with each passing decade.

The entire thrust of the scientific mind has been to free itself of any need to explain the workings of nature other than by chemical and physical interactions. There is no clearer description of the approach by which researchers attempt to wrest nature's secrets from her than the one offered by William Harvey, the seventeenth-century English physician who discovered the circulation of the blood. Harvey wrote of his certainty that the mechanistic answers sought by an inquiring mind were to be found by the process of close observation, devoid of prejudgment: "Nature herself must be our advisor," he wrote in 1651. "The path she chalks must be our walk. For as long as we confer with our own eyes, and make our ascent from lesser things to higher, we shall be at length received into her closet secrets." The simple phrase "from lesser things to higher," written almost three and a half centuries ago, is as direct a definition of the principle of inductive reasoning, the logic pattern of scientific research, as has ever been expressed.

Nevertheless, I do not hesitate to propose that man is in some as-yet-undiscovered way more than the sum of his biological parts, that a thing greater than the innate has somehow been crafted from the innate—that we have taken nature's endowment and made with it the stuff of spirit and all that is implied by my use of that word. How indeed, it might justifiably be asked, can such a formulation escape the charge of being just another form of vitalism?

Actually, vitalism has nothing to do with the case. My formulation does not require the existence of any "energy"—nor any substance, either—beyond what is already well known from the study of physics or chemistry. I propose a human body each of whose constituents follows, as do all living things, the well-researched principles of biophysical systems that are amenable to study by standard scientific methods. I am not postulating a uniqueness within human tissue that presupposes the need to discover new natural laws to understand it. The human spirit of which I speak depends for its elucidation on principles already explicated. In the battle of vitalists versus mechanists, I stand squarely in the center of the mechanist camp. My concept does not involve a "nonmaterial" source.

The quality I call spirit is in its very essence the product of the organization and integration of the multiplicity of physical and chemical phenomena that is us. It has to do with the way the various parts of the human body communicate and are coordinated with one another under the control of the evolved masterwork that is the human brain. That such a complex multicellular organism functions with a unity of purpose is the result of a myriad of messages of various sorts, and of integrative capacities that in themselves originate in *nature's* purpose, which is merely to keep the organism from dying, at least until its reproductive capacity is spent.

Reason, conscience, and morality cannot be studied in the same way as can the speed of an impulse along a nerve fiber or the quantity of adrenaline necessary to constrict an arteriole—they are immeasurable qualities. But still, an understanding of such abstract and ultimately indefinable factors in our lives depends on familiarity with their biological underpinnings—for they do rest on a biological foundation. They are examples of the ways in which our inherent protoplasmic workings have been fortunately exploited by us to develop a philosophical edifice that far transcends the requirements of survival and reproduction—achieving a magnitude unconfined by the restrictions of instinct alone or the mere necessities of ongoing cellular life.

If I am at all close to correct in claiming that the human spirit is the developed product of the potentialities inherent in human biology, then the key to exploring it is to begin by learning as much as possible of what is at present known about the physical aspects of our bodies, from atoms to organs.

To those who believe that biology is destiny, each of us is neither more nor less than the stuff of his cells. To such convinced reductionists, I would answer that biology has given us our ingredients, but we have given biology the glory of using its gifts to cook up a miracle. We have found, or perhaps made, potentialities that could have led us along all manner of misdirected pathways, or even nowhere. By bountiful experiment with the raw materials of survival, we have invented humankind.

In the gradual process of concocting the ingredients that are biology into the cassoulet that is man, the sauces of instinct have simmered into the motivations and yearnings that give flavor to human life. As our ever-exploring species has by trial and error responded over eons to that fluxing of the world around us and the world within us, we have slowly—imperceptibly—created the phenomenon that is today's civilization.

Man is to be understood only if his biology is understood. Organic structure and function are the essential starting points for any exploration of humanity and spirit. Conjecture about the human condition can have no validity unless it begins with observable, reproducible evidence that can be validated by repeated challenges to its accuracy. And so if we are to attempt to elucidate the elusive quality that is the human spirit, we should launch our efforts from the stable foundation that is the study of nature. In this, we take up the work of Aristotle, and perhaps with the same perspective. Quite early in his great biological opus *Parts of Animals*, Aristotle did, after all, articulate in words that by now have a familiar ring his fascination with observing the phenomena of life: "In all things of nature, there is something of the marvellous."

By letting nature be our adviser, as William Harvey advised—by researchers depending on the evidence of their senses—it is possible to identify the qualities that are characteristic of living things:

RESPIRATION
CIRCULATION
RESPONSIVENESS
ADAPTABILITY
DIGESTION
ABSORPTION
ASSIMILATION
EXCRETION
MOVEMENT
GROWTH
REPRODUCTION

This is not an attempt to list the characteristics in order of importance, because every one of them is absolutely necessary if the organism is to continue to live. (An exception must be made for circulation, which does not take place in the ordinary sense within single-celled organisms.) The order I have chosen, in fact, is constructed specifically from the viewpoint of how the listing would be prioritized for a human being were an emergency attempt being made to preserve life in a case of respiratory or cardiac arrest. When that kind of urgency arises, first consideration must be given to maintaining an airway for the entry of oxygen and then to restoring circulation of the blood. Such other life-sustaining functions as digestion, absorption, assimilation, excretion, and movement can come later; ultimately, a human cannot survive unless all of them are intact.

Growth and reproduction, the other qualities inherent in living things, are not absolutely necessary to prevent death of the individual, but they are crucial to perpetuating its cells.

Although the description that follows will pertain generally to characteristics of all life, I shall restrict specifics to you and me, vertebrates of the class Mammalia, order Primates, family Hominidae, genus *Homo*, species *sapiens*.

Respiration refers to a sequence of events by which an organism obtains oxygen from its surroundings, uses it, and rids itself of the carbon dioxide produced as a byproduct of the process.

The purpose of *circulation* within a multicellular organism like us is to carry oxygen, nutrients, and various cellular and chemically active substances to tissues and individual cells, and to remove the waste products of cellular activity. In humans, as in all other vertebrates, circulation is propelled by the pumping action of the heartbeat.

Responsiveness, which in earlier generations was called *irritability*, refers to that quality of living things that enables them to appreciate changes in their environment in such a way that they react appropriately. A foreign body imbedded in the skin, for example, produces the pain, redness, local heat, and swelling characteristic of the body's means of responding to intruders by mounting an inflammatory reaction whose purpose is to isolate the foreign material and destroy or extrude it. This is a simple example of an enormous range of the propensity of living things to sense changing circumstances, merging into their *adaptability*, which comes into play in sustaining life and protecting the DNA so well that further generations are produced. Among the adaptabilities is the much more complex gift of using our nurture to influence our nature. The existence of responsiveness and adaptation gives the lie to those who focus on inherited characteristics as the sole or necessarily the predominant influence on behavior and health, not to mention such a cluster of qualities as we package together under the amorphous heading of intelligence.

Digestion, *absorption*, and *assimilation* are closely related to one another. In humans, digestion converts food to simpler substances that can be absorbed into the thin-walled blood vessels and certain other channels running within the layers of tissue lining the stomach and intestine. The circulation next carries those substances to the organs of the entire body for assimilation into molecular structures usable in cellular functioning. Waste products of such functioning must then be returned into the circulation and brought to the kidneys for *excretion*. Some cellular by-products or noxious circulating chemicals are dealt with in other ways, one example of which is detoxification by the liver.

Human *movement* refers not only to such grossly apparent actions as walking or reaching for a drink but also to the spontaneous internal movements required for ongoing life, such as the heartbeat and the undulating wavelike motion of the gut, called peristalsis.

To increase body size is to exhibit *growth*, but growth of less than the total body can take place in individual organs or tissues as required by the need to respond to changed circumstances. The liver, for example, is capable of regenerating great pieces of itself when enough of its substance has been lost to injury or sickness; the lymph tissue enlarges to combat either local or generalized infection.

Reproduction, although not necessary to sustain my life or yours, is a potentiality of all living things, and is crucial to the survival of a species. After self-preservation, protection of the DNA for the next generation is the organism's highest priority. The ability to reproduce is dependent on efficient functioning of every one of the other life characteristics, but there is not one of them that cannot proceed perfectly well even should the individual as a whole never produce progeny. And even here there is a *but*. If your cells or mine should suddenly lose the ability to reproduce by division into two offspring, we would soon die with them. All growth and all healing depend on the ability of cells to divide and thereby reproduce themselves. And so, at the microscopic level at least, reproduction is obligatory even for us as individuals. As noted in an earlier chapter, some biologists put the matter in very blunt terms: The only purpose of any organism is to pass its DNA on to the next generation—that in itself is the function of all living things. In this sense, then, reproduction of multicellular organisms like us is of the lowest priority in preserving our individual lives, but of the highest in preserving our kind.

As much as this list provides a perfectly adequate accounting of all the biological functions that characterize living things, it should not be mistaken for an attempt to define such an abstruse and immensely complex word as *life*. Even the most detailed mechanistic descriptions of our nuts-and-bolts physicochemistry are missing something, and that something is immense. More than one scientist has expressed frustration with how little is even now known about the *organization* of our biological faculties, functioning with such an integrated degree of coordination that we are capable of what would seem to be mental and physical miracles, not to say spiritual ones, which would appear to transcend the mere interaction of molecules.

Otto Loewi was such a scientist. Loewi was the physiologist who first demonstrated that impulses pass across the gap between the fibers of one nerve cell and the next by means of chemical substances, rather than via an electrical wave, as had been previously supposed. Loewi's friend Henry Dale later showed that acetylcholine is the chemical that enables the crossing of the open junction between a nerve ending and the muscle it activates. By the work of Loewi and Dale—who shared a Nobel Prize for their contribution, in 1936—the method of conveyance of messages by chemicals (which later came to be called neurotransmitters) was finally clarified. There could not have been a more mechanistic laboratory-produced explanation of a phenomenon that for millennia had been presumed a miracle of inscrutable mystery.

But even in the face of his discoveries, something in Loewi refused to give up his fascination with that mystery. He never lost his sparkling ability to wonder what else there might be. As a scientist he devoted his life to proving that the totality was no more than the sum of its physicochemical parts, but there were playful elves in his soul, and once in a while he would smilingly say the kind of thing reported by his nephew: "The beauty of the Budapest String Quartet can never be explained by a little acetylcholine in the nerves and muscles." After his death in 1961 at the age of eighty-eight, Loewi was quoted by his nephew as having believed that "the life sciences contain spiritual values which can never be explained by the materialistic attitude of present-day science."

Not all science, of course, is materialistic—and very few scientists, in case you have wondered, are cynical about the sources of life. Not infrequently, the selfsame researchers whose career-long work it is to seek out materialistic or mechanistic answers are precisely those most urgently questing after the secret sources of the mysteries; the very wonderment that sustains the ceaseless curiosity of so many researchers is precisely their wonderment at the marvels of nature. Otto Loewi has never been alone in his awe. To search for a biochemical or physical basis for those marvels is not to lessen that awe. In some ways, the awe is greater, in fact, when it can be shown that no magic or vital force is needed to elucidate one or another of nature's closet secrets. That enormously complex biological interactions are so flawlessly coordinated as to result in such obvious manifestations as human thought or the electrical activity that drives the heartbeat is as exciting to me—actually more exciting—than such phenomena were when I was a small boy and thought them divinely (in the supernatural sense) driven. And the knowledge that even the coordination is directed by factors increasingly accessible to experiment—well, I revel in it.

What I find most exhilarating is not even the freedom from invoking magic that modern science provides. As paradoxical as it may at first sound coming out of the mouth of a skeptic like me, the ultimate exhilaration derives from my conviction that the whole *is* greater than the sum of its parts. The physicochemical and genetic may have provided us the basic ability to integrate our mental and somatic functioning, and they may also have given us the inborn ability to adapt, but the very way in which we have made use of that adaptability is the real secret of how our species has transcended what might have been the limitations of mere survival. By going beyond the basic needs fulfilled by our molecular function, we have never ceased making use of our unique biology in constantly expanding ways.

The anatomy of the intricate circuitry in our brains and the balance of hormonal controls in our physiology can take us only so far, although they do provide the physicochemical means to go further. Millennium by millennium, *Homo sapiens* has, I believe, built up patterns of anticipatory thought and evolved a culture that could not have been predicted by relying only on knowledge of our anatomy and physiology. It is necessary to take other factors into account.

By the way we have gradually brought ourselves to utilize our inborn and developed neurological connections, chemical messengers, and cerebral centers, we piecemeal over eons discovered and formed the pathways and linkages within us that produced the qualities of abstract thinking which are the hallmark of our species. The achievement of this discovery has been us—gradually emerging, continually adapting, ceaselessly striving—as we journey from *Homo* to humanity.

To a degree beyond the abilities of any other animal, we have put our enhanced cerebral capacities into the service of coordinating our responses to the outside world. By increasing our level of independence from the dangers of the environment, we have also heightened our opportunity to enjoy the fruits of that independence. Because it is not necessary to focus constantly on mere survival, our species is free to turn our attention to the quality of life and the development of personal relationships beyond those required for mutual safety.

One of the ways in which we have created some of the qualities we call human has been to make use of certain structures and capabilities that we share with all vertebrates and many lower animals. We use these structures and capabilities to serve as go-betweens, as it were, by which our conscious thought may sometimes interact with and affect some of our automatic internal functions (such as circulation and digestion, for example) that would seem to be beyond the control of will. The automatic, inward-acting entity is a separate and distinct subdivision of the nervous system—called the autonomic nervous system, consisting of a widespread web of nerve cells, circuitry, and chemicals whose function is to preserve the constancy within us. While watching over the inside, it keeps careful track of what is happening on the outside and responds to it. The autonomic nervous system bridges the gap between conscious knowledge of our actions and control over them on the one hand and cellular activity on the other. It is doing this while all other parts of the nervous system act in relation to the outer world that surrounds us.

The autonomic nervous system existed among lower animal forms long before conscious control existed—it is the oldest and most primitive part of the nervous system's structure. Only much later in evolution did animals develop any elaborate sense of real awareness of their surroundings, as well as the ability to change their behavior voluntarily and knowingly based on comprehended input from the outside world. The worm, for example, has no brain or spinal cord. Brain, cord, and nerves appear first in vertebrates; animals below that level do not have them.

Among vertebrates (like us), the totality of the nervous system is divided into what are known as its *central* and *peripheral* regions. The central nervous system consists of the brain and spinal cord, while the peripheral nervous system consists of the nerves that carry messages into and out of the brain and cord, traveling to and from the internal tissues and periphery of the body—the autonomic is one of its two subdivisions and the so-called somatic (*soma* is Greek for "body") is the other.

The peripheral system—both autonomic and somatic—has receiving structures that pick up entering messages from inside the body and from its surface, which it then transmits along its nerves to the brain and cord. These receiving structures are located in the skin, muscle, and the various internal organs and are attuned to events happening in both the outside world and the inner world of the body's constant functioning. When the brain and cord have processed the information thus brought to them, they send their responses back to the body via outgoing nerves of the peripheral system.

To reiterate: The peripheral nervous system has two parts, of which the autonomic is one and the somatic—carrying messages about sensations and movements of the head, trunk, and limbs—is the other. It is the incoming messages of the somatic system that originate in receiving structures in the skin, tendons, and voluntary muscles; it is the incoming messages of the autonomic system that originate in areas of involuntary function such as the blood vessels, glands, and internal organs.

The relationships of the various parts of the nervous system may be diagrammed as follows:

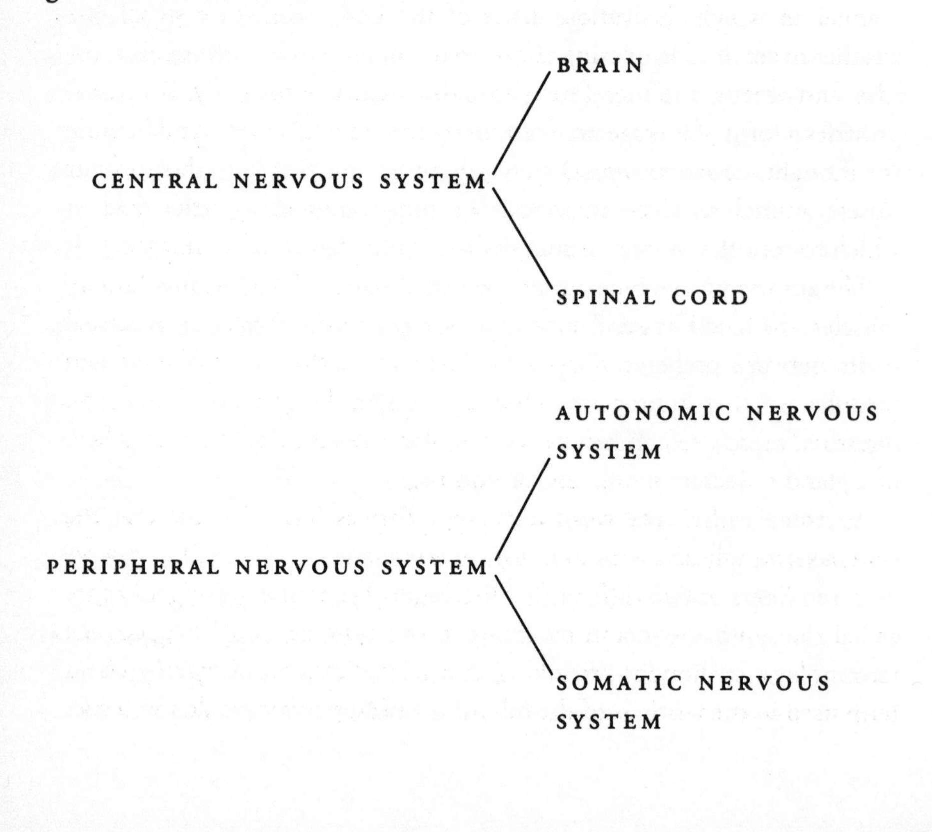

The autonomic mechanisms work in coordination with the rest of the nervous system's structures. In fact, parts of it travel along with or are anatomically within them. Not only that, but the autonomic and somatic systems use some of the same receptor structures in various tissues and they also share certain processing centers in the central nervous system. The autonomic activity is controlled by centers in the brain stem, particularly the two called the hypothalamus and medulla. As will be described in more detail below, the hypothalamus exerts control primarily over the responses that require sudden action, and it also integrates autonomic responses with those of other brain areas, including those involving consciousness and the emotions. The medulla deals more with routine (so-called vegetative) matters, such as digestion, respiration, and normal circulatory dynamics.

Accordingly, although autonomic activity is in most ways distinct from the rest of the nervous system, it dovetails with the entire spectrum between an ability as cerebral as deliberate thought or emotion and an ability as mindlessly chemical as molecular interactions inside of cells. In all higher animals, the autonomic nervous system is one of the keys to the manner in which the various parts of the body constantly signal one another to act in coordinating efforts to maintain stability within cells, tissues, and organs, and therefore within the totality of the body. Because it provides a form of linkage between thoughts and molecules—and because the thoughts of *Homo sapiens* are vastly more complex than those of our closest animal relative—its wondrous functioning is a fertile field in which to seek the sources of some of the extras that make us human.

The autonomic nervous system governs the activities of the involuntary muscles, the blood vessels, and the glands. Sensory information is carried to its network of nerve fibers and cells and to the hypothalamus and medulla, so that appropriate instructions may be returned, telling an intestinal muscle to contract or relax, a blood vessel to constrict or dilate, or a gland to secrete more, less, or nothing.

As noted earlier, the word *autonomic* derives from the fact that the system seemingly acts autonomously of conscious will. It consists of nerve cells and fibers of two differently functioning types, comprising what are called the sympathetic and parasympathetic systems, and here, too, the terminology is thought-provoking in its implications. *Sympathy* was a term used in the writings of the followers of Hippocrates to denote a special accord between various organs, in the sense of one responding or demonstrating some abnormality when the other is injured. The uterus and the breast were thought to have such a relationship, as though they were some kind of protoplasmic Corsican Brothers, or sisters. Over the centuries, the term came to be understood to refer to situations in which organs were affected by events or treatments occurring elsewhere in the body. When some of the anatomical parts of the autonomic system were identified in the seventeenth century, and then later as more and more of its activity was elucidated, the terms *sympathetic* and *parasympathetic* were applied by physicians influenced by classical thought, because these nerves quite obviously were responsible for some of the effects that for more than two millennia had been called sympathy.

The two parts of the autonomic nervous system balance each other's effects in the interest of maintaining the body's homeostasis. The parasympathetic system keeps things slow and steady; in general, it takes care of the regular housekeeping functions of humdrum life, like normal intestinal movements and the dependable beat of the heart. In general, these activities are governed by centers in the medulla. The sympathetic system, on the other hand, is the one designed to leap into action when certain urgent demands appear, as in excitement. It tends to inhibit the general upkeep function of the parasympathetic system while it takes over to respond to sudden dangerous or exhilarating stimuli. The term *fight or flight* has been used to encompass the range of responses of sympathetic activity. Centers in the hypothalamus govern it.

As a rule, a structure will receive nerve fibers from both systems, but often in greater abundance from one than from the other, to the point where one or the other is dominant. Different parts or functions of an organ may be dominated by fibers of either kind. An example of this occurs in the intestine, whose blood vessels are sympathetically controlled but whose peristaltic movements are under the domination of the parasympathetic system.

A few illustrations will suffice to demonstrate how the two systems do their balancing. Sympathetic stimulation dilates the pupil, and parasympathetic contracts it; sympathetic stimulation decreases intestinal peristalsis, and parasympathetic increases it; sympathetic stimulation decreases secretion of the salivary glands, and parasympathetic increases it. In a few situations, the two systems act to accomplish some common end result, as when erection, which is due to parasympathetic stimulation, is followed by ejaculation, which is sympathetic. Both erection and ejaculation, incidentally, are examples of functions that are hardly (pun unavoidable) as autonomous as they are in other animals.

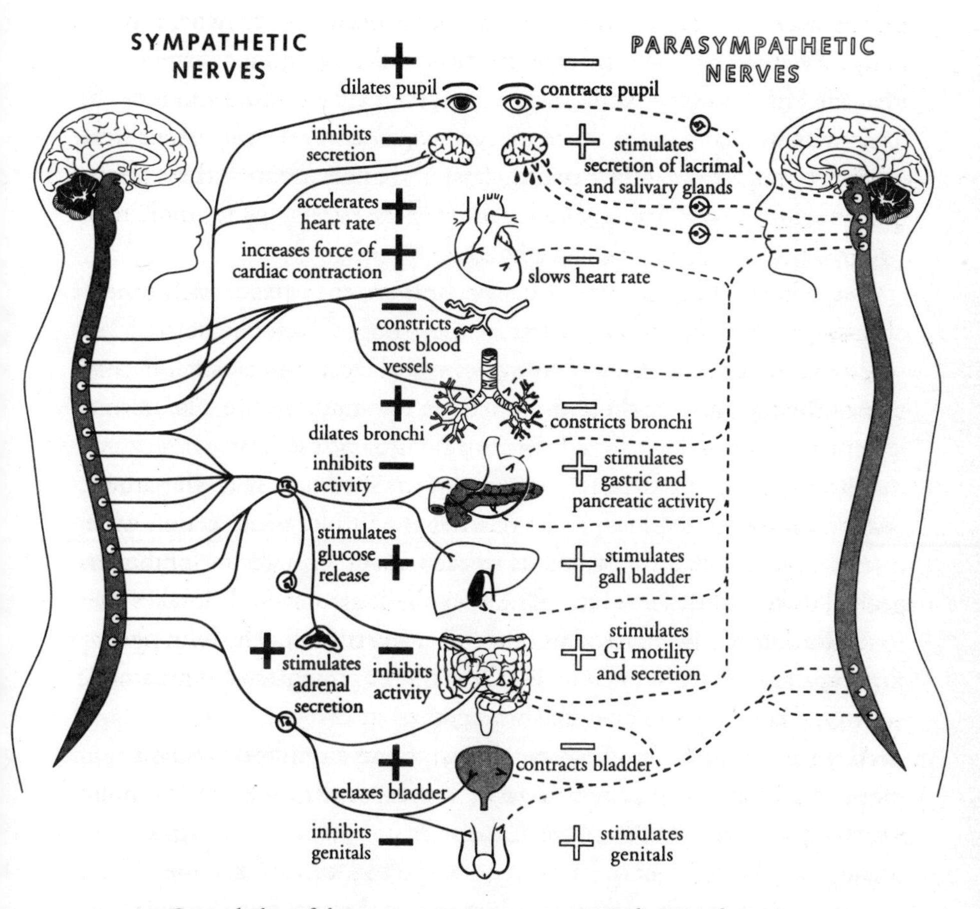

General plan of the autonomic nervous system, indicating the actions governed by sympathetic and parasympathetic impulses.

Because it can be confusing to remember all of the actions controlled by each of the two systems, it is best simply to keep in mind that in general the effect of the sympathetic system is to prepare us for urgent or dangerous situations. It constricts blood vessels, raises the speed and force of the contraction of the heart, widens air passages in the lungs, releases glucose into the bloodstream for quick energy availability, dilates the pupil, presides over the shifting of blood from the intestinal tract to the heart and voluntary muscle—all of these are needed in an emergency, for confrontation, action, or escape. The parasympathetic system, on the other hand, superintends functions of a more long-haul nature, the so-called vegetative functions. It slows the heart, increases peristalsis and the tone of the muscle in the intestinal wall, stimulates secretion of saliva and gastric juice, and contracts the bladder to permit urination. It can be seen that the two systems work together to assure the maintenance of the body's stability and the safety of our lives. In one way or another, whether together or separately, they assure that we can respond to whatever exists in our environment that is inimical to life, whether it appears suddenly or is a condition that must be constantly dealt with.

There is one more notable difference between the two systems. In order to achieve their effects on the tissues they supply, they liberate different neurotransmitters at their nerve endings. The parasympathetic nerve fibers liberate acetylcholine, the same neurotransmitter we have already encountered at the nerve endings to voluntary muscle. With a few exceptions (like fibers to sweat glands, to blood vessels, and to the adrenal gland), the sympathetic neurotransmitter is a variant of adrenaline called noradrenaline. Like adrenaline, noradrenaline is also secreted by the central portion of the adrenal gland.

So the adrenal gland secretes into the bloodstream not only the neurotransmitter noradrenaline but also adrenaline, a closely related compound with similar effects. It does this when stimulated by acetylcholine liberated (in an exception to the general rule) at the terminals of sympathetic fibers that enter the gland. Clearly, a sympathetic, or fight-or-flight, response has plenty of backup. Clearly also, there is considerable dovetailing and crossover between the endocrine and nervous systems.

There is an anatomic relationship between the autonomic and the rest of the nervous system. The sympathetic fibers travel with the somatic nerves leaving the spinal cord at the level of the chest and abdomen, the so-called spinal nerves of the somatic nervous system. The parasympathetic fibers travel with several nerves (called cranial nerves) that come down from the brain stem, as well as some somatic nerves leaving the spinal cord at the level of the lowest part of the bony vertebral column, called the sacrum. The most important parasympathetic nerve is a large trunk called the vagus (in Latin, *vagus* means "wandering"; the nerve is so called because it comes down through the neck from the medulla and enters the chest and then the abdomen, where its ramifications wander everywhere), which carries parasympathetic fibers to the heart, lungs, and digestive tract.

In considering just how it is that autonomic nerves exert their effects on tissues and organs, it must be appreciated that everything that happens in any structure of the body is the result of the interactions of molecules within the cells themselves. Cells of the affected tissues carry an assortment of so-called receptors (not to be confused with the receiving structures that pick up stimuli in the nervous system, described on page 90) in their enveloping membranes, responsive to acetylcholine, to noradrenaline, or to adrenaline. Receptors are highly specialized protein molecules; they combine with the neurotransmitter molecule either to excite or to inhibit some action of the cell. The interaction between noradrenaline, for example, and certain noradrenaline receptors in the membrane of a muscle cell in the wall of an arteriole (except if the vessel supplies a voluntary muscle) is what sets up a chain of chemical reactions that excite the muscle to contract. Its contraction, in combination with the contraction of all its nearby kin by the same mechanism, results in constriction of the arteriole and a consequent decrease in blood flow to the tissue it supplies.

The foregoing several pages have not been particularly easy to follow, but perhaps matters can be clarified by an example of the kind of thing that happens in real life.

As I thoughtlessly step off a curb, I am frightened by the screeching tires of a car whose brakes have just been slammed down to avoid hitting me. My sudden perception of the danger initiates a series of events. The sensory impulses from my ears are carried by the auditory nerve to a hearing center in the lower part of my brain, the so-called brain stem. Instantaneously processed in the brain stem's hearing center, the impulses then find access to my hypothalamus as part of a general fear reaction, of which the thinking part of my brain is instantly made aware. The hypothalamus, owing to its control over sympathetic activity, instigates the transmission of messages down the appropriate sympathetic nerve fibers, resulting in the liberation of noradrenaline at the nerve endings in the walls of arterioles supplying my intestine. At the same time, sympathetic fibers to the central portion of my adrenal gland carry impulses instigating the outpouring of adrenaline that goes directly into my bloodstream to add to the total effect. The noradrenaline combines with receptors in the membranes of smooth muscle cells in the arteriolar walls, and a series of intracellular chemical reactions is accordingly initiated, as a result of which the muscle cells contract, narrowing the arteriole and thereby decreasing the blood supply to my gut, among other organs. This part of the fright reaction has deprived my gut of a significant share of my total blood volume, but where does it go?

It goes directly where it is instantly needed, to the leg muscles that will help me jump out of the way. Because the sympathetic fibers to the blood vessels in my voluntary muscles paradoxically liberate acetylcholine, the fright reaction makes the arterioles dilate, so that more blood pours through them. In essence, the instantaneous response of my sympathetic nervous system has shifted blood from my gut to my legs. Were I fleeing a saber-toothed tiger, I would need every drop of that extra blood, as I would also need the other elements of the response, such as the dilated pupils that allow more light to enter my eye, the added output of my heart provided by the accelerated rate and forcefulness of contractions, the widening of my air passages so that I can exchange more oxygen and carbon dioxide, the supplementary energy-rich glucose put out by my stimulated liver—all changes that help me cope with the threat to my life and allow me to escape to reproduce offspring with the same valuable traits of responsiveness to stress.

At the same time, I have done a few things that are strictly voluntary—or at least they *seem* to be voluntary, in the sense that my voluntary muscles make them happen. Once processed in the hearing center of my brain stem, the signals that originated in my ear are transmitted not only to the hypothalamus but elsewhere, as well. They are also fired down to a motor pathway in my spinal cord and from there out through the appropriate somatic nerves to the leg and trunk muscles that make me jump. Other networks carry a message up to the thinking part of my brain, the cortex, to apprise it of the "voluntary" parts of the response and to bring about a conscious sense of danger. Thus, both higher and lower centers in my brain will be involved in my response to the frightening sound of screeching tires.

The message to the higher center allows me to recognize the significance of the sound and evaluate the situation so as to take willed further action should I choose. In fact, much of the way my leap to safety plays itself out—like the direction I take, the place I land—is determined by choice; once I knowingly perceive my peril and the elements of the fright reaction have set the subsequent events in motion, I make lightninglike decisions about which purposeful movements will achieve my objective.

As soon as the very earliest response begins, I know it. Not only do I observe my overt reaction, but also, incoming autonomic signals make me acutely conscious of some of the sympathetic response that has been generated. I feel the pounding sensation that my racing heart is by now producing in the center of my chest; my breath comes fast, and I have a tight feeling in my gut.

In all these ways—conscious and unconscious, willed and reflexive, higher brain and lower brain, central and peripheral, somatic and autonomic—the various elements of us combine their efforts to maintain stability and life. The boundaries are vague, and no one knows where molecules and mind come together to achieve each other's intentions.

The part of the fright reaction that travels from hearing center to spinal cord to somatic nerves to voluntary muscles is known as the startle reflex, a relatively commonly experienced phenomenon in daily life and one with considerable survival value. With conscious knowledge but without total conscious control, we leap out of the path of an oncoming car or in some other way avert imminent danger.

Of course, in today's world, I am not done with my sympathetic outburst just because I have escaped injury. It starts up again as I see the driver swearing at me and extending the middle finger of his right hand in an obscene salute. Without thinking, I shake my fist at him in rage and frustration, but he is away in a cloud of high-octane exhaust. Every time I think about the son of a bitch during the rest of the day, my autonomic system pitches in with a brief flurry of hyperactivity, not as much as when I needed it to survive my own carelessness, but sufficiently evident that I am aware of it.

Somewhere between the conscious brain and the workings of the hypothalamus and medulla, or perhaps elsewhere, influences of will can, under certain still-unexplained circumstances, sometimes be exerted on the autonomic nervous system. Although there is as yet only minimal understanding of these things, it is empirically demonstrable that such behaviors as relaxation techniques and autohypnosis can sometimes change the responsiveness of sympathetic and parasympathetic nerves. Masters of meditation are even known to have developed the ability to control such seemingly autonomous functions as heart rate, blood pressure, and intestinal peristalsis. Although still a long way off, it is in the ramifications of the autonomic nervous system that the linkages will be found by which conscious thoughts may one day reliably be put in the service of homeostasis and emotional tranquillity.

The preceding description of the fright response is only one example of a characteristic of living things that is probably the key to our survival: the organized way in which our bodies respond to stimuli, which enables life to exist and organisms to flourish. In the scale of animals, the integration of responses is first seen at the level of single-celled organisms and hugely increases in magnitude and sophistication as forms become more complex. The wondrously coordinated interdependence of every part of the human body with every other part depends on various kinds of what might, at least metaphorically, be called awareness—an awareness universally distributed through the agencies of the nervous, endocrine, and circulatory systems, which in one form or another are in contact with every cell. Of the many characteristics of our humanity, none would exist without the intricacies of our multiply interwoven signaling systems.

Yet what I am describing here as awareness is distinct from the rational thought processes that constitute what is commonly thought of as a state of *literal* awareness. The awareness to which I refer is a meld of such qualities as instinctual sensibility, subconscious perception, and almost certainly levels of knowing not (at least not yet) accessible to any methods of studying transmission of information at its deepest internal levels. It is at least partially on the basis of this kind of awareness, I postulate, that much of the integration and coordination of the organism is made possible.

I conceive of such an awareness as part of a continuum along which may be found qualities of sentience seamlessly linking the simplest to the most complex components of the entire organism, thus allowing information to pass between them. I have become convinced that some overall impression, an awareness, of the general pattern of cellular events and organization is transmitted upward by means of this continuum through gradually increasing levels, until it imprints itself on our very patterns of interpreting external events and responding to them. It is by this means that our lives, I contend, and even our culture come to be influenced by, and are the reflection of, the conflict that exists within cells, between forces that would break them down and forces that would build them up. In our cells and in our daily existence, that conflict and the inherent instability that we employ in our constant striving for equilibrium and homeostasis are the underlying principles, I would propose, that form the basis of everything we are.

The response to sudden danger represents examples of instinctual behavior—we survive in large part by instinct. Although a word variously interpreted by the philosopher, the psychologic theorist, and the laboratory scientist, there would seem to be a general agreement on a basic formulation provided by the lexicographer. My *Webster's* puts it this way: "INSTINCT: an inborn tendency to behave in a way characteristic of a species; natural unacquired mode of response to stimuli; as, suckling is an *instinct* in mammals."

The word is derived from the Latin compound verb *instinguere*, "to urge onward or impel" (*instigate* obviously has the same origin), constructed from the prefix *in-* ("on") and *stinguere* ("goad," "sting," or "prick"). The involuntary aspect of the word's implications is clear, especially when *stinguere* is traced to its Indo-European root, which is *steig*, meaning "prick," or "pointed," or "sharp." In the mid-sixteenth century, instinct acquired its current usage as "an innate impulse," or an inborn tendency specific to a particular species. Interconnecting circuits within the nervous system and biochemical processes such as those regulated throughout the body by hormones mediate the instincts that have permitted humankind to survive the omnipresent dangers of the environment.

In other words, though we may or may not be mindful that they are functioning, instinctual behaviors are inherent and driven—in a more basic sense, they are the result of goading and pricking. Whether we consciously want to or not, our innate drives goad and prick us toward staying alive and reproducing ourselves. Instincts and the form of awareness in which I claim instincts to be a participant require an integrated action of every factor in the body's responsiveness. Those elements that involve any part of the nervous system are dependent on the receiving, processing, and transmitting capabilities of a remarkably specialized cell that is a marvel of nature's biochemical and physical engineering. To understand fully the response mechanisms of the body, it is necessary to know something about the way this cell does its job.

The nerve cell, or neuron, is the fundamental unit of the nervous system's functioning. It is a cell so highly developed and dedicated to such narrowly specific tasks that it cannot reproduce. Although the various structures and subunits within it are constantly being repaired and replaced, a lost neuron is gone forever.

It is the neuron's job to detect changes in its environment and then to transmit appropriate messages in order that action can be taken. The cells of the nervous system compose a massive organization of processing stations and wires that extend its influence into every protoplasmic nook and cranny to perceive, integrate, and coordinate events in the furthest reaches of its domain. In some ways, it shares this kind of responsibility with the blood and endocrine systems, but its speed of transmission is vastly more rapid than theirs. The average speed of nerve impulses is 50 meters per second, but they range between 0.5 and 100 meters per second. The faster times are generally those involved in rapid reflex activities (such as response to pain), while the slower are likely to be found in the autonomic nerves. Impulses in the brain travel at about 20 meters per second; a meter being 39.37 inches, 20 of them equal about 65 feet.

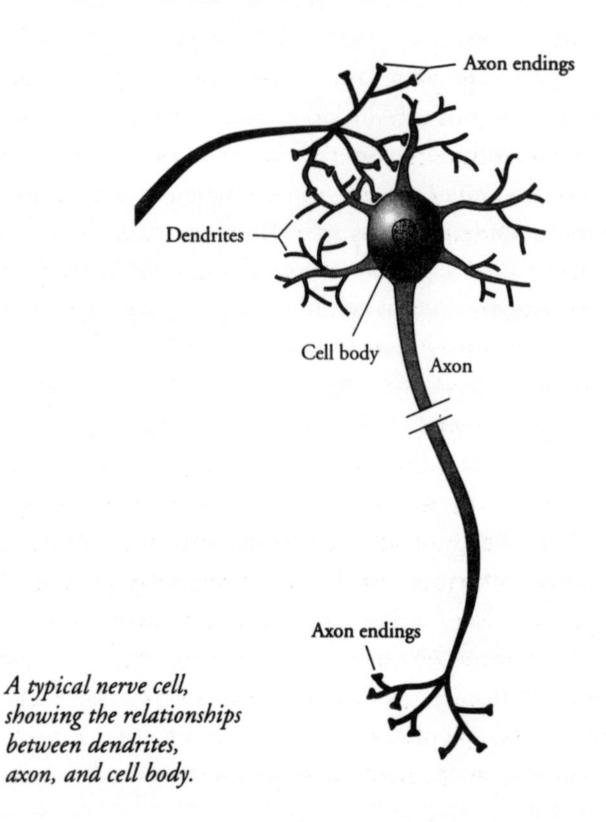

A typical nerve cell, showing the relationships between dendrites, axon, and cell body.

Neurons come in many varieties, but they all have the same basic structure, consisting of the cell body and two different kinds of extensions. The first is the *dendrite*, which receives stimuli from its immediate surroundings, from other neurons, or from sensory structures in the skin, muscle, or internal organs. Each cell has many dendrites and each dendrite has many branches, which explains the origin of the word from the Greek *dendron*, meaning "a tree." Because these trees have such a large number of receiving branches, the body of any one neuron is connected to many sources of stimuli. A single long extension called the *axon* (Greek for "axle") carries the impulse away from the cell body and toward other neurons and their dendrites, or directly to muscle and gland cells. Because the axon is, like the dendrite, also multibranched, it, too, connects with a number of receiving structures. All of this arborization serves to enhance markedly the number of connections that a single neuron can have to other parts of the nervous system. A nerve cell may have as many as hundreds of dendrites, but it can have only one axon.

The junction at which the impulse is transmitted from one nerve cell to another is called the *synapse*, from the Greek *synapto*, "I join together." It is here that the message passes from one of the branchings of an axon to one of the branchings of a dendrite, or, alternately, directly to the cell body. The axon-to-dendrite relationship is reminiscent of Michelangelo's magnificent *The Creation of Adam* on the ceiling of the Vatican's Sistine Chapel; here, the extended right hand of God is reaching toward the outstretched left hand of Adam, but their adjacent fingers are not quite touching. There remains the smallest of interspaces between them, which the essence to be transmitted must cross.

Michelangelo must have believed that some divine spark of energy leaped across the gap between God and man, and much the same was at first thought about the transmission of nerve impulses through the synapse. But as logical as it seemed, this hypothesis was not supported by later study. When the impulse reaches the tiny intervening space, a chemical substance—a neurotransmitter—is released from the axon into the junction. In addition to acetylcholine and noradrenaline, there are about fifty other neurotransmitters. The molecules of a neurotransmitter bind to receptor molecules on the dendrite or cell membrane. The binding changes the shape of the receptor; this, in turn, opens up pathways, called channels, in the membrane that envelops the receiving cell.

The consequent flow of chemical substances through the membrane passes the impulse into the receiving cell. The neurotransmitter will either excite the receiving cell or inhibit it, depending on what kind of channels it opens up. A given neurotransmitter can do either, and specific circumstances determine which will occur.

No neuron acts alone. It is always part of a collective response to some event in its environment, which involves not one but large numbers of nerve cells sending impulses to various destinations through their arborizing axons and dendrites. Entire nets and circuits of neurons interact with one another, involving as few as two or as many as millions, averaging about a thousand and connecting across excitatory or inhibitory synapses. Because axons travel together in distinct groups of commonality which after a bit may divide and go off in separate directions, these messages may be carried across prodigious distances and to several centers virtually at once.

Nerve cells sharing a common function are often grouped together in one specific location within the brain or spinal cord—called a nucleus (not to be confused with the nucleus within each cell). Neurologists use the term *nucleus* to refer to an assemblage of neurons that originate, modify, or relay information in concert with one another. A nucleus is a focus upon which a brain center is based, such as the respiratory center in the medulla or the temperature center in the hypothalamus.

It may in general be said that there are three classes of neurons: sensory neurons detect stimuli and relay them to the spinal cord and brain; intermediate neurons, called interneurons, in the cord and brain integrate and coordinate the messages—because they transmit messages only locally, their axons are short; and motor neurons transmit information to the cells that can act on it, such as those in muscles or glands.

What we dignify by calling a nerve impulse or a signal or even a message is in fact nothing more than a tiny charge of electricity created when certain chemical and physical changes take place in a nerve cell or fiber in response to stimulation. The charge amounts to about ninety millivolts, and it speeds along the axon like a hurtling express train on a wide-open track. When it reaches a synapse, it causes a neurotransmitter to be released there. A nerve cell can fire as many as 50,000 times per minute if necessary.

In this way, the impulse acts (at least metaphorically) as a kind of information to be carried from one part of the nervous system to another.

Because of the extensive arborization of axons and dendrites, an impulse originating in a small group of sensory cells can be transmitted widely and into far-flung parts of the entire system. An example of this is the way in which the loud and sudden screech of those automobile tires caused almost simultaneous autonomic, reflex, and conscious responses at multiple sites in my body, even though the stimulus got its start in one tiny focus.

The sound wave from the tire's screech strikes my eardrum and sets it vibrating. The vibration is transmitted internally to a lineup of three small bones in a narrow canal called the middle ear. The first bone is in contact with the inner side of the drum and the third with a fine membrane called the oval window, situated at the entrance to my inner ear. The oscillations of the bones amplify the vibrations of the drum some twenty times and pass them along to the oval window. In turn, the vibrations of the oval window are transmitted further internally to the cochlea (from the Greek *kochlias*, "a small spiral shell"), a snail-shaped structure containing cells capable of transforming (biologists prefer to call it "transducing") the mechanical energy of the vibration into electrical energy in the form of an impulse. The impulse thus produced is chemically communicated across synapses to the end branches of fibers of the auditory nerve, and from there to hearing centers in my brain. Once recognized as a sudden sound and interpreted as dangerous, it is simultaneously transmitted along several pathways to other centers, resulting in the variety of responses described earlier.

(Incidentally, the inner ear has a double function. In addition to the cochlea, it also houses three fluid-filled canals of a semicircular shape, which can detect alterations in the position of a person's head relative to the earth's gravity and transmit them to the brain. The canals, therefore, are part of the body's mechanism for maintaining positional equilibrium.)

Large numbers of axons traveling together in the peripheral nervous system are ensheathed in an encapsulating envelope—the entire package is the cablelike structure we call a nerve. Typically, a nerve contains thousands of axons. Those that carry messages *to* the spinal cord from the periphery are sensory fibers; those that carry messages *from* the cord or brain to the periphery are motor fibers; autonomic fibers travel within the nerves. The three varieties—sensory, motor, and autonomic—are each clustered into a distinct bundle within a major nerve trunk. These bundles are also compartmentalized by the destination of the motor stimulus or source of the sensory stimulus. This means that although a major nerve trunk may contain a mixture of all three elements, the elements travel together in distinct groups within it, organized by destination or source and the nature of the specific kind of information they carry. The further from the cord, the more separate are the various subdivisions of each major trunk. The analogy to a multipurpose cable or telephone trunk line is inescapable.

Tracing a major nerve trunk containing all three elements from the periphery to the center, it is seen to divide into two parts, or roots, just before reaching the cord. One of them (called the dorsal root) enters toward the back of the cord, while the other (the ventral root) is in communication toward the front. The dorsal root contains the sensory components, bringing in signals to the cord from the skin or muscle or internal organs; the ventral root is the motor component, carrying signals heading out to structures that are to obey the transmitted commands, such as gland or muscle.

As noted on page 76, sensory signals originate in specialized cells or little multicelled receiving bodies in the skin, muscle, and internal tissues. These receivers have the ability to detect a stimulus and transduce it into a nerve impulse. They pick the signals up in dendritelike projections that carry them to the cell body. The specialized cells in the cochlea are an example.

Any individual receiving cell or body responds to a particular kind of stimulus and no other, whether it be from a source that is mechanical (like pressure, or the vibration of sound waves), chemical, heat-related, painful, or light-generated. So specialized are these receivers that a sensation such as touch, for example, has different representatives for light touch, coarse touch, and pressure; an individual taste receiver on the tongue responds either to sweet, bitter, sour, or salty materials, but not to the other three. (In spite of being limited to four basic kinds of receivers, our food sensation is widely varied, because the brain coordinates the signals from taste receivers with those from an extensive assortment of smell receivers. The actual taste of the food in our mouths is the result of the blending of the two. This is why a stuffy nose detracts from the enjoyment of a meal.)

Transducing is akin to translating. The receiving cell or body picks up a message in one language (mechanical or chemical or one of the others) and converts it into another, the electrical impulse. The electrical impulse is the universal commonspeak, the lingua franca, of nerve transmission.

The cells of any given receiver are able to translate into the commonspeak from only one kind of stimulus. Imagine a group of interpreters at some ideal United Nations, each of whom can understand only one language. Each of them listens to that language and no other as it comes in through headphones. He interprets what he hears and turns it into Morse code for transmission.

Once the sensory stimulus coming in via a peripheral nerve reaches the cord through the nerve's dorsal root, it is processed in either or both of two ways. It can pass via the intermediary of interneurons directly to motor neurons in the cord and then out their axons, thus activating a reflex without any need for the involvement of higher connections; and it can also be sent on up to the brain. These, actually, are the two functions of the spinal cord: to act as a nerve center that takes in signals and causes reflex responses; and to be the intermediary between the periphery and the brain.

When a nerve impulse ascends to the brain from a sensory neuron in the cord, it does so along an axon that is joined into a specific bundle or tract along with other axons carrying the same message. There are a number of distinctive ascending tracts, each of which is composed of axons originating in neurons involved in some specific sensation. Each sensory filament carries a single type of message, whether it is pain, touch, heat, or any other. This means that the sensation of touch has a tract of its own carrying it up to the brain, as does pressure, pain, or temperature, for example. Each of these pathways conveys its distinctive sensory message to a specific position in the thalamus, a structure in the lower part of the brain, lying above the hypothalamus. In the thalamus, the information is fine-tuned and interpreted, then sent higher to the thinking part of the brain, the cortex. The cortex recognizes the source of the signal and acts on it. Meanwhile, the ascending tracts have also sent collateral branches to assorted parts of the lower brain that affect autonomic and other involuntary responses.

In these ways, a single sensation, such as the one caused by sitting on a thumbtack, goes off in various directions after it has reached the cord through the dorsal sensory root. Its ability to take multiple pathways results in a variety of almost simultaneous responses, ranging from the reflex act of leaping up from the chair (mediated only through the cord), to a slight increase in heart rate and other minor manifestations of the fight-or-flight phenomenon (mediated autonomically through the hypothalamus), to swearing at whichever harebrained idiot has left the goddamn tack in such of a helluva stupid place (mediated through the cortex).

Other bodily responses are set off at the same time, requiring the integrated activity of a variety of mechanisms that coordinate with one another to protect the thumbtack-sitter from lasting harm. A drop of blood appears at the tiny puncture wound on the surface of the injured buttock, and the clotting and inflammatory mechanisms go instantaneously into action, their purpose being to limit the damage to the immediate locale of the wound; the arterioles feeding the site go into spasm so that only that tiny bit of blood escapes; within minutes, a little clot is forming and protective white blood cells and plasma rich in the proteins that will begin the healing process have found their way to the area, having passed with ease through the walls of the local capillaries, which have increased their permeability to allow these substances to move through. Within instants, the body has called up a company of forces against a sudden threat to its integrity.

All of the neurological part of the response is activated by the passage of the impulse along nerve fibers. Though we metaphorically call it a message, the impulse is really only an electrical charge traveling at high speed, having been set off at its origin when the initiating stimulus caused certain chemical and physical changes along the membrane enveloping the nerve cell. In synapses, within neurons, and through the centers called nuclei, all sorts of filtering, coordination, and integrating occur, but the basic mechanism that transmits the signal remains the same—quite ordinary chemical and physical changes in the neuron, resulting in the production of a tiny electrical charge.

There is one additional wrinkle to all of this. As noted earlier, an individual impulse entering a neuron in the central nervous system can have one of two effects: It may excite the neuron or inhibit it. Since a neuron receives many signals at once, their total effect is determined by adding those that are excitatory and subtracting those that are inhibitory. A neuron will not fire unless its threshold is reached. Thus, when many thousands of neurons are acting together, the summation of their effect may vary anywhere along a continuum from very weak to very strong. The variability of the strength of the total signal transmitted along a large bundle of axons adds yet another dimension to the variability of possible responses.

Though all of us are aware that messages about such stimuli as pain or temperature are ceaselessly being conveyed brainward, we pay much less heed to certain other sensations that are nevertheless of considerable importance to the constancy of our lives. Most people, for example, are blissfully ignorant of proprioception, the ability to detect position and movements of our extremities, head, and trunk. Almost every evening from early spring to late autumn, my wife and I stroll through the streets of our neighborhood, deeply engrossed in conversation that is sometimes serious and sometimes silly, neither of us ever paying an eye blink of attention to the position of our feet at any given instant, or to how we are coordinating the large number of completely voluntary movements that go into taking a single step. And yet if asked to close our eyes and describe the position of a foot, a hand, or even a fingertip, each of us would do it unerringly. We can identify the specific location of a moving extremity because of a combination of proprioception and the mental image our brain carries of the body. With our eyes still closed, we could just as unerringly touch a finger to the tip of the nose. When Muhammad Ali floated like a butterfly and stung like a bee, he was using a wide array of his body's physiological control mechanisms to guarantee the perfection of his movements and the timing of each punch's arrival at its target, but the one of which he was least likely to be conscious was proprioception. The same might be said of Vladimir Horowitz on the evening when Otto Loewi sat in the audience at Carnegie Hall, or of me when I dissected the delicate tissues around Marge Hansen's pancreas.

The ability to know where our parts are and to control and direct voluntary muscular movements starts with those microscopic receiving structures that exist in large numbers in our joints, tendons, and muscles. Stimuli arising within cells in these receivers are carried by sensory nerves to the dorsal roots and into the cord, where they are chemically conveyed across synapses to become part of a reflex arc. At the same time, they cross other synapses that enable them to be transmitted up their own distinctive tracts to the medulla, located in the lowermost part of the brain, which is called the brain stem. From there, they are relayed across the midline to the thalamus (explaining why the right brain deals with stimuli from the left side, and vice versa), and thence onward to the sensory cortex. In all three—medulla, thalamus, and cortex—coordination and integration of incoming messages assure that motor stimuli are sent back down the cord, which in turn coordinates and integrates the movements of appropriate muscles and any necessary autonomic responses. All of this happens without our having to plan a single aspect of it. Because of this perfectly synchronized series of responses, we walk, play the piano, and punch one another in the nose with the freedom that comes from unthinking certainty that everything will happen precisely as we expect it to.

I grew up in close contact with a man who had lost a great deal of his proprioceptive sense. Max Tailor was, like so many of the middle-aged men I knew as a boy, a Jewish immigrant who had arrived in the United States in the first decade of this century as part of the vast influx from the impoverished ghettos of Eastern Europe, seeking a better life in America. For him, the better life was slow in coming, if indeed it ever came at all. Although descended from a succession of tailors (hence the choice he made when he discarded his jawbreaker of a Yiddish-Slavic name in favor of something that sounded more American), he had never been trained in the family tradition, and had to be satisfied with a semiskilled job as a needleworker in one of the dress-manufacturing lofts of New York's garment industry. The Tailors lived in a walk-up apartment on the fifth floor of the same Bronx tenement where I grew up. Max earned just barely enough to support his family of a wife and two sons. Or it would have been enough had his proprioception not begun to desert him in his late thirties, when his boys were not yet teenagers. As it was, the family just scraped by.

The neurological process was slow but inexorable, as Max's condition gradually deteriorated over a period of fifteen to twenty years. No reason for his increasing disability was ever discovered during his lifetime, and Max went to his grave without a diagnosis. Which is to say that he went to his grave without a clinical diagnosis from the doctors—there was plenty to diagnose, or at least to arouse the pained compassion of those who watched not only Max himself but also every member of his family descend into the chronic state of anxiety, frustration, and finally hopelessness that overwhelms those who are witness to the physical and emotional deterioration of someone they love. There was great tension in the household, arising from so little comprehension of the ever-worsening tragedy of their lives that Max's sons sometimes sought explanations in nonorganic causes, like lack of will. Herb, the boy who was my age, more than once resentfully complained that his father seemed trapped in his own refusal to help himself. We didn't know of things psychosomatic in those days, or at least in that place, but his meaning was clear even without the words to express its anguished bitterness. Herb raged at the unknowable illness, even as his heart broke to see his father's relentless decline.

And yet the four of them soldiered on. Max and Fannie Tailor had come to America to fulfill a dream, but they soon found that their dream would have to be brought to reality by their children. Like the sons and daughters of immigrants everywhere, Herb and his brother, Joe, were their parents' guides through the bewildering chaos of the New World—which never lost its newness or its strangeness even after decades. There was an unspoken principium in that small tragedy-struck family that—as long as good health prevailed—all obstacles yielded to hard work and to education. And so they did. Both boys struggled, but both boys succeeded. Through all his difficulties, Max retained a highly developed sense of irony, which in time became a prevailing theme of his conversations, bringing at least some grim humor to his worsening days. He would have had many a sardonic chuckle had he lived to see what would become of his boys: Joe manages money, which Max and Fannie never had; Herb is a professor of English literature, which Max and Fannie never learned to read.

All through his illness, Max was a charity patient of one or another of the hospitals supported by the city of New York. As his condition worsened, hospitalizations became more frequent, and his family felt increasingly beholden to the doctors who cared for him on the public wards. When he died, his two sons agreed with one of the neurologists that an autopsy should be done. Their one proviso was that the report be sent to an old friend who was by then training as a surgeon in New Haven. For almost his whole life, the old friend had known Max well, they told the doctors, and he would understand exactly how to interpret the findings for them. And that is how I came to comprehend just what it was that I had been witnessing for two decades.

The first thing I had noticed, when I was perhaps eight or nine, was that my friend's father was beginning to walk in a peculiar way. In those early years, he seemed just a bit unsure of his balance, and after a while he took to planting his foot down watchfully at each individual painstaking step, with what appeared to be a conscious deliberation. In time, he had acquired what I can only describe as a slapping and somewhat rolling gait, with the soles of his widespread feet smacking down one after the other onto the pavement. He seemed always to be gazing directly downward as he walked, as though to be sure his legs were going where he wanted them to.

By the time I was fifteen, it was a real trial for me to eat a meal at the Tailors'. It was not that Max's hands shook, but something equally ruinous had begun to affect his attempts to bring a spoonful of food to his mouth, and it took me awhile to figure out what it was: Max had little certainty of where his hands were at any given instant, unless he paid constant attention to them. Unlike the rest of us, who could bring a cup or a forkful to our lips without needing to think about it, he seemed to plan every part of the arc that allowed him to complete the entire cycle. Yet even with the greatest of care, soup or even solid food was always falling from his utensils and onto his lap or the worn linoleum of the kitchen floor.

Even when he finally achieved his goal, there was something uncontrolled about Max's chewing. I never ate a meal in his presence without noticing partially chewed bits around his lips. By then, his shaving was erratic, and glistening wet particles of food would catch in the areas of stubble he had missed. I reached the point where I could never look directly across the table at him without being overtaken by a wave of nausea. No matter how his sons tried to reassure him otherwise, he must have known what an embarrassment he was to them. By my late teens, I could no longer bring myself to eat at their apartment.

There were other problems, some of which I knew about only because Herb described them to me. Every few weeks, Max would have an episode of severe pain in his shins, striking with such lightninglike suddenness that he would sometimes shout loudly with the discomfort, which always spontaneously abated after a few minutes. Also, he was subject to prolonged attacks of agonizing abdominal pain of no apparent cause. Doctors examining him at those times would shrug their shoulders and say something like, "It's just one of Max's episodes," without having any real idea of what they meant. In later years, whenever a patient told me of some symptom that I knew was meaningless even though unexplainable, I would find myself allaying their unnecessary anxieties by pronouncing, "We see this—yes we do." Every time I uttered those words, they made me think of Max.

In addition to everything else, there was the smell. Max always had about him the faint odor of urine, which I never understood until Herb told me that his father seemed to lack complete control over his bladder function. He would often dribble into his pants and not know it. In time, his bladder lost its capacity to empty properly, leading to recurrent episodes of urinary infection, one of the precipitating factors in his eventual death.

During his last decade, Max was given several hypothetical diagnoses, all of diseases in the general category of multiple sclerosis, in which the spinal cord is damaged in scattered areas, giving rise to a range of symptoms caused by the involvement of various parts of the central nervous system. But none of the diagnostic alternatives fit the entirety of Max's symptoms. It was only after his death that the whole picture became explicable.

When Max's autopsied spinal cord was examined under the microscope, it was found that the dorsal roots had undergone significant degeneration and so had the ascending tracts in the back part of the cord, no doubt from some forgotten long-gone spinal infection. These sites were shrunken, and the area of decreased fibers was partially replaced with scar-like tissue. The destroyed tracts were exactly those that carry proprioceptive stimuli up to the brain. The leg and abdominal pain had obviously been caused by involvement of the dorsal sensory roots, sending jangled messages up to the brain. This was also why Max had lost coordination of his bladder function.

I reported the findings to my friend Herb and described how the microscopic pathology explained each symptom. I was not surprised at his response. He wept softly for a few minutes, and when he finally spoke, it was with regret and guilt for all the years he had truly believed that his father was victimizing the family with an imagined illness, "all in his head," meant to bring him sympathy and control. He had never been able to understand, Herb said, why Max could not manage things about his body that the rest of us just take for granted. And now, when it was too late, he finally knew.

5 | THE FUNDAMENTAL UNIT OF LIFE: THE CELL

Man's knowledge of his body has not been acquired in a smooth continuum. It took its first sustained leap forward during the golden age of Greece, not only via the observations and speculative thinking of Aristotle but also in the more pragmatic clinical studies of physicians trained under the influence of the Hippocratic school that arose in the fifth century B.C. The Hippocratic tradition flourished most prominently in the person of Galen, a physician from Asia Minor who practiced in Rome during the second century A.D. Although his studies shed new light on human biology, so convincing was his authority, and so imperiously did he enunciate his message of having learned everything worth knowing, that new discovery was stifled for the next fourteen hundred years.

It would take the full cultural force of the late Renaissance to loosen the stranglehold on knowledge bequeathed by Galen, thus permitting unbiased studies of the human body to be done once more. From the sixteenth century onward, dispassionate observation was the goal and method of medical investigations, even though physicians and researchers remained limited by the absence of instrumentation adequate to supplement their five senses and objectify their findings. Despite the bounds thus placed on them, within a century the experimental method and inductive reasoning became the means of achieving major advances in science. The art of clinical medicine was the art of the individual doctor interpreting what he saw and heard, in light of an ever-expanding comprehension of the processes of disease.

The introduction of three crucial instruments in the nineteenth century did much to redirect the path of medical progress away from clinical artistry and toward the goal of scientific objectivity. In 1819, the stethoscope was introduced and the very first distancing took place, both literal and symbolic, between doctor and patient; in 1830, a newly identified law of optics permitted the correction of an aberration in lenses, thereby making the microscope a useful tool for the first time since its invention in the late seventeenth century—in the process giving doctors the opportunity to study ever tinier and more disembodied bits of their patients; and in 1895, Wilhelm Konrad Roentgen discovered X rays, and the need for the presence of even minute samples of patient lessened, to be replaced by a shadowed image imprinted on a photographic emulsion.

Of the three innovations, the new microscope would produce the most far-reaching effects. Research done with these amazingly revelatory optical systems led directly to the realization that the fundamental unit of all life is neither organ nor tissue but a tiny structure of previously unappreciated significance, invisible to the naked eye. For almost two hundred years, this smidgen of architectural curiosity had been called the cell. The improved lenses soon showed that the cell is the smallest bit of living thing that can, when the appropriate environment is provided, live independently. It is with the cell, therefore, that the story of human structure and function—and eventually of humanity and spirit—most properly begins.

The existence of cells had been known long before the more accurate means of magnification made it possible to study them meticulously. In 1665, the English polymath Robert Hooke published a book called *Micrographia,* in which he described studies he had conducted of various animal and vegetable forms by using a rudimentary microscope. When he examined a thin slice of cork, he found that it "consisted of a great many little Boxes," which he called "pores or cells," from the Latin *cellulae*, meaning "small rooms." Making a rough estimate of their size, he calculated that a cubic inch of cork contained "about twelve hundred Millions, or 1259712000 [cells], a thing almost incredible, did not our microscope assure us of it by ocular demonstration." Hooke, ever a man to appreciate the significance of a new finding (and possessing the timeless urge to establish priority of discovery), wrote:

> *I no sooner discerned these (which were indeed the first* microscopical *pores I ever saw, and perhaps, that were ever seen, for I had not met with any Writer or Person, that had made any mention of them before this) but me thought I had with the discovery of them, presently hinted to me the true and intelligible reason of all the* Phaenomena *of Cork.*

Hooke had made a discovery that was far more consequential than even he realized. But almost two centuries had to pass before it could be shown that the cells seen by Hooke were examples, albeit dead ones, of the building blocks from which the entire architecture of each living thing is constructed. From time to time, a reference to the existence of cells would appear in a scientific treatise, but no one knew quite what to make of the little boxes. Finally, as a direct result of the 1830 lens revolution, a rapid sequence of new findings was made possible, leading to the proposition that came to be known as the cell theory.

The first step in the process of cellular explication was the 1831 report of a Scottish botanist, Robert Brown, that every plant cell he examined with his new microscope contained an object that appeared to be a spherical core structure, to which he gave the name "nucleus," from the Latin *nucula*, "a kernel" or "little nut." Seven years later, another botanist, the German Matthias Schleiden, published what would prove to be a landmark paper, stating his conviction that all plant tissues, without exception, are composed of cells. Within a year, Schleiden's friend Theodor Schwann put forth the proposition that the same was true of animal tissues and, in fact, of all living structures.

Although Schwann supported his thesis with plenty of direct evidence, it was considered by many to be a revolutionary and perhaps heretical idea. Having no wish to offend anyone or to violate the precepts of his church, the devoutly Catholic Schwann obtained the approval of his bishop prior to publication of a monograph so seminal to the history of biology that students more than a century later would be required to recite its title from memory—in the original German. Having been one of that put-upon legion of young victims of misdirected pedagogy, I can still do it on demand: *Mikroskopische Untersuchungen über die Uebereinstimmung in der Struktur und dem Wachsthum der Thiere und Pflanzen—Microscopical Investigations Concerning the Correspondence in the Structure and the Growth of Animals and Plants.*

With the appearance of Schwann's book, all but one of the basic elements of the cell theory were in place. The only problem remaining to be solved was to identify the source of any given cell—from what organic material does a cell arise? Within a decade and a half, the microscope would provide the solution to that riddle, too. The answer came in 1855, in the form of a single resounding Latin phrase uttered by Rudolf Virchow, then professor of pathology at the University of Würzburg, and the most prominent medical researcher of his time—*omnis cellula e cellula*—"every cell originates from a previously existing cell." With that one statement and the evidence to support it, Virchow all but wiped the scientific slate clean of the dozens of speculations purporting to explain the existence of living things on the basis of the spontaneous appearance or generation of viable tissue from unformed or even inorganic matter. Within a few years thereafter, Louis Pasteur, in a series of papers published in a French scientific journal, would write the final epitaph for spontaneous generation. Henceforth, the continuity of life would be recognized as being dependent on the cellular reproduction of life already present. The key to the continuity was shown to be the equal division into two offspring cells from one parent. In this way, nature ceaselessly duplicates the smallest of its units that are endowed with the potentiality of independent existence. The establishment of the primacy of cellular life has proven to be the most important structural concept in the entire field of biology.

For a century afterward, the overarching thrust of physiological research emphasized the paramount importance to the cell that the stability of its surroundings be maintained. Under the influence of Claude Bernard and several generations of progressively sophisticated studies of the *milieu intérieur*, a great deal was learned about the local environmental conditions that must be met in order to ensure continued cellular existence and the optimal performance of tissues and organs. But in recent decades (and owing largely to progress in biochemistry and to technological innovations like the electron microscope), it is the cell itself that has become the focus, quite literally, of attention. Knowledge of its structural and molecular components has multiplied rapidly, hurtling forward at such an astonishing rate that it has gotten beyond the reach of most of us who attempt to follow it. An ever-more-abstruse body of knowledge increasingly eludes the comprehension of the vast majority of those ordinary citizens who do not happen to be specialists in one or another of its arcane branches of study.

What follows is an attempt to provide a "core curriculum" of sorts on the biology of the cell. I have tried to avoid counterproductive detail while at the same time deliberately going over some of the ground several times here and there when it has seemed necessary in order to be certain of its comprehensibility. Though the result may fall short of being a masterpiece of inclusiveness, it is written with the assumption that we are all starting from the same point—namely, scratch.

The approximately 75 trillion cells of the human body, each measuring between 0.0002 and 0.0008 inch in diameter, make up about two-thirds of its weight. The structure of any cell is characterized by three predominant features: an enveloping membrane that, like a casing or skin, demarcates the cell from everything around it; the nucleus, containing the cell's DNA; and the cytoplasm, which includes all the cell's contents other than its nucleus.

The membrane is a thin, all-encompassing barrier having the capability of selecting exactly what it will and will not allow to enter or leave the cell. It might on first thought be assumed that such a barrier should be a solid structure, but in fact the cell membrane is neither completely fluid nor completely solid; it has characteristics somewhere between the two, for which the term *semisolid* is most appropriate. Although a single continuous layer, it is not smooth—cells come in a vast variety of shapes and geometric forms, depending, among other determinants, on their function. Because the membrane surface area of each must be maximized for optimal efficiency, nature has used the stratagem of providing all manner of infoldings and outpouchings in certain cells, so their surface contours are likely to be quite irregular. Even fingerlike outgrowths are sometimes seen in special circumstances, as in the cell membranes of certain tissues whose main function is absorption, such as those of the intestine and gallbladder.

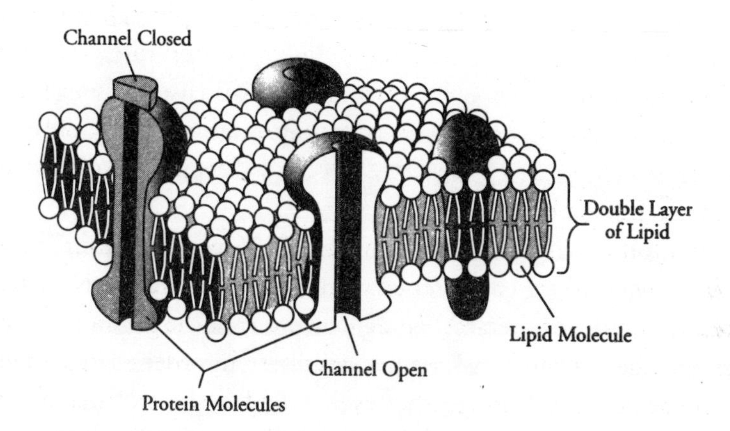

A representative area of a typical cell membrane, showing its double layer of lipid, or fatty material, and several of the protein molecules that function as channels.

Irrespective of its contours, a cell membrane is always composed of a flexible double layer of an oleaginous kind of fatty material, or lipid, whose consistency has been compared to that of a light grade of machine oil. Protein molecules float in the membrane in such a way that they are in effect embedded in it. Some of the embedded proteins project outside the cell, some into its inside, and some traverse its entire thickness of approximately 0.0000002 inch, so that one end of the molecule is in and the other is out. Individual proteins or groups of them have distinct jobs to do, such as regulating the rate of in-and-out traffic by controlling what they allow to pass through them—these function as pores, or selective channels through the membrane; other proteins function as receiving foci, to which specific nearby molecules attach themselves, thereby bringing signals from neighboring cells or the extracellular fluid—such proteins are called receptors; others function as enzymes that are catalysts for chemical reactions; others function to provide linkages between cells.

Molecules may either traverse the cell membrane passively or require more active means to effect their movement in or out. For example, when the concentration or pressure of particles on one side of the membrane is higher than on the other, the two sides tend to become equalized by such passive means as diffusion or osmosis. But when the needs of the cell demand that the substance enter or leave *against* the pressure or concentration gradient, an active form of transport requiring energy must take place through the embedded protein channels. Active transport demands that physical work be done by the cell. The power for the work is made available by being released during chemical reactions that break down certain energy-rich molecules within the cell. Active transport through the membranes uses up to 40 percent of a cell's energy supply.

Under certain circumstances, yet other mechanisms are employed to enable substances to enter or leave the cell. Two of the more common of these are engulfment (this process, which biologists call endocytosis, is the common method by which bacteria or other large particles are taken in) and expulsion (called exocytosis) of blisterlike projections pushed out through the membrane.

In these several ways, its membrane determines the contents of the cell and controls the intake of nutrients and the release of wastes and other harmful materials. It is selectively permeable. The receptors on its surface are ever on the alert to monitor the never-ending changes in the cell's environment—they initiate the process by which appropriate response is made possible. By such means, the membrane of each individual cell is a crucial participant in the series of tactics by which entire organisms—you and I—adapt to the constant flux of conditions around them and within them.

And now to the nucleus. The kernel of a thing is its essence, or its most essential part. As early as 1831, Robert Brown seems to have intuited the dominance of the newly discovered object he named the nucleus. At the very least, he recognized it to be the cell's dominant anatomic structure. What he could not have known was that it is also the center of the cell's activities. The reason for its authority is that it contains the material that regulates cellular function and determines heredity. That material is a chemical compound called DNA—deoxyribonucleic acid. No wonder biologists have said of the cell that it is "the sphere of influence of its nucleus."

The nucleus is enclosed within a double membrane, or envelope, perforated at intervals by numerous pores that allow various substances to enter and leave. Its major component is a large number of loosely coiled threadlike fibers of a material called chromatin, which, during the process of reproduction, coil more tightly and are then called chromosomes, forty-six in number for humans. Each chromosome is a single DNA molecule plus a considerable amount of protein material that acts as scaffolding and is approximately equal in mass to the DNA. A DNA molecule is an enormously long chain containing genes. Because it is a member of a class of chemical compounds called nucleic acids, the fiber that results when it combines with protein is called a nucleoprotein.

In its normal state within the nucleus, the DNA is not stretched out to its full length, but is complexly folded into a package. When a chromosome is observed under a standard microscope, therefore, what is actually seen is a package of folded-up DNA and its protein scaffolding. Were it to be unfolded, stretched out, and pinned end to end, all the DNA in a human cell's forty-six chromosomes would measure six feet in length.

This is as good a point as any at which to introduce the concept of genes. Genes are biological elements that contain information that can be transmitted from one generation to the next. A gene, of which each of us has some 50,000 to 100,000, is therefore a unit of heredity; it is made of a bit of DNA. Each gene is a localized region on the length of the DNA molecule. In fact, only a small fraction of a cell's DNA is composed of genes, probably less than 10 percent. And so, on the face of it, a gene is neither more nor less than a chemical structure lying somewhere along the DNA molecule. If you believe there is no more to life than that, you're ready to be sold the Brooklyn Bridge.

And exactly what is a DNA molecule?

A molecule of any nucleic acid is a long chain made of structural units called nucleotides, of which DNA has millions strung out one after the other. Some nucleic acids (DNA being one) are actually not one but two strands of such a chain, coiled around each other to make a double spiral (being occasional devotees of the recondite, scientists have chosen less pedestrian verbiage, thus calling it a double helix), like the twin serpents of a caduceus.

Each nucleotide in a strand of DNA is itself composed of three molecules, all joined together into a single unit: a sugar named deoxyribose, a nitrogen-containing molecule called a base, and a phosphorus-and-oxygen-containing molecule called a phosphate. Because there are four different kinds of bases—guanine, cytosine, thymine, and adenine—there are also four different kinds of nucleotides—respectively, guanine nucleotide, cytosine nucleotide, thymine nucleotide, and adenine nucleotide.

The DNA molecule is organized in such a way that the base of the nucleotide contributed by one strand of the double helix is joined to the base of the adjacent nucleotide on the other, much like rungs on a ladder whose uprights have somehow become twisted around each other, or a spiral staircase with a handrail on each side. Each rung, or stair, therefore consists of two bases attached to each other across the double helix. Scientists call this arrangement the principle of base pairing. Guanine can pair only with cytosine, and thymine can pair only with adenine. If one were to trace the pairs down the length of the double helix, it would be seen that any pair of bases may follow any other, which means that the number of possible different sequences is huge. Some idea of just *how* huge comes into focus when it is pointed out that there are some 3 billion base pairs in the DNA of a single cell. Any one of our forty-six chromosomes may have as many as 100 million base pairs.

The DNA of all animals and plants is the same, one species differing from another only in the sequence in which the base pairs are aligned one after the other. It is the sequence—the order of bases down the length of the DNA molecule—that determines the genetic message that is to be transcribed and transmitted from one generation of cells to the next. So similar are these nucleotide sequences that those of *Homo sapiens* and his closest animal relative, the chimpanzee, differ by only about 1 percent—but what a 1 percent!

Small alterations that sometimes arise in sequences of base pairings of a species result in genetic changes called mutations, and those mutations that can withstand the rigors of the earth's environment survive. Those that withstand it best come to dominate less hardy varieties. In that brief statement are to be found the secret of natural selection, the secret of evolution, and the secret of continuing life on this planet.

To understand how DNA duplicates (scientists prefer to say "replicates") itself, picture the two strands unwinding from one another and separating. There are now two individual strands, each of which is a single chain of millions of nucleotides. The base, whether guanine, cytosine, thymine, or adenine, of every nucleotide is projecting out like the broken-off half rung of an immensely long ladder split up the middle. There are plenty of available free nucleotides figuratively hanging around in the nucleus, the jutting-out bases of which can each find their complementary jutting-out base on the DNA strand. Because a guanine nucleotide is capable of attaching only to a cytosine nucleotide and a thymine nucleotide only to an adenine nucleotide, the new strand formed in this way is precisely like the one it is replacing. The end result is another double helix, exactly like the one that existed prior to unwinding.

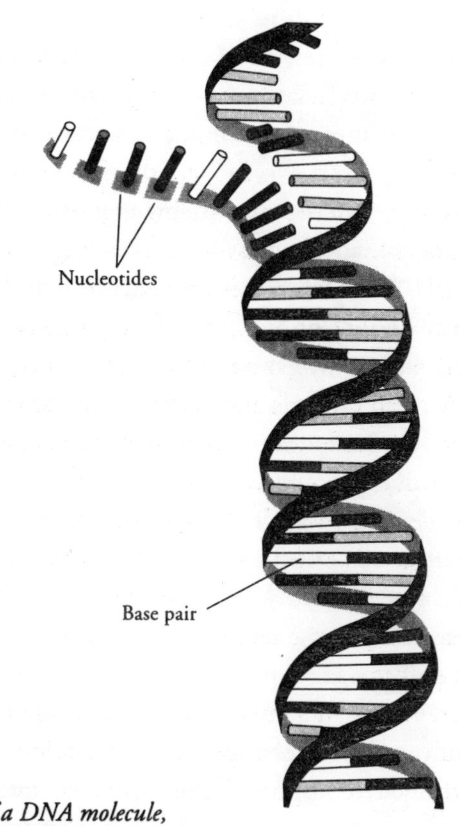

Short segment of a DNA molecule, showing the unwinding of the two strands that begins the process of replication.

This entire process is initiated and controlled by specific enzymes (called replicating enzymes) in the nucleus. An enzyme is a protein that speeds up a chemical reaction. It does this by lowering the amount of energy required for the reaction to occur. Accordingly, it will sometimes initiate a process that might not take place in its absence. Enzymes are catalysts, and each of the thousand or more in our bodies catalyzes a distinct kind of chemical reaction.

Not only do enzymes exist that cause unwinding, recoupling, and proofreading of the replication process but also there are others that repair damage that may have been done to the DNA by such noxious influences as chemical agents and radiation. Groups of enzymes are at all times moving along the many strands of DNA, patrolling them for damaged or abnormal areas, which they fix immediately on discovery. One method is for a so-called excision repair enzyme to snip out the faulty bit, which is then reconstituted properly with molecules from the immediate vicinity. In their own intracellular way, these mechanisms are yet additional examples of the entire organism's constant determination to maintain homeostasis.

All the enzymes and other proteins that are part of a cell are built of constituent parts made under the direction of the chemical information contained in DNA. The nature of the information is determined by the order and type of nucleotide sequences in the enormously long molecule. At the risk of bludgeoning my readers over the head, I'll repeat statements made earlier: A DNA molecule is a long string of paired nucleotides; a sequence of these nucleotides makes a gene; a gene is therefore one distinct sequence of nucleotides along the DNA—it is one short segment on the DNA molecule. Not all DNA is made up of genes; in fact, by far the major part of it is not. In sum, a DNA molecule is a long string containing genes interspersed with areas that seem to be nonfunctional, or silent. What, if anything, these silent areas do is not at present known. The function of the great majority of genes is to direct the synthesis of proteins.

Accordingly, when scientists use the term *genetic code*, they are referring specifically to the relationship between the nucleotide sequences in DNA and the sequences of building blocks (called amino acids) in the protein molecules that will ultimately be synthesized using the information contained in the DNA—the relationship, in other words, between the genes and the proteins whose manufacture they determine.

Proteins are the most diverse of all biological molecules, and are equipped to carry out a wide variety of tasks. In addition to the catalytic proteins known as enzymes, previously mentioned, there are structural proteins, such as those that make up muscle and bone; regulatory proteins, such as hormones and the many proteins that regulate the activity of genes; channel proteins, which act like pores or conduits to move molecules in and out of cells; receptor proteins, which receive signals; and proteins called antibodies, which act as weapons to defend the body against foreign substances.

A protein molecule is made of one or more chains of polypeptides; a polypeptide is a chain of three or more amino acids; each gene (in other words, each short stretch of DNA) carries the code for one distinct kind of polypeptide chain. It must be decoded and transmitted in such a way that the message, telling the cell how to make the gene's own unique kind of protein, can be used. Here it must be pointed out that the word *short* is relative when used to describe the length of the bit of DNA that constitutes a gene. Actually, the average length of a sequence on the DNA that codes for a protein is roughly several thousand base pairs.

When scientists break the code of a particular gene, they then know which sequence of nucleotides (aka gene) produces which polypeptide chain and therefore which protein. No wonder the excitement in the community of genetic researchers that such a thing might be possible for all human genes within the next decade. Imagine the consequences for our species of being able to focus attention on every one of our genes individually, and to identify its role in the twin objectives of self-preservation and reproduction.

Like so many biochemical pathways, the one by which genes accomplish their mission of creating polypeptide chains is an indirect one. It requires an intermediary, in the form of another nucleic acid called RNA, ribonucleic acid. RNA can, in a sense, be thought of as the translator for DNA. Like DNA, it is a linear chain of nucleotides. It differs in being a single strand and in using ribose as its sugar instead of deoxyribose, and a base called uracil in place of thymine. Like thymine, uracil can bind with adenine.

RNA is synthesized in a way analogous to the series of events by which DNA is replicated. A piece of DNA first "unzips," but only in a particular area—that is, the two strands locally uncoil somewhere along the molecule. Following this, nearby available nucleotides containing ribose and uracil bind themselves base to base, under the direction of enzymes, onto the DNA so that a sequence of nucleotides is assembled to make the new molecule of RNA. The RNA is then released from the DNA, which "zips" itself up again. The DNA can thus be considered a template for making RNA. The process by which this happens is called *transcription.*

The term *translation* is used for the method by which RNA is in turn used as a template to synthesize a polypeptide chain. In summary: A specific message encoded in a short length of DNA (a gene) is transcribed into a specific message encoded in RNA, then translated into a specific polypeptide that becomes part or all of a molecule of a specific protein. The codiscoverer of DNA's double-helix structure, Francis Crick, has put the whole matter into the trenchant and pithy statement that DNA makes RNA and RNA makes protein.

Translation, then, is the process in which specific amino acids available in the cytoplasm become attached and positioned in the proper locations along the RNA molecule and thus build up a polypeptide chain. This occurs in the cell's cytoplasm, the RNA having left the nucleus via the pores in its membrane.

Looking back over the total process of synthesizing a protein, it is easily seen that several elements are required: a source of information—which is the nucleotide sequence (the gene) in DNA; a way of transmitting the information in an appropriately decodable form—which is by transcription, the synthesis of RNA; and a way of decoding the message and using it to make new protein—which is by translation, the process by which RNA functions as a template for synthesis of a polypeptide chain. The entire undertaking is overseen by enzymes, themselves proteins manufactured by the very same processes. As a catalyst, an enzyme enormously speeds up chemical events that take place in living things, with each enzyme being involved in a specific type of reaction. Enzymes are also involved in the construction and metabolism of the four major kinds of molecules required and used by cells: proteins, fats, carbohydrates, and nucleic acids. This means that the DNA, by creating enzymes and other proteins, is the basis on which all the structure and function of a cell are dependent.

Accordingly, a specific abnormality in some tiny part of the DNA molecule will show itself in an aberration of the protein it produces. Even a single defective inherited gene is capable of causing pathology of major consequence. Among the most crippling of such pathologies is an abnormality of hemoglobin synthesis called sickle-cell disease.

Although also found in low frequency among Saudi Arabians, Indians, and those whose origins are around the Mediterranean basin, sickle-cell disease is primarily encountered among black persons of African descent.

In the United States, it is almost always a malady of African Americans. Because it is due to a so-called recessive gene, it must be inherited from both parents to result in the full-fledged form of the disease, which affects 75,000 Americans. If the gene comes from only one parent, it exists in the form of what is known as sickle-cell trait, a condition without symptoms, except under extremely rare circumstances.

In people with sickle-cell disease, a single defective gene is guilty of synthesizing one wrong polypeptide in the hemoglobin molecule, resulting in abnormal hemoglobin production. It does this by placing one incorrect amino acid into the polypeptide chain. Under certain circumstances, the abnormal hemoglobin molecules form aggregates in the red cell, which align themselves into crystal-like structures capable of stretching the cell membrane and distorting the entire cell into a shape resembling a sickle. Among the circumstances that cause sickle formation, or sickling, is a decreased oxygen level in the blood. Such sickling is exaggerated when the blood becomes too concentrated (as in dehydration) or if its acidity should increase (as can happen in serious infections).

People with sickle-cell disease are anemic, often severely so, because the changes in the red cells result in loss of their natural deformability as they are squeezed through capillaries, thereby shortening their lives considerably. Their distorted shape and irregular membrane cause them to clump together as though in a logjam, in this way obstructing tiny blood vessels. The resultant decrease in flow and the delivery of oxygen to the tissues supplied by those vessels results in pain that sometimes becomes very severe—even excruciating. Such an event is called a sickle-cell crisis. If enough vessels are shut off, the tissue simply dies, causing such problems as small areas of bone death, skin ulceration, or even destruction of an entire organ. The organ most likely to be destroyed is the spleen, which gradually loses its blood supply over a period of years and usually shrinks down to a nonfunctioning nubbin about the size of a plum by the time the patient enters adolescence. The loss of splenic activity has a deleterious effect on the immune system, resulting in increased susceptibility to infections.

As a surgeon, I have not been called upon to treat patients with sickle-cell disease for any but problems arising secondary to their primary sickness. For example, such people do frequently need to have their gallbladders removed because the chronic breakdown of hemoglobin from the disintegrating red blood cells releases substances that agglomerate to form stones. But I have been curious enough about how this singular genetic abnormality actually affects one's daily existence that I asked one of my hematologist colleagues to arrange a meeting for me with Mural Penn (called Kip, for his middle name, Kiplyn), a thirty-five-year-old African-American man whose personal history is so paradigmatic of some of the hazards associated with the vexing illness that he has on one occasion been asked to speak to a group of medical students about it. Before Kip told me his story, I could not have imagined the havoc wrought on a man's life by a single incorrect amino acid placed in error by one infinitesimal defective gene.

For most patients afflicted with it, sickle-cell disease is ceaselessly and interminably the dominating fact of each hour lived. No moment is free of the apprehension that an excruciating crisis may suddenly strike. Any illness, any hot summer's day, any overexertion can bring on a grueling agony of pain that lasts for days and requires astonishingly large amounts of narcotic to ease it, necessitating increases in dosage over the years as tolerance rises. Kip Penn has visited the emergency room of Yale–New Haven Hospital literally hundreds of times for pain medication, and he has spent so many days as an inpatient that he long ago stopped counting the numbers of nightmarish hospitalizations he has endured. His admissions for pneumonia alone, a disease to which sicklers are particularly susceptible, number almost twenty.

Even when not in crisis, Kip is always very anemic. Measured as the percentage of blood that is composed of red cells (called the hematocrit), his count is usually at about 24 percent, though the level drops to somewhere around 18 percent when he is actively sickling and in crisis. The normal value is in the neighborhood of 42 percent. This means not only that he has no reserve for exertion but also that his tissues are always getting less oxygen than they need. As with so many sicklers, the low oxygenation has resulted in a degree of impaired growth. Kip is five seven and weighs 131 pounds. His small, spare frame conforms to the pattern that medical textbooks call "aesthenic." He has shared yet another characteristic with many sicklers: For twelve years, he was addicted to narcotics.

I asked Kip what his childhood and teen years were like.

> *A kid with sickle cell always feels different—deprived. I couldn't do the activities the other kids did, especially sports. So I tried to*

excel in things like drawing and artwork, where there was no exertion. I was mostly a loner because I couldn't run with the crowd.

I had some crises before, but the first time I remember getting pain medication was when I was eleven. The introduction of the pain medicine into my system—man, it felt so good for the pain to stop, you know. It was Demerol they gave me, and after that I had a lot of hope that I'd be able to stop the pain when it came. From then on, I knew I wouldn't have to be in pain for a long time—when I had an attack, all I had to do was go down to the hospital and get some medicine and it would stop the pain.

A lot of people never heard of sickle-cell disease, and when I told them I had it, their first reaction was they'd do things like flip their fingers like a cross—like you're trying to make a vampire go away—like, Oh no, get away from me! People would ask me why my eyes were yellow [the breakdown products of hemoglobin accumulate in the blood and often cause jaundice]. I was embarrassed to talk about it because I was different—other people's eyes didn't turn yellow and things like that. It was hard for me to keep up with classes when I got to high school because I was sick so much. I was very thin and it was a bad thing to be real skinny. I couldn't gain weight. So I withdrew from people and that's why I became a loner.

Kip would have most of his crises in the winter, either from the pneumonia or because "the cold gets right in the joints"; less often he would have them in the heat of summer when he perspired enough to become dehydrated. The attacks brought him to the emergency room again and again, and they often kept him in the hospital for days or weeks. Being, like all sicklers, markedly prone to infections, he on one occasion developed a pus-filled abscess under his jaw, requiring surgery under general anesthesia to drain it through a large incision. The infection had caused so much swelling of the skin and underlying tissues of his throat that a temporary opening had to be made into his windpipe (a tracheostomy), for fear he would suffocate. The scars, easily visible above his open-necked shirt, are a constant reminder of the many times he has narrowly escaped death.

Although Kip has had crises during which the pain is primarily in his upper and lower extremities, its greatest severity is almost always centered in his abdomen. His hematologist tells me that this is because the small vessels supplying the front of his vertebral column and those to his belly muscles become obstructed by the clumps of deformed red cells. The net effect is like having a horrible charley horse that lasts for days.

> *It feels like someone has a handful of my intestines and he's squeezing it like at a pulsating kind of rate thing—just like a pulse. It can be to the very mild or to the very extreme. I've had pain so bad—you know, I just wanted someone to kill me and get it over with, it hurt that bad. I've had mostly the regular things with sickle cell, like my gallbladder operation and the pneumonias and that abscess in my neck when I was in my early twenties. I had hepatitis B, too, but that was from the intravenous street drugs. I was so sick that time that the doctors were telling my parents they'd help with the funeral arrangements and everything. I lied to the doctors and told them I got it from sexual contact, because I didn't want them to know my drug use was at the point it was at.*

Kip's most severe experiences with sickling occurred when he was in his early to mid-twenties. For a period of two and a half years, he was turning up in the emergency room an average of three times a month; each attack would last three to four days. It was during that period that he had the abscess, the hepatitis, and a succession of pneumonias. But none of these episodes was as bad as the one he considers the worst of his life.

> *I was sleeping over at my parents' house [Kip comes from a stable working-class family. His father was a truck driver and his mother intermittently took a variety of jobs, from clothing presser to domestic. An older sister is married, with two children], and I woke up in excruciating pain. It started out as abdominal, but soon I had it from head to toe. Very high fever—103. I called to my father. He came in and picked me up off the floor because I was in a knot. I had rolled out of bed and fell on the floor because I couldn't even stand up. He took me to the emergency room. I had been a frequent visitor to the emergency service—in fact, I was at a point where I was abusing the emergency room to get drugs just to feel good.*

There was a certain doctor on call that night who I'd seen pretty regularly there and she was the kind of doctor who was really adamant against people abusing the system. I was really having a real hard time this night and she was, like, not too enthused to see me first of all, and was giving me very minimal amounts of medication for pain, 'cause she thought it was all an act, you know. Sometimes when I would come to the hospital, I'd be in a little pain, but not as much as I pretended to be.

She gave me 150 milligrams of Demerol [twice the standard dose for a person of Kip's size] when I first came in the door, but it didn't touch the pain. Then she let me sit there for, like, three hours. I was rolling around and crying out in agony. My father was really getting upset, so the doctor must have figured, "I might as well give him enough to knock him out so I won't have to hear him anymore," you know. She gave me ten milligrams of morphine i.m. [into the muscle of the buttock], five milligrams at a time, separated by about ten minutes, but it had no effect on the pain whatsoever. By this time, she had my blood work back from the lab and it was really low, so she knew I wasn't pretending at all. So twenty minutes after the first morphine, she gave me another ten milligrams intravenously, and that lessened the pain to the degree where I could sit up. I was in the hospital for two weeks after that, getting 150 milligrams of Demerol every two or three hours.

Kip had by that time turned to street drugs. But even in this he abused the system, feigning pain to manipulate an overly compassionate and trusting resident physician into writing him weekly prescriptions for absolutely enormous amounts of Demerol and methadone. The scam worked remarkably well for a few months, until the misguided resident was confronted by her colleagues, who pointed out just how far she had misplaced her compassion. But until then, Kip was able to maintain an arrangement with one of his drug-dealing accomplices who would meet him on a nearby street corner after Kip had obtained his weekly quota of 100 hundred-milligram tablets of Demerol and an equal number of ten-milligram methadone tablets. The dealer paid nine hundred dollars for the lot, a 10 percent discount from the street value—as an incentive for buying from no one else. Kip would use the money for cocaine, which he "purified" into the crack he craved. Although he had started his habit on straight cocaine, he turned to crack within a few years, during the time it was becoming increasingly popular.

Kip had begun using drugs in response to a situation quite common in this soul-destroying disease.

> *It's to the point where you go through a phase where you're having a lot of realistic pain and you express this to your doctors. But somehow, somewhere down the line, they stop believing that it's all real. So you can't get as much medicine as you need, and you turn to other sources to kill the pain, and a lot of times that's street drugs. I started with cocaine. I was introduced to it by my own uncle. At times I also used Percocet, Percodan, Tylox—I'd tried heroin, but I didn't like it because it was too strong for me. After a while, crack was my real drug of choice. I smoked it for six or seven years. Before that, I was injecting the cocaine powder mixed up in water that was 90 percent of the time not sterile.*
>
> *Actually, crack is purified with heavy chemicals like ether and ammonia. I never smoked real crack—what I smoked is called base. I just got it back to its natural base by boiling it in water after mixing it with baking powder. The impurities come out and the cocaine congeals into a little ball of oil that solidifies into, like, a rock form when the water cools. Then you smoke it in a pipe. It makes a crackling sound, so it's called crack, but what I smoked was really base, except I call it crack so people get the idea.*

Kip supported his habit by becoming a dealer. "It was a kind of status symbol thing, to sell drugs and have money. I made so much money, I'd send a lot of it to relatives far away so I couldn't get at it. They saved it for me, because I knew I'd need it when I quit all this. I always wanted to be able to quit some day." But his addiction only kept getting worse.

Kip led a perilous life. For one period of four years, from 1991 to July of 1995, he seemed barely alive, sleeping in alleys and abandoned houses, becoming involved in fights, shooting and being shot at, mugging nurses leaving the 3:00 P.M. to 11:00 P.M. shift at the Hospital of St. Raphael. His world shrank to an area of four square blocks in the most drug-infested part of New Haven. He has a long scar on his forehead, sustained in a brawl during which, high and fallaciously sure of himself on cocaine, he fought off three larger men trying to club him into insensibility in a robbery attempt. At one point, he slashed the throat of his girlfriend during a fit of rage—fortunately not deeply enough to injure her seriously. As he now puts it, "I was the stereotype." And all this time, he was having one sickle-cell crisis after another. Again and again, he found himself in the emergency room and often hospitalized.

And yet in the midst of all the self-debasement and the misery, Kip knew he had to quit. Things had become so bad that even his close-knit family did not want to see him coming around. His anguished mother would say things like, "I don't know who it is behind those eyes. When it's my son again, you can come back."

And then, in February of 1995, Kip's father had a massive heart attack. When Mrs. Penn found her son and told him the news, he was in the midst of getting high. Even today, he feels guilty for not having been able to pull himself sufficiently together to rush down to the hospital.

At first, Mr. Penn seemed to be doing well. But somehow he must have known that he would not survive. A few days after the attack he spoke to his son, appealing to whatever sense of responsibility had not yet been burned out by dope: "Kip, you make sure you take care of your mother." When his father died two weeks later, something changed for Kip.

The narcotics addiction had begun when Kip's uncle introduced him to cocaine. Uncle Jerome Perry had always been the younger man's role model—"He was my idol"—a fast-stepping, smooth-talking charmer who was irresistible to women and seemed to know just how to stay perpetually one step ahead of self-destruction. But even he had his limits. When he reached the very edge, he one day stepped back from it and determined to break his habit. On the morning of his brother-in-law's funeral, he had been clean of narcotics for about a year, living with others like himself in a group home in Willimantic, Connecticut, halfway across the state from New Haven. He offered his nephew a place there. And so it happened that the selfsame man who had started Kip on street drugs became his rescuer.

When I first met Kip (incidentally, on the anniversary of his father's death), he had been drug-free for five months. He had overcome his addiction without therapy or counseling or any form of detoxification program. He did it on his own—it was a triumph of willpower.

There have been payoffs beyond the obvious. Most important, not a single sickle-cell crisis has occurred, almost certainly because Kip has been taking much better care to avoid precipitating causes, such as dehydration, chilling cold, pneumonia, cocaine, and physical overexertion. The day after we spoke, he had an interview scheduled for a job as an automobile salesman with a dealership in Willimantic. "I'll reach for the moon," he told me. "If I miss, I'm still among the stars." Besides being not very accurate astronomy, it's a platitude—hearing it made me wonder whether the trap still lies in wait for the unwary, into which that gullible drug-prescribing resident fell. But coming out of the mouth of Kip Penn, after everything he has shown of himself in the past twelve months, the words somehow sound like a vow.

When I asked Kip where he would like to be in ten years, he answered without hesitation. "I want to be a doctor." At the age of thirty-five, with only a high school diploma, and burdened by his disease and its almost certain limitation of life expectancy, such a goal would seem beyond reach. But for a man with the determination to walk away from a major drug habit after twelve years while at the same time fighting the debilitating ravages of sickle-cell disease—well, it is not impossible. If he fails, I suspect he will still be among the stars.

It should be apparent that the genes and proteins determining our essence are neither more nor less than chemical compounds. No matter their complexity, they are still only molecules, and a molecule is neither more nor less than a group of atoms linked together by bonds of energy. The word *molecule* is derived from the Latin for "small mass," which makes it clear that it is a small mass of atoms. The atoms of which you and I are made are some of the atoms that compose the most elementary living and nonliving bits of matter in our universe. And yet it is my contention that the richness and variety of human life are the products of our molecular and cellular selves: The one grows out of the other.

Analysis of rocks that are 3.6 billion years old, before life existed, has shown them to contain all of the atoms found in the molecules of which living things are composed. Accordingly, it would seem certain that the elemental ingredients of today's man were readily available at that early time, not only in the earth's structure but in its atmosphere as well. All that was required to create life was a source of energy to bring these atoms together into simple molecules and then into molecules of gradually increasing complexity. There is evidence indicating that this is precisely what happened.

In an experiment that has since been repeated numerous times in multiple variations, Stanley Miller, a graduate student at the University of Chicago in 1953, actually created amino acids from quite ordinary gases that must have existed in the earth's atmosphere during those primeval days of the earth's childhood. His adviser, Harold Urey, had been the discoverer of heavy hydrogen (known also as deuterium), an accomplishment for which he was awarded the Nobel Prize for chemistry in 1934. Urey's deuterium research and further work with radioactive atoms led him to studies on the atmosphere and origins of the planets in our solar system. The work convinced him that the earth's primitive atmosphere was much like that now present on Jupiter, a mixture of water vapor and simple gases—hydrogen, methane, and ammonia—none of which is composed of more than four atoms. Neither water nor any of these gases contains atoms of any element other than carbon, oxygen, hydrogen, or nitrogen.

Miller re-created this atmosphere in a glass chamber and bombarded it for a week with repeated electrical discharges to simulate lightning. At the end of that time, analysis of the mixture revealed that it contained a number of the complex compounds made by living things, including certain amino acids.

Miller's discovery gave rise to a new subdivision of research, called abiotic ("without life") chemistry, whose purpose is to study the events out of which life is thought to have emerged approximately three and a half billion years ago. By providing an energy source and an appropriate environment, abiotic chemists have demonstrated that many of the complex molecules made by living things can also be made from atoms and simple atmospheric compounds. Since the original experiments, Miller and his associates, now at the University of California at San Diego, have created thirteen of the twenty known amino acids. Not only amino acids, but sugars, the fatty materials called lipids, and even nucleotides (some of which are components of DNA and RNA) can form under abiotic conditions.

In the past few years, Miller's group and others have carried out experiments that provide reasonable explanations for the origins of a number of the building blocks of such substances. Admittedly, some chemists have reservations about the design of Miller's experiments and his interpretation of their results, and so do not regard them as definitive evidence that life arose as he claims; and there are several alternate theories. But each of them—with the exception of one that supposes the earliest organic molecules to have arrived on Earth by traveling on comets, meteorites, or interplanetary dust particles—is based on the probability that life on our planet evolved from early molecules that in one way or another were formed from inert inorganic substances.

There is nothing miraculous about any of this—given enough hundreds of millions of years and proper conditions for the random coming together of atoms, the addition of an appropriate source of energy will result in the binding of atoms into ever-more-complex forms. Enough combinations will eventually occur among the relatively few kinds of atoms found in organic matter that it would be inevitable by the laws of chance that nucleotides will be among the vast multitude of molecules generated. Once that crucial point is passed, it would seem only a matter of a few more eons until the appearance of DNA and then of life. The stage is then set for mutations and the survival of only those emerging forms that can best adapt to the earth's atmosphere. Everything else follows from that, and finally *Homo sapiens* makes his first hesitant entrance. Man is a walking hulk of molecules ultimately made up of atoms of carbon (the chemical symbol is C), oxygen (O), hydrogen (H), nitrogen (N), sulfur (S), phosphorus (P), magnesium (Mg), potassium (K), calcium (Ca), iron (Fe), and a few others. In high school, I learned a mnemonic for the most significant of the relatively few elements whose atoms are found in living things: S. P. COHN'S MgK (pronounced *magic*) CaFe. In those days, the whole pile of chemicals in a man was said to be worth ninety-eight cents, the price of nineteen New York subway rides from the Bronx to the Battery, with enough left over for three sticks of gum.

If it is true, as the vast majority of scientists believe, that life on earth arose as (1) the virtually inevitable outcome of straightforward chemical reactions governed by the laws of thermodynamics, and (2) the development of multitudes of combinations governed by the laws of chance, and (3) the survival of some of these new structures governed by the laws of natural selection, is it not also possible that some form of life is present elsewhere in the universe, where similar concatenations of atmosphere and energy are available? Not only possible but probable? Not only probable but virtually certain?

These questions have been addressed by many of our era's most profound thinkers. I am much taken with a few sentences recently written by Christian de Duve, a Belgian biochemist who won a Nobel Prize in 1974 for his studies of the structure and functioning of the cell. Referring to the abiotic processes that produce the constituents of living cells, he writes:

> *These processes were inevitable under the conditions that existed on the prebiotic [before life] earth. Furthermore, these processes are bound to occur similarly wherever and whenever similar conditions obtain. . . .*
>
> *All of which leads me to conclude that life is an obligatory manifestation of matter, bound to arise where conditions are appropriate. . . . [T]here should be plenty of such sites, perhaps as many as one million per galaxy. . . . Life is a cosmic imperative. The universe is awash with life.*

De Duve's point of view reflects a thesis shared by many scientists who have studied life's abiotic origins. It may not be chance in the usual sense, they point out, that brought together the various molecular forms into the complex structures that can build proteins and guide the phenomena of life. The very chemical and physical conditions present on earth those billions of years ago may have made the beginnings of life inevitable. Given those particular atoms and the characteristics of this particular energy source, the outcome was predetermined by the laws of physics and chemistry. This is a concept that philosophers, particularly philosophers of science, call determinism. Determinism is the notion that, given certain conditions, the outcome observed is the only one possible. For any given phenomenon, there is, of course, no way to prove such a proposition.

Does the existence of inviolable laws of physics, chemistry, energy, chance, probability, and even natural selection mean that what is predetermined by existing conditions is also predestined?

The very word *predestined* should give pause to those who would rule out the role of God. As opposed to agnostics, who say simply that such things are at present unknowable, those who would call themselves atheists might consider the reality of our profound ignorance (even in an age when some speak with facility of big bangs, superstrings, quanta, and quarks) of first principles, of the origins of what would appear to be natural laws, and of the ultimate sources of energy and matter. I believe these mysteries may one day be solved by the same methods of hypothesis, experiment, and reason that have taught us about abiotic chemistry and every other scientific field we study. Just as there will always be new forms of inquiry, new theories and new principles will be discovered. Still, despite my confidence that science is the path toward eventually explaining all, our present ignorance leaves me with an emptiness within that matches the void in our knowledge, a kind of sickness in what some would call the soul. It may be precisely that sickness which mankind seeks to cure by adding, ever adding, to the life-affirming, life-expanding edifice of qualities that are to me the human spirit.

If, following from the statement of the seventeenth-century philosopher and mathematician Gottfried Leibniz, "there are no souls without bodies," it must certainly be true that there is no spirit without cells. It is necessary to return to them. Thus far, only the membrane and nucleus have been discussed. It remains to describe everything else that is in a cell, collectively called the cytoplasm. Having derived *cell* from Latin, our wordsmith forebears, for their own inscrutable reasons, turned to Greek for a word to describe its contents: *kytos* is "cell" and *plasma* is "a thing formed," making cytoplasm the formed material within a cell, although it is actually more than that. It consists of a semifluid medium called cytosol, in which exist many thousands of structures that do the cell's work and provide its energy.

The cell is given its shape by an irregular arrangement of protein filaments of various kinds, appropriately called the cytoskeleton. Within the cytosol are many specialized compartments called organelles, which, like the nucleus, are enveloped by membranes. Each type of organelle has its own function, among which are digestion; synthesis; modification and distribution of proteins and other materials; disposal of wastes; and extraction of energy for fueling various kinds of cellular activity. Because they sometimes enter the more esoteric dialects of biospeak, it would be remiss not to list the names of some of these organelles, although I have no intention of muddying the waters by describing them further. In no particular order, some of the more important ones are Golgi bodies, lysosomes, mitochondria, endoplasmic reticulum, and ribosomes. Actually, the nucleus is itself an organelle.

In short, all of the functions listed in the previous chapter as characteristics of living things are being carried on in each of our cells during every moment that we exist, much of it within or on the organelles. At any given instant, in any given cell, millions of molecular interactions are taking place. Were they not noiseless, the din emanating from the center of a cell's ceaseless tempest of surveillance, commands, and determined activity would be painful to the ear of some imaginary creature infinitesimally small enough to listen to it. The sound made by one of the body's organs would be intolerable, and as for a whole man—well, he could be heard from the next county. The atmosphere of chaotic busyness is comparable to the cacophonous disorder of a tumultuous factory whose every straining machine and shouting worker is situated in the same room, all united in the dynamism of furious, churning intensity that is required to guarantee, under circumstances of constant, instantaneous change, the stability of a single product. That product is life.

A detailed knowledge of cellular physiology is not required to make a man's jaw drop with wonderment when he contemplates all that he is. Ralph Waldo Emerson knew nothing of such ultramicroscopic finery as endoplasmic reticulum and ribosomes when he wrote, "One moment of a man's life is a fact so stupendous as to take the lustre out of all fiction." Our vast knowledge of molecular biology has only deepened the jaw's drop and widened the wonderment. The unheard din of living is the symphony before which the chorale of the spirit soars in song.

Like the engines of any industrial undertaking, the entire basis on which a cell functions is the existence of readily available sources of energy, energy being defined as the capacity to do work. An atom can exist only because the particles of which it is made (electrons, protons, and neutrons) are held to one another by bonds of energy. Similarly, molecules exist because the atoms of which they are composed are held to one another by bonds of energy. It is energy, therefore, that keeps a human being in one piece. Ultimately, virtually all energy on our earth comes from sunlight. The sun's energy is absorbed by plants and animals and in this way becomes part of the food we eat. It is incorporated into the bonds that hold the food's molecules together.

Accordingly, all molecules contain energy, but some contain a great deal more than others, because of the powerful nature of the bonds required to hold them together. Adenosine triphosphate (called ATP) is the name of the most important energy-carrying molecule existing in all living things. It takes a great deal of power to keep its components from separating from one another, and even at that it is not very stable—not only does it come apart with relative ease but it does so with the release of considerable energy.

The function of ATP is to carry energy from one site to another within the cell, releasing it as needed for the various chemical reactions to take place. Since cells are constantly involved in so many countless such reactions and interactions, the crucial role played by ATP is obvious. The energy in ATP is, in turn, supplied by glucose, which is released when this simple sugar is broken down in the cell into carbon dioxide and water. The sequence of events is as follows: Glucose resulting from the digestion of food is absorbed into the tiny vessels in the intestinal wall and then carried in the bloodstream to every living cell in the body. Once in the cell, it is broken down into carbon dioxide and water, releasing energy in the process. This energy is picked up into new molecules of ATP, which are constantly being built, to function as the bonds that hold each molecule together. The ATP delivers the energy to sites within the cell where chemical reactions are taking place, at which point it is released by breakdown of the ATP. Thus, ATP's role is essentially to transfer energy from glucose to those chemicals within the cell that need it in order to carry out the interactions that are essential for normal functioning; it does this by ceaselessly being broken down and reconstituted. Glucose is, therefore, the body's primary source of energy. Ultimately, its origin is always the food we eat.

So dependent are our cells on ATP that every day they break down a quantity of it that is more than the total of our body weight. Because we keep eating and breathing, we are constantly taking in new sources of energy and the means to rebuild and thereby recharge the ATP. Every molecule of the power-packed stuff is like a tiny battery that is used up and reconstituted thousands upon thousands of times a day, and sometimes millions.

There is fascination in all of this, and a great deal of remaining mystery. But there is no miracle to be found in the atoms and molecules and the exchanges of energy. Those who seek it, seek it in vain. The miracle is not in our flesh but in our uses of it. In the earliest stages of animal life, flesh and instinct began the long process of pointing the direction in which organisms were impelled to adapt so that they might survive and make more of their kind. As flesh and instinct developed ever-more-complex strategies to deal with the flood of constantly changing stimuli in their surroundings, higher animal forms evolved, culminating in *Homo sapiens*. But always, in the beginning, is the cell.

The three principles of the doctrine that historically minded scientists still call the cell theory are these: The cell is the basic unit of all living things; all organisms are composed of one or more cells; every cell arises from a preexisting cell.

Reproduction by division into two is the means by which cellular heredity is made possible. Each of the two offspring (traditionally called daughters) must be given an exact copy of its parent's DNA as well as a complete set of cytoplasmic structures. To make this happen it is necessary prior to cell division for DNA to replicate, as described earlier, and for cytoplasmic contents like organelles, enzymes, and RNA molecules to be stockpiled and positioned within the cell in such a way that they can be properly apportioned to the two daughter cells.

Earlier, DNA replication was described as being straightforward, and it is. But this straightforward event requires an intricate organization: a huge number—some 3 billion in each cell—of base pairs must be copied no more than once; this has to occur at exactly the right instant in the life of the cell, and it must be done with perfect accuracy; to complicate the issue further, the DNA is scattered through the forty-six chromosomes, and various parts of it must be triggered to replicate at various times in the process of cell division. Is it surprising that scientists now trying to discover just how this intricacy is organized approach the many parts of it with the same awe and wonderment that the religious have for millennia felt when they contemplate the miracles they attribute to God?

Science is a system of belief. Accepting it may affect one's conception of God's role, but it need not erase it. One's decision in these matters is far more likely to be made on the basis of deep personal commitments and needs than it is on the facts of the case, and this is as much true for agnostics as it is for believers. Though many scientists have chosen to place their belief in such things as thermodynamics, others have simply rethought their philosophy and discovered a new strengthening of faith.

With the exception of sperm and ovum, the cells of human beings are diploid (from the Greek *diploos*, "twofold"), which means that every one of them has two complete sets of chromosomes, one contributed by each parent. There are twenty-three from the father and an equivalent twenty-three from the mother, for a total of forty-six. Because genes for the same characteristic lie in the same position in each of the two equivalent parental strands of DNA, the members of the pair are called homologous chromosomes. The word *haploid* is used to indicate that the sex cell (sperm or ovum) has only one set of twenty-three chromosomes (from the Greek *haploos*, "single").

The DNA in every cell of any given species is characteristic of that species. The geography of that DNA—its sequence of genes—is what is meant by the word *genome*. The genome of a species is therefore its complete set of hereditary factors. When scientists speak of mapping the human genome, they refer to identifying each of our approximately 50,000 to 100,000 genes and its location on one of the twenty-three enormously long molecules that is haploid human DNA. The job is scheduled for completion in 2005. At that time, the entire catalogue of genes will be available, like the guide to titles and locations of books in a library.

Mitosis is the name given to the sequence of steps by which cell division results in each daughter cell being given the same number of chromosomes and an equal complement of cytoplasm as its parent. But the sperm and ovum are different from all of their colleagues in the body. Because they are haploid and arise each from a predecessor type of cell that is diploid (called *spermatogonium* and *oogonium*, respectively), the predecessor's division must be modified in such a way that only twenty-three chromosomes are handed on, the haploid number. This modification of mitosis is called *meiosis*. If not for meiosis, the new cell resulting from the sperm and ovum coming together (the new cell is called a *zygote*) would result in a zygote with ninety-two chromosomes.

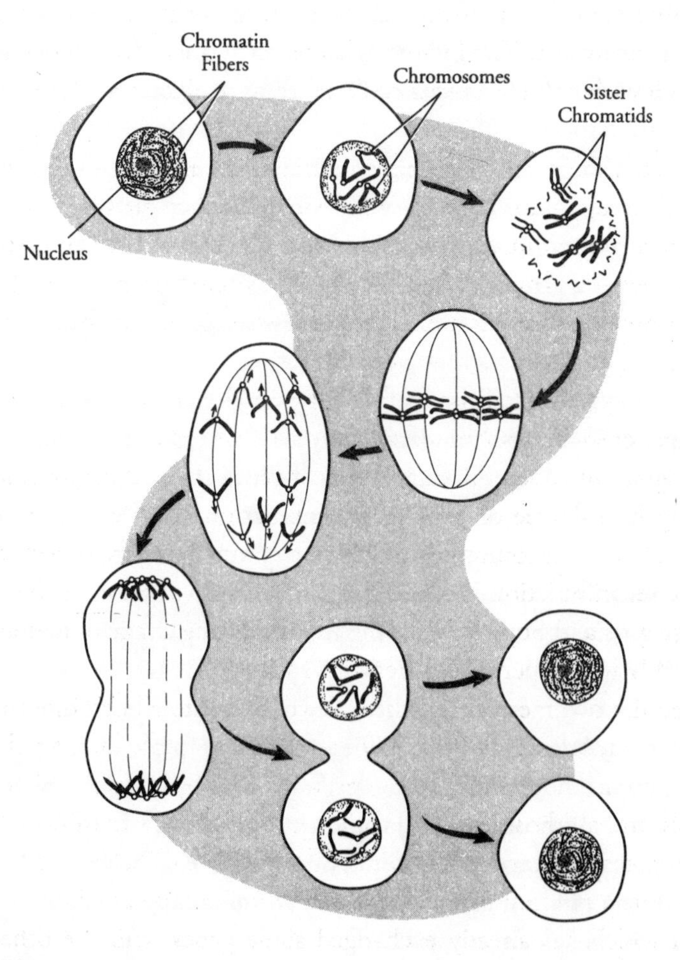

The process of mitosis.

When mitosis is about to begin, the chromatin in the nucleus condenses itself into forty-six recognizable chromosomes, which when seen through the microscope look like so many threads of various sizes (the Greek word for "thread" is *mitos*, hence mitosis). The DNA in each chromosome then replicates as described in chapter 5. The two resultant chromosomes remain attached to each other for a brief period, during which they are called *sister chromatids*. Soon afterward, the sister chromatids separate, and each goes off to an opposite pole of the cell. They can do this because the enveloping membrane surrounding the nucleus has meantime degenerated, allowing them to enter the cytoplasm. Once the chromosomes are lined up across from each other, a furrow develops around the cell's circumference, as if the cell membrane were being pinched at its middle into a waist. The furrow deepens until the cell has separated into two, each with half its cytoplasm and a copy of all the genes that were in the parent.

Meiosis is somewhat more complicated because its purpose is to result in a spermatogonium or oogonium with half the original chromosome number, a process in which each resultant sperm or ovum (a sperm or ovum is called a gamete, or sex cell) acquires one member of each pair of chromosomes. This requires two successive divisions, which may be referred to as meiosis I and meiosis II.

These divisions occur in a sequence of steps during which there is an exchange, or crossing over, of small amounts of DNA between the two homologous members of each of the twenty-three chromosome pairs, resulting in a degree of genetic recombination. This creates essentially new DNA, each chromosome of which contains genes from both parents. Genetic recombination is one of the reasons for variation in the traits of the little girls and boys who are the intended long-range outcome of this remarkably well-supervised series of events.

When the two members of these newly constructed chromosome pairs separate from one another in meiosis I, each can go to either pole of the cell; it is not as though all those originally from the person's mother go to one pole and all those from the father to the other. Accordingly, each of the two poles will have a complete set of twenty-three chromosomes, but they will be a random mixture from maternal and paternal contributors, each of which has already exchanged some genes with the other. This brings about a situation in which the postdivision cells contain genes from each parent in any of a large variety of combinations. This factor of random alignment plus the effects of crossing over between two homologues mean that each sperm or ovum produced by any of us will have one of some 8.4 million possible combinations of maternal and paternal genetic material. Some estimates are even higher. When soft music and moonlight intervene and sexual intercourse takes place, sperms made in this meiotic manner try to find their way to a similarly derived ovum. Should one of them succeed, the diploid number of chromosomes will be restored, resulting in the random admixture of characteristics donated by countless generations of each parental line.

In humans and other animals, one additional wrinkle takes place during the process of meiosis, but only in females. In each of the two divisions, one of the daughter cells receives nearly all of the cytoplasm. This means that one of the eventual four daughter cells is very large and the other three are very small. Only the large one can function as an ovum, because the others don't have enough cytoplasm to carry out the functions necessary to stay alive. Not surprisingly, all three shrivel and die. The result of meiosis in the oogonium, then, is not four equal cells but a single large one, the ovum.

The number of variations that occur when the millions of possible combinations in an ovum meet up with a similar number in a sperm is so high as to be virtually incalculable, especially by me. From this vastness of possibilities arises the notion of the gene pool. A tribe or any other group whose male and female members cohabit has a large pool of genetic variations from which an offspring may potentially draw its hereditary qualities.

How that heredity determines the characteristics of the offspring turns out to depend on more than the genes themselves. The combination of genes present within a fertilized ovum or within the adult who develops from it is called that individual's genotype. But simply knowing the genotype is only a beginning toward knowing what that individual will be when he is born. What the person actually looks like, inside and out, is called his phenotype, from the Greek *phaino*, "a thing brought to light or observed." The phenotype results from the interplay between the genotype and the surroundings in which it finds itself; it is determined by the environment around the gene, the environment around the cell, and the environment around the entire person. It depends also on the ways in which the individual genes of the genotype interact with one another.

An example of the latter factor is to be found in eye color. Eye color is determined by the cumulative effect of many different genes, all contributing to the production of the pigment called melanin. Depending on how their various contributing genes express themselves and interact, two blue-eyed parents can produce a child whose eyes present any of a range of varieties of blue. The notion of gene expression, yet another element contributing to the outcome, will be discussed in the next chapter.

For a (very, very) simple illustration of the effect of pure *external* environment—the environment around the entire person—think of two boys who have quite opposite hereditary predispositions to build up muscle mass; we'll call them Sonny Strong and Sonny String. Should String take up body-building in a serious way while his friend Strong spends his days cultivating violets, String may well grow up stronger than Strong, while Strong may stay a string bean. Or take a young girl with the genetic probability of becoming a tall adult. She will not fulfill that potential if she chances to fall into a situation that leaves her malnourished. Earlier in this century, such environmental effects on heredity were commonplace, as the children of suboptimally nourished immigrants consistently grew taller than their parents.

Internal environment plays an even more significant role in making a phenotype of a genotype. The deepening of a boy's voice at puberty is dependent on genes that control the development of the larynx, the open boxlike container of the vocal chords. But those genes require certain minimal levels of the male hormone testosterone in order to be expressed. Should anything occur to decrease the amount of testosterone circulating in the blood, the larynx does not mature normally and the boy's voice remains at a higher pitch than would be expected from a familiarity with his genotype.

For an example of internal and external environment combining to affect phenotype, I'll use my own attempt to escape what would appear to be a genetically destined rendezvous with malignancy. Being a man whose mother and brother died of cancer of the large intestine, or colon, I find myself thinking, perhaps more often than most, about the relationship between genotype and phenotype. My statistician colleagues tell me that I stand at least twice the chance of dying of the disease than does some other fellow without such a worrisome history. This one organ accounts for some 100,000 new malignancies each year in the United States, and the disease kills approximately 56,000 people, making the colon second only to the lung as an origin of cancer deaths. It gives me the willies to find myself in a high-risk category for a disease like that.

To develop a colonic cancer requires the cumulative effect of several genetic abnormalities, occuring over a period of years. They are thought to come about because of a combination of factors, including the deletion of fragments of chromosome and the ineffectiveness of the enzymes responsible for proofreading and repair of this particular error. The cancers begin as proliferations of cells in the intestinal lining, and they gradually form themselves into polyps as new genetic errors occur. Polyps are growths that resemble tiny but luxuriant trees with short trunks, projecting into the cavity of the intestine.

Because ten years ago my colon was found to contain several polyps, including one that was quite large, I have good reason to assume that I carry the genotype for my family's cancer, and that it seems well on the way toward doing its intended job on me. (A gene having the potential to produce cancerous changes is called an oncogene, from the Greek *onkos*, meaning "mass" or "tumor.") But its past behavior doesn't necessarily mean that the polypoid and cancerous phenotype will continue to appear. I have some ways to decrease the likelihood of that happening.

There are plenty of data indicating that factors in the external environments of our lives influence the development of colon cancer, and the most important of them is dietary. When large populations are studied, diets high in animal fat and low in roughage are consistently shown to increase the incidence of this form of malignancy. The disease is some ten times more common in meat-eating, industrialized societies than in the less well-stuffed peoples who depend for sustenance on plant foods high in fiber. The exact mechanism by which the cancer is initiated is uncertain, but it is theorized that fats increase the colon's content of bile acids and certain bacteria, which in turn create other chemical products that result in genetic changes which encourage proliferation of the lining cells of the intestine. The rapid multiplication and heaping up of cells produce the polyps and then the cancer.

Fat's precise effect on gene expression (or whether altering the behavior of genes is, in fact, its modus operandi at all) is not yet known, but what does seem virtually certain is that decreasing its dietary intake will lower the probability that a genetic predisposition will flower into a cancer. After the polyps were discovered, I drastically cut down my consumption of steaks, chops, and all of those other forms of meat known for their greasy excess.

To prevent my polyps from developing into cancers, a gastroenterologist removed them through a long serpentine tube called a colonoscope. On reexamination a year later, he excised a few small new ones. Repeated colonoscopies haven't turned up any polyps since 1989. Two years ago, I added a low dose of aspirin to my program, based on several large studies indicating that this further step would provide additional protection in some as-yet-obscure way, but almost certainly relating to modification of gene expression.

By cutting down on fats, I altered the *milieu intérieur* of my body and probably the environment within my colonic cells as well, thereby in all likelihood affecting the risk that my genetic predisposition would be expressed. Colonoscopy allows me to alter my external environment by removing polyps that may have eluded my attempt to influence phenotype; it reduces by approximately 90 percent the odds of my developing a cancer, according to a recent report from the National Polyp Study Workshop headquartered at Memorial Sloan-Kettering Cancer Center in New York.

In view of all this, the notion that biology is destiny takes on an entirely new meaning. Even those who insist that the proposition is true recognize perforce that what would seem to be the most preprogrammed of our biology-driven qualities—namely, our genetic predispositions—are not only subject to change but actually *must* change with changing circumstances. What is even more damaging to the argument for an inflexible biology is the way in which certain perfectly conscious and willed behavior patterns are capable of taking the starch out of biology's seemingly stiff collar of inflexibility.

There appears to be a wide spectrum of holds in which the physics and chemistry of each person's genetic inheritance grip the evolving patterns of his life. Some are enormously strong, and some are yielding, to the point of pliability. At the one extreme of possible outcomes are those physical and mental qualities over which a predetermined internal set of circumstances—genetic and internal environmental—does exert unwavering control. For such characteristics, the dictum of Horace holds true: "You may drive Nature out with a pitchfork, yet she still will hurry back." At the other extreme are responses to conscious and even unconscious decisions about how we want to live. As gamblers have been telling us since long before modern science was weaned from the breast of philosophy: It ain't just the cards you're dealt; it's also the way you play 'em. In the seventeenth century, these things were said more elegantly and were not infrequently rendered especially authoritative by being adorned in clerical garb. The Anglican divine Jeremy Taylor, writing in 1650 about what he called "holy living," made much the same observation, but in better prose: "We are in the world like men playing at tables; the chance is not in our power, but to play it is."

The fact that so much of the playing is beyond conscious control should not be allowed to obscure the influence of the world around us and our relationship to it, as well as the ways in which what is external can affect what is internal, including an internal factor as seemingly beyond reach as the functioning of genes. Even were environment not an ingredient of outcome, left to their own devices our inherited genes do things we could not predict simply by knowing their original purpose. The final product, a child with its own unique characteristics, is the result of much more than biology's equivalent of mix and match.

Because diploid organisms like us carry two genes for each characteristic, one from either parent, it was long thought that the phenotype was simply a matter of one member of this pair being dominant over the other. If this were true, each of us would be a mix of pure characteristics of the two parents, but simple observation of our own children or ourselves shows this not to be the case. No matter how close the approximation, no one has exactly his father's nose, body-hair distribution, buttock curvature, and tendency to develop hardening of the arteries, along with exactly his mother's long earlobes, thumb shape, broad foot, and high waist. This is the kind of situation that would exist if one of a gene pair was always expressed (because it was dominant) while the other was not (because it was recessive). Instead, various levels exist, in the degree to which one of the pair is dominant over its colleague on the homologous chromosome. Accordingly, the phenotype of any particular characteristic does not commonly turn out to be precisely that of either the mother or the father. Another of the factors that makes a child's phenotype often quite different from either parent is that many of our body's characteristics require the interdependent action of more than one gene, with the resultant phenotype being the outcome of such relationships as well as the recombination described earlier. As noted above, eye color is one of these qualities.

But even an accurate knowledge of the resultant phenotype does not guarantee any certainty in predicting overall outcome, since that seemingly unalterable aggregate of qualities is not born into a vacuum. Once out of the womb and into the world, it confronts an immense ambience of things and events to which it must adapt itself.

For each of these externals, the array of possible responses may be large or it may be very limited. Our cerebral capacity having evolved into its present abundance, we humans have a remarkably good chance to do something about these surroundings, and a remarkably great deal to say about which of the range of responses we will choose. Memory, vision, altruism, and a rational free will, all characteristics highly developed in (I would argue *by*) our species, are the resources which under proper circumstances may permit the guiding of what would appear to be a fixed biological destiny into the mutable composition that is humankind. Sometimes, what begins as unalterably grim phenotypical certainty proves to be the malleable stuff from which can be molded a living reality of hope and fulfillment.

Added to these various influences and effects is the possibility of *mutation* of one or more genes. The word itself comes from the Latin *mutare*, "to move or alter." As used in biology, it refers to small inheritable changes that occur from time to time in individual genes of the DNA. Such changes are molecular alterations of the nucleotides, and they can take place in any of a number of ways. Typical examples of the mechanism by which it happens are incorrect pairing of a base during DNA replication, or movement of sections of DNA strands into other parts of the molecule. Also, certain molecules called mutagens have the ability to change the bases. Cells have enzymes specialized to repair such damage, and they are usually effective. But sometimes they fail. The average mutation rate for any gene is one in a million replications, making them very rare. Nevertheless, there are so many of us in any species that the total number that elude the repair enzymes in a whole population of people is quite large over time.

Mutations may be spontaneous and random or they may occur as the result of some factor in the external or internal environment, such as radiation or chemicals. Based on their effect on the organism, the DNA changes fall into one of three categories: those that may help it to survive and reproduce, those that harm it, and those that do neither.

Harm ordinarily leads to death of the individual or of his progeny. Those mutations that are neutral have no effect on survival, either because they result in new proteins that are relatively unimportant to the cell; because they occur so late in development that they will be surrounded by normal cells that negate their effect; or because they do not change the protein in such a way as to alter its function. Obviously, a harmful mutation taking place in the cell of an adult is usually of far less consequence than one in the embryo.

Across the vast in-between land that separates harmful from neutral mutations, there is a range of possibilities. In this middle ground are to be found cells whose functions may be adversely affected but not sufficiently to cause death. When enough of these mutated cells have made their appearance, the total organism or person is handicapped by their presence, but the situation is still consistent with life. When the sick embryo develops into a sick fetus, the stage is set for a sick child to be born. There are more than three thousand human diseases (of which the vast majority are fortunately rare) in which this situation exists, most of them involving the cell's inability to produce one or another enzyme. Among the best-known of such conditions is phenylketonuria (PKU), in which the absence of a specific enzyme results in mental retardation. PKU, incidentally, is an example of a genetic disease that can be corrected by adjustments in the environment—namely, the institution of a diet low in the animal acid phenylalanine.

From time to time, a mutation occurs that improves an organism's ability to cope with its environment. Such a change increases the likelihood that the plant or animal will be more successful and will win out in competing for such essentials as food and mates against forms that are without the new advantage. In time, the new form will predominate and produce more surviving offspring than the old, adding to the advantage. Commonly, this results in the eventual dying out of the old. This is the process known as natural selection, and it is the key to explaining how modern plants and animals have evolved from less advantaged, or lower, forms.

Charles Darwin had an optimistic view of the eventual outcome of nature's enterprise. In the penultimate paragraph of *The Origin of Species*, he wrote, "And as natural selection works solely by and for the good of each being, all corporeal and mental endowments will tend to progress towards perfection."

Having described the ways in which a set of genes gives rise to developed characteristics that may be quite different from those predicted, we can now return to meiosis and some unfinished business. In addition to the normative influences on development, some quite aberrant events occasionally take place. It need not be pointed out that such a complex process as meiosis lends itself to the possibility of mishaps anywhere along the way, although it is nothing less than astonishing that they occur less frequently than once in every one hundred live births. Among the more common of the few mishaps, however, are improper separation of one of the new chromosome pairs, an abnormality called *nondisjunction.* When this happens, one of the resultant sperms or ova is left with two of a particular chromosome when it should have one; the other has none and so dies. Should an ovum with a diploid chromosome chance to be fertilized by a normal sperm, the resulting zygote (the zygote is the fertilized ovum, from a Greek word meaning "yoking together") is said to exhibit *trisomy.* Trisomy, therefore, is the condition in which the zygote and all the cells of the individual fetus created from that zygote have three of a particular chromosome instead of two, for a total of forty-seven.

Chromosomes come in a range of sizes and have been assigned numbers largest to smallest from one to twenty-two, with the exception of twenty-one, which is actually smaller than twenty-two. The numbering ends at twenty-two because the twenty-third is the chromosome that determines the sex of the child. It is called either X or Y to distinguish it from the others. Females carry two X chromosomes, while males have an X and a Y. The twenty-two nonsex chromosomes are called *autosomes.*

Trisomy 21, the genetic defect in which the fertilized ovum has three chromosomes twenty-one, was named (in terminology characteristic of that medically ingenuous time) mongolian idiocy when it was first identified in the middle of the nineteenth century, but later it came to be called Down syndrome, after the man who described it. In every case of this condition, the extra chromosome has come from the mother.

Children with Down syndrome illustrate as well as anything I can think of some of the ways in which environmental factors can define how genetic or chromosomal abnormalities actually affect the pattern of a person's life.

Down syndrome is the diagnosis for approximately 20 percent of all patients in institutions for the retarded in the United States. In 95 percent of cases, its cause is pure trisomy 21 due to nondisjunction; in the remainder, the chromosomal abnormality is more complex. Although it is certain that the imbalance that trisomy creates in how a gene expresses itself is the cause of the clinical findings in Down syndrome, the way this comes about is just beginning to be elucidated.

Prior to the widespread use of amniocentesis, about one in twelve hundred Americans was born with Down syndrome, amounting to some three thousand each year. About a quarter of these were (and still are) the children of women beyond the age of forty. Although mothers of all ages give birth to babies with Down syndrome, there is a gradual increase in frequency as the mothers become older. Several theories have been proposed in an attempt to explain why the possibility of nondisjunction increases with age, but they are too speculative for any certainty.

An experienced clinician can usually make the diagnosis with precision, but it is common these days for amniocentesis, which has an accuracy of 99.4 percent, to have been done prenatally. In this procedure, a needle is inserted through the wall of the mother's uterus to draw off a sample of the fluid surrounding the fetus; this sample can then be subjected to any of several studies, including microscopy, for congenital abnormalities. Should the parents desire it, early aborting of the pregnancy is available when a serious departure from normal is detected.

In general, the IQ of children and adults with Down syndrome is symmetrically distributed around the mid-40's, which puts them in the moderately retarded category. But they tend to have other problems, some of which, such as the 40 percent incidence of congenital heart disease, can be life-threatening. Another problem of major concern throughout the life of a person with this condition is the predisposition to infections of various sorts, due to congenital disturbances of immunity. Low thyroid function, hearing problems, and cataracts are also commonly seen. For uncertain reasons, muscular tone is poor. The frequency of Alzheimer's disease among patients with Down syndrome is six to ten times what it is in the general population, and it develops at a much younger age. The result of all of these associated problems is that far less than 50 percent of people born with Down syndrome survive beyond the age of thirty and less than 3 percent beyond fifty. However, in the absence of congenital heart disease, nearly 80 percent will reach thirty.

No matter Down's intention when he decided to call his patients mongols, the name did provide a striking image of certain aspects of their appearance, tending to keep it fixed in the minds of physicians. Much of what he attributed to racial characteristics is the result of abnormalities in the bony structure and skin of the skull and face. The head is short and the forehead is broad. The general effect of facial flatness is accentuated by a poorly developed upper jaw and the small nose with its lack of a prominent bridge. The most specifically "mongoloid" qualities are two distinctive features around the widely set eyes: an upslanting and narrowing of the palpebral fissures, which are the slits between upper and lower lids, and a vertical fold in the skin of the nose at the inner angle of the eye. Actually, even this is somewhat in error. The fold in an Asian eye is a continuous one around the upper and inner aspects of the eye, while the Down fold is restricted to the inner angle.

Certain other visible features will often alert a pediatrician to the diagnosis of trisomy 21 immediately following the infant's delivery. The most prominent of these are widening of the space between first and second toes and the presence of only a single crease across the palm. The child's ears are small, he will have an increased amount of neck skin, and the bones of his extremities will be shorter than normal.

It has been my good fortune to befriend a young man with trisomy 21 whose achievements reflect, if not an undiluted triumph over biology, then at least a strikingly successful thwarting of its attempts to stand in the way of his independence. His name is Kirk Selden, and he was the co-captain of Team Connecticut, my state's entry in the 1995 Special Olympic World Games. As if to turn on its head the well-documented symptom of poor muscle tone and strength in trisomy 21, Kirk's Olympic event was the power lift.

At the age of twenty-three, Kirk Selden has upside-downed far more than merely an inborn set of handicaps. If an unpardonable pun may be pardoned: It would not be inappropriate to point out that his life thus far has represented the epitome, in fact, of the upside of Down. When reasons are sought for Kirk's successful adjustment, the search leads, as it almost always does in such situations, directly to home and hearth. It leads even more directly to an attitude that pervades the Selden household, put into virtually aphoristic form by Kirk's mother, Debbie, when we spoke about it not long ago. Referring to the aim with which she and her husband, Van, brought up Kirk and his younger brother, Christopher (who does not have Down syndrome), she said, "We wanted them to grow up loving themselves."

In what seems a simple sentence, Debbie Selden was articulating the basis on which all devoted parents attempt to raise the young people entrusted to their care, whether handicapped or not. Among the several reasons for the success that crowns at least some of such efforts is the fortunate ability to identify, whether or not we are conscious of it, with our children as they are growing up. Our responses to their behavior (including its least admirable qualities) are informed by our recall of what our lives were like at their age, and at the same time we try to instruct ourselves about their journey toward adulthood. We strive to create an atmosphere for them that fulfills the universal human yearning to be understood.

When a home is steeped in the sensibility that everyone in it is committed to an understanding, even an appreciation, of everyone else in it, a child is likely to remain emotionally rooted within that nurturing soil, even during those sometimes prolonged periods of inner turbulence when a wish to be understood by parents or anyone else seems furthest from his mind. When it is time to go out into the world, the very fact that the roots of understanding are so strong and deeply implanted in the soil of home makes them flexible and capable of growing across long distances. The most firmly rooted of our children travel confidently to the farthest destinations, metaphorically and sometimes physically, from our hearths. When they put down new roots with new people in new places, the old ones remain undisrupted. They never cease being conduits for sustenance, nor do they stop deepening. If we are to be successful parents, each year should enhance our children's ability to be sustained by the nurturing roots whose source is home. The search for stability most properly begins with the search for understanding.

It seems a commonplace to observe that understanding another person requires the ability to imagine oneself in his place, to see the world through his eyes. Something of the sort has been expressed in so many religions and philosophies that the principle has become virtually axiomatic. With minor variations, one culture after another has weighed in with its own version of the doctrine that the basis of morality is to be found in imagining oneself to feel the emotions of another, to go a step beyond empathy.

The notion of projecting oneself into another's place is the basis of what Christians have called the Golden Rule since the seventeenth century, it having been promulgated sixteen hundred years earlier in the gospels of both Matthew (7:12) and Luke (6:31). But Confucius had already beaten both New Testament authors to it by some five centuries, Aristotle by four, and Rabbi Hillel by the length of a lifetime. Hillel, of course, derived his dictum by a gloss on the injunction in Leviticus to "love thy neighbor as thyself."

But nowhere is the principle of imagining oneself in another's place more sublimely expressed than in the writings of a young English poet. Paradoxically, the poet turned to prose as he instructed his readers that "The great instrument of moral good is the imagination." It is imagination, he pointed out, that enables us to understand one another, and through the understanding to feel love. Love, in turn, is the basis of morality. Here is the twenty-nine-year-old Percy Bysshe Shelley, in *A Defence of Poetry.*

> *The great secret of morals is love; or a going out of our own nature, and an identification of ourselves with the beautiful which exists in thought, action, or person, not our own. A man to be greatly good, must imagine intensely and comprehensively; he must put himself in the place of another and of many others; the pains and pleasures of his species must become his own.*

Shelley was writing here of the use of imagination to absorb from others all that has beauty around us, by appreciating their comprehension of it; of learning to experience the beauty through the eyes and minds of those who have already discovered it, or perhaps created it. And yet he writes about something else, as well. He is telling us that without imagination of another's mind there can be no understanding of that other and therefore no love, and without love, there can be no morality. He tells us further, I believe, that the imagination to see ourselves and the world around us through the eyes of another must precede all else, because without it, there can be no depth of knowledge of those we would nurture. Our gift of imagining allows us to understand what is within them, and thereby to perceive what is lacking that only we can provide.

I would add that not only must we see with the eyes of others but we must also look *into* the eyes of others and see ourselves reflected there.

"The pains and pleasures" of those who need us must become our own if we are to live up to the expectations they and we ourselves have of us.

But to "imagine intensely and comprehensively" the consciousness of another, we need some similarity of experience in which to base our imaginings. Without it, we lack a path of insight into the perceptions and interpretations of those whose world we seek to enter. Shelley, and the theologians and philosophers, addressed an audience whose orientation was toward people fundamentally like themselves, sharing a commonality of what we call human nature. How can moral imagination be attained when the subject is someone whose interpretation of his surroundings is clouded by the gauzy veil of mental retardation?

The answer may prove to be far simpler than might be supposed. Much of our imagining takes place without conscious effort or even awareness. Our behavior toward others is predicated with virtual automaticity on the probability that they are much like us. When we treat people with kindness, we expect a certain response, and when we treat them with hostility, we expect another—we usually get what we expect. Our problems in relating to people do not as a rule arise from a lack of predictability of the other so much as they do from a lack of our own conscious awareness of the motivations and messages we transmit to him. By and large we find it much easier to understand the other fellow than to understand ourselves. Most of the time, an unexpected response from someone is not caused by his mishearing of the message, but by his hearing very clearly a message we did not realize we were sending. In many ways, we interpret others far better than we interpret ourselves.

The retarded are not beings from another planet. There is no complex code to be cracked that permits the reading of mystical texts within their minds, no key that would unlock the sealed gate that prevents entry into the circuitous alleyways of what we conceive to be their impenetrable minds. When retardation is not profound and when it is not complicated by such mountainous challenges as severe autism, for example, the basic rules of human interaction apply just as they do for the rest of us. The vast majority of the retarded tacitly and without being aware of it ask of us that we look to Shelley's principle, to our moral imagination, as the key to understanding them.

The thesis that the retarded follow the same core development stages as their fellows with normal intelligence underlies the principles of current medical and psychological thinking about their upbringing. Even if biological constraints hinder improvement in those intellectual capacities that are determined by strictly organic factors, emphasis on social, emotional, and adaptive development fosters circumstances that maximize the possibility of an enjoyable, useful life. Full inclusion into a family's daily doings is the beginning of a process that leads toward participation in the mainstream of a community.

By this thesis, the vast majority of the retarded are not to be treated as though they have some unique psychology that excludes them from sharing the same needs as the rest of us, or being affected by similar experiences. Their social and emotional development are keys to adaptation in the same way as for people of average intelligence. When a family is able to overcome its initial grief at the realization of their infant's disability and then to go on to the point not only of acceptance but of actually taking pride in his accomplishments, that growing child will amply reward their confidence in him, and in themselves.

If I have absorbed nothing else from my contact with the Seldens, observation of their interactions with one another has substantiated my belief in the importance of everyday things. It is in the accumulation of attitudes, moods, small occurrences, bits of conversation, shared pains and pleasures—all of it with the automatic ease that comes of long custom—that we live our lives, not in the big events. There is no question that the big events, particularly those unexpected, have the potentiality to change the direction of lives, but the background of accumulated years of daily living against which they occur is the crucial determinant of the extent of that change. As the man said, it's the way you play the cards. This, then, is the sum total of child rearing—every moment of every day, we are teaching life to our children. There is no such thing as quality time and there is no such thing as downtime—it is all just time, and none of it goes unnoticed by a developing mind.

Steadiness is, in fact, among the most difficult of all the problems in raising a child with Down syndrome. "Many parents," Van Selden points out, "have trouble dealing with the day-to-day push, because it's so draining." Careful, prolonged study of the child is required, and especially imagination.

> *I'd like to think I've been able to identify with his situation, not because he has Down's but because he's a person going through life cycles. Although he's twenty-three now, I have to remember that in*

> *some ways he's ten, and I've got to switch gears by remembering ten. It hasn't been easy, but I've come to know him well enough to be able to do that. For example, we sometimes almost forget he's retarded, and just then he says he has to get to the TV to turn on* The Power Rangers *or some other show for children.*
>
> *Like all these kids, Kirk has his own strengths and weaknesses. We've worked to find the strengths and support them, at the same time we minimize the weaknesses.*

The emphasis placed on identifying areas of strength and competency finds full expression in the philosophy behind the Special Olympics. By the time a young retarded person takes part in these games, he has trained for years and has been concentrating not only on his abilities but on his sense of purpose, as well. Even one-man events demand teamwork with coaches and other individuals working on the same skills. Systematic studies carried out at the Yale Child Study Center have shown that "sustained, active participation in the Special Olympics is associated with remarkable improvement in social competencies, such as the ability to form friendships, communicate with others and participate more fully in the community. Indeed, the participants' level of work and friendship may far outpace their tested intellectual level and approach that of individuals with normal intelligence." The director of the studies, Dr. Elisabeth Dykens, has said, "The opportunity for engagement in Special Olympics not only strengthens athletic skills and motivation but it also helps enhance the adaptive skills necessary for participation in the workplace and full responsible sharing in family and community life." Considering that mental retardation is defined not only on the basis of an IQ below 75 but also as involving demonstrable deficits in adaptive functioning, such a statement says a great deal about what can be accomplished when expectations are high. A retarded mind needs to develop socially, and it sometimes can also develop intellectually up to a point. But none of this is likely to take place unless expectations are had of it and demands are made on it.

The Seldens had expectations of themselves and of their son almost from the moment of his birth. Kirk was Debbie and Van's first child. She was a twenty-five-year-old speech pathologist and he, at thirty-one, was beginning his career in public administration, working for the state of Connecticut. They had done everything right—dutifully attending Lamaze classes, learning all they could about pregnancy and the care of the newborn, and preparing themselves for the wonder of parenthood. As in the first pregnancy of all couples whose lives are going well and whose future seems bright, it was a very exciting time. For the full nine months, there wasn't a cloud in the bright blue sky.

At precisely the right instant in the early period of labor, Van took Debbie to the hospital. The labor and delivery were as uncomplicated as the pregnancy had been. Kirk's was a well-orchestrated and quite perfect natural childbirth. The baby boy was gently placed in Debbie's waiting arms for that first glorious glance so joyously anticipated by human parents, that first moment that is to begin a lifetime of memories. The word *beatific* has few uses in these fast-paced and often cynical times, but beatific is the only way to describe the smile that appears on a mother's face when she first gazes raptly down on her newborn babe.

> *But as soon as I looked at him, I knew. I told myself, There's something wrong—this child has Down syndrome. I had been working in a classroom with preschoolers and two of them had it, so I was very familiar with it from that and from some other children I had worked with. The obstetrician said I was mistaken, but I knew better. But then he had me sedated, and I didn't know why. A few hours later, they came in and told me I was right.*
>
> *You go from the highest of highs, and then within seconds to the lowest of lows—anticipating the birth of this child, absolutely delighted to be a mother—and then realizing that the child you expected is not the child you have.*
>
> *I was devastated, but Van wasn't. He had his good cry and then he said, "Well, does the fact that this baby isn't perfect change the reason we decided to have a child? It doesn't, does it?" He had been an artillery officer in Vietnam not long before that, and it somehow put things into perspective for him in a way I was incapable of at that moment. I'm not sure I could have done all of this as easily as I finally did if it hadn't been for Van. Not only was he incredibly understanding of how I was feeling but he was also absolutely delighted with his son.*

I asked Van to explain. The near-perfection of Debbie's speech therapist's diction is modulated by the lilting softness of a few barely perceptible whisps of the South, lightly carried on the very surface of her tongue since girlhood as a minister's daughter in Mississippi. Van, on the other hand, speaks with the authoritative precision of the military and with the crisp directness of an administrator who is firmly in charge of himself and perpetually on the ready for the next challenge. He has what professional soldiers call "command presence." His belly is flat, his back is straight, and he turns his eye directly on whomever he is addressing. The no-nonsense image is betrayed, fortunately, by a twinkling smile and regular bursts of deliberately silly humor. Like his wife, he enjoys—savors, in fact—the company of people who approach serious matters without solemnity.

> *Life is a series of adjustments. I've known that for a long time, but being in Southeast Asia made it even more clear. So you adjust and go forward, and that's that. It's my philosophy.*

Van's adjustment and determination to go forward were almost instantaneous, but Debbie took longer. For several weeks, she felt depressed and withdrawn, but it would have been unlike her to persist in pessimism. Finally, an intellectual acceptance was followed by her complete emotional emergence from the somber aftermath of her baby's delivery. The whole process took only two weeks. "In the end, I came to peace with the situation, in my heart and not only in my head."

So much had she and Van "come to peace" with the fact of their child's disability that the next decision was easy. The doctors told the new parents they had three options.

> *They said we could take Kirk home and grow attached to him, and learn to love him—and he would ruin our lives. Or we could take him home, try to stay detached, and institutionalize him when he was five or six; the other option was to institutionalize him right away.*

They never hesitated. Kirk was brought home, and the hard work began in earnest. It was Debbie's experience in special education that made the Seldens determine to push intensively with the development of their child's verbal and motor skills. As she puts it today, "I always felt there was more we could—should—be doing." Kirk was taken to Boston

Children's Hospital within a few months for advice about early intervention. The Seldens returned to Connecticut with the plans for an entire program. At fourteen months, Kirk was entered in New Haven's Open Door Nursery School for children with special needs. It says a great deal about his parents' outlook that he was the youngest child ever to have been enrolled there. This was to be the start of an educational process that would last until he was twenty-one.

Kirk did well at Open Door, but in the meantime another misfortune had struck the family. When Kirk was a year old, Debbie gave birth to a little girl, normal in every way except that she had severe congenital heart disease. The child died after cardiac surgery when she was four months old. None of the anguish around that time was permitted to interfere with the attention paid to Kirk's development.

When he was three, the directors of Open Door recommended that Kirk be enrolled in a regular nursery school, because they felt he would benefit from being with normal peers on whom to model himself. It was difficult to find a school willing to take a child with Down syndrome, but Debbie persisted until she found a very good one. When Kirk reached the kindergarten–first-grade level, he entered a nearby public school with facilities devoted to special education. He remained there until he was twenty-one. During all of those years, Kirk was treated as just a neighborhood kid by all the children in the area. Debbie says, "He's always been accepted by the young people in our area, even though their lives and activities separated from his as they grew toward adolescence. To this day, when they come home from wherever they are, they see Kirk in the street and stand around and talk with him and treat him no differently from anyone else, even though their lives have taken such different paths."

From the very beginning, both adult Seldens became active participants in group efforts to aid the retarded. As Van told me one evening, "If you're going to have children, you have to get involved in their lives—that's true whether or not they're retarded. So very early on, we both became involved in ancillary activities that dealt with mental retardation." When Kirk was very young, Debbie founded a support group called Parents of Young Retarded Children, which still meets occasionally, even though the kids have now grown up. She and Van have taken an active part in the Association for Retarded Citizens, and Van has been a major participant, including board president, of an organization called the Vantage Group, whose purpose is to establish group homes for young retarded adults. Vantage now owns five such homes, including the one Kirk lives in. Van has served on the advisory board of Connecticut's Department of Mental Retardation. Debbie speaks of these activities with a sense of mission.

It's been important for us to make the lives of people with mental retardation easier and better, and to try to change some of the societal mores and values—to create a system that was more open and willing to accept all people. Actually, my whole life has been channeled in that direction, even before we had Kirk, and I've continued professionally to pursue a career toward meeting that goal.

We've always wanted to be sure that no one suffered the trauma we did when Kirk was born at the hospital in Hartford. We were faced with a child with special needs and not having our questions answered. We did in-service training for pediatric residents at Yale–New Haven Hospital, trying to do what we could to educate them about these things. We'd speak to new parents of Down syndrome kids whenever it was needed. We were on twenty-four-hour call, and sometimes we went in the middle of the night because that's when parents wanted to talk.

You know, when parents can't adjust, it's because of an inability to accept the child, and a rage that never goes away, of "Why did this happen to me?" We have friends who've never fully accepted it.

When Kirk was two, his parents adopted an infant of mixed race, white and black, whom they named Christopher. The closeness that has developed between the two boys is among the greatest rewards of the sense of solidarity that has been fostered within their family. "Our greatest delight is the joy they take in each other. The two of them are our dreams."

A transformation took place in the hierarchy of their relationship when Kirk was about ten and Christopher was eight. The younger boy gradually began to watch out for Kirk, as though he were the older brother. It was at Christopher's insistence one day, for example, that the training wheels on the twelve-year-old Kirk's bicycle were taken off. The two of them worked together all afternoon and when Van came home in the early evening, they were ready—Kirk rode out under his brother's proprietary eye.

Kirk became involved in athletics in 1980, when he was eight and playing Wiffle ball. Gradually, he became interested in several events—basketball, swimming, track and field, skiing, baseball. He cannot think fast enough to do especially well at team sports, so his concentration has been on the individual events, particularly power lifting.

Van has been very active in the Special Olympics. He was the Commissioner of the 1995 World Games held in New Haven—a series of competitions lasting nine days—which brought 7,200 retarded athletes into town, representing some 140 countries around the world. It was a proud week for the Seldens, in many ways the culmination of what all four of them had striven toward day by day during the previous two decades and longer. Kirk competed in two events he had been working on for seven years, the dead lift and the bench press. With his whole family and a crew of hard-rooting neighbors cheering him on from the stands, he lifted 305 pounds and pressed 152 pounds, being awarded a Bronze Medal for each and a Silver for both combined. At the end, his broad victory smile was the most disarming look of unabashed triumph I have ever seen on an athlete. Even the high fives were anticlimactic after that wide-open sunny-faced grin of achievement.

There is a sweetness about Kirk that is almost physically palpable. The evening I met him, he was dressed in dark trousers, a scrupulously clean shirt, and a neatly knotted tie. As we were introduced to each other by Van, he looked me directly in the eye in the manner of his father, swallowed hard, extended his stubby arm, and soberly shook hands with me, just a trifle tentatively but with the firmness of a well-schooled gentleman. He is five feet six inches tall, weighs 140 pounds, and looks rounded and soft, as do most people with trisomy 21. But the impression of softness, other than softness of the hands and face, is an illusion. In fact, Kirk is a squat, muscular fireplug of a man.

We sat at the kitchen table later and talked about his parents and his brother and his home.

> *I think I love them very much. They're peaceful. If you have problems, just go to Debbie and Van. My father is peaceful. He's kind to everyone. He never hurt me or my brother or my mom. He's very*

comfortable—same thing with my mom. I feel happy for Debbie and Van, because they raised very good sons.

Chris, my brother, he's kind and peaceful, too. If you know him, he's there for you. If you have problems with him, he doesn't say anything. We go outside and play some basketball, come in and watch TV. Sometimes we play some pool.

Kirk told me of meeting President Clinton at the Special Olympics and pointing out to him that the First Lady is much better-looking (*beautiful* was the word he used) in person than on TV. He had good things to say about his meetings with Jean-Claude Van Damme and Arnold Schwarzenegger. Schwarzenegger and his in-laws, the Shriver family, have been a driving force behind the success of the Special Olympics, which were originally founded by them and are now led by the actor's brother-in-law Tim. When I asked what the muscleman is like in person, Kirk responded with a knowing smile. "Arnold," he said, laboriously drawing out the vowels in a way that is characteristic of his speech patterns, "likes to talk."

With me at least, Kirk speaks very slowly and with a quality of deliberation, as though picking over his words before choosing them one by one. Every half dozen or so words, there is a bit of a stutter, manifested by a slight hesitancy between the first consonant and the vowel following it. I've noticed that he is more likely to go somewhat faster when speaking to his parents or friends, and from time to time, he will even fire off a short rapid-sequence sentence. The reason for his careful delivery seems to be more complicated than would appear—it's not just that he needs extra time to think. When at one point I asked him what he considered to be the biggest problem of growing up with Down syndrome, he replied, as though describing strangers, "Well, they have speech problems. That's what they have with Down syndrome, speech problems." When I pressed him, he bit by bit made it clear that his own speech problem is caused by the difficulty of exhaling in coordination with his vocalizing of sounds, or at least that's the way it seems to him.

Like virtually all people with trisomy 21, Kirk's round, pudgy features give him a baby-faced look, and the impression is enhanced by a minimal but definite liquidization of his *l*'s. His general sweetness of disposition, the easy smile so often accompanied by the twinkle of some apparent inner joke, and that baby face combine to produce an appearance that can only be described as irresistible. He seems literally to radiate unself-conscious pleasure at being with others. As we sat at the kitchen table the first evening, I more than once had to fight off an urge to get up, throw my arms around him, and plant a kiss on his chubby cheek, as an adoring uncle might do to his lovable nephew. Had I done that, I feel sure he would have understood and laughed at the fun of the thing. He is a young man accustomed to the warmth of spontaneous affection.

The certainty of others' feelings toward him is only part of a general sense of self-confidence that Van and Debbie had prepared me for. Kirk unabashedly accepts compliments as though they are the most natural thing in the world, and he points out his accomplishments with well-justified pride. When given a verbal pat on the back, he is likely to say, "That's right." He has no doubt that when the next Special Olympics rolls around he'll be power-lifting far more than his present 305 pounds. The way he puts it is, "I got more in me."

For the past two years, Kirk has lived with two other retarded young men in a small house about two miles from his parents' home. It is owned and furnished by the Vantage Group and is supervised by a manager who spends a good part of every day there. The three residents keep the place in scrupulous order. They take turns shopping, cooking, and cleaning, and they do a much better job of it than I am accustomed to seeing in homes shared by Yale graduate students. Kirk's large, well-ordered room is adorned with the memorabilia of his life, as any young man's would be. In addition to several large pictures of his family, athletic trophies, plaques, photos (including one with the loquacious Arnold), and certificates are prominently displayed as though an impressionable young woman might pay a surprise visit at any moment. And of course, such a thing is not only possible but frequent. Kirk has had several girlfriends in the past few years, and has maintained a wide circle of friendships with many of the young retarded people with whom he went to school. As far as Debbie and Van have been able to ascertain, there have been no sexual relationships, at least not yet.

The three young men get along well together. The other two, both named Jeff, also have trisomy 21, complicated in one by a minor degree of autism. They seemed joyous in one another's company on the evening I visited, teasing and horsing around like a trio of high-spirited fraternity brothers. Kirk introduced me to Jeff and Jeff, and they immediately included me in their high jinks. There is something about being in the company of lively young men that always puts me into a frame of mind similar to their own, and I couldn't help but interject a wisecrack or two. As I stood in the doorway preparing to leave a bit later, Kirk wrapped his powerful right arm around my shoulder and pulled me to him with such unexpected suddenness that my feet were yanked from their planted position. With a broad grin on his face, he exclaimed to one of the Jeffs as I stood there awkwardly immobilized in the vise of his unyielding grip, "I like this guy!"

Kirk did something even more remarkable that evening. Although I had been introduced to him as Dr. Nuland and tried to convince him to call me by my nickname of Shep, he insisted on referring to me in quite another way—three separate times he called me Dr. Nudelman, a remarkable error considering that "Nudelman" was precisely my family's name before my cousin Sol invented "Nuland" some time in the 1930s and thereby accelerated the process of making real Yankees out of all of us.

Kirk has a job, and it is not his first. He works three hours a day at a local restaurant which is part of the Red Lobster chain. He polishes woodwork there and takes care of the silverware, which he enjoys far more than he does his volunteer work at a nearby nursing home. The old folks with whom he tries to play checkers and other games do not respond with any kind of enthusiasm, and that takes his enjoyment away.

The process of learning to read has been laborious for Kirk, and was made even more difficult than it might have been by an element of dyslexia. It is only in the past few years that Debbie's persistence and skill have borne fruit, and he can now read handwritten notes and similar printed material with comprehension. His favorite reading for pleasure, he told me, is *The Little Engine That Could.* Justifiably, Debbie calls the conquest of reading the mountain she is most proud of having climbed with her son.

Debbie and Van Selden knew something about themselves, and intuited a thing too, while their son was still in his cradle. What they consciously knew about themselves was that they were incapable of turning their child over to an institution or to anonymous caretakers. What they intuited was another incapacity: They did not have it within themselves to love their child less or to place him at an emotional distance just because he was born retarded. Well before it was generally appreciated that integration into a family and community would provide the optimal developmental perspective, their hearts told them what the experts were only beginning to say. Though he had special educational and social needs, there was nothing special about Kirk's need to be understood. It has been their moral imagination and their inability to withhold the devotion of their hearts that has brought about the interweaving of experience and biology that is personified in the adaptation of their son to the world in which he lives. Less well-informed people might consider Kirk's accomplishments a miracle. For Debbie and Van, it has been only what they expected.

7 | THE ACT OF LOVE

If, as I have claimed, human beings are somehow greater than might be inferred by the study of only their physical and biochemical aspects, is it possible to discover the basis of the as-yet-obscurely-defined qualities that go beyond mere biology? Do we have the tools to investigate without either religious or atheistic bias the essence of humanity that fills the gap between what we have been given by nature and what we have made of it?

In order to answer these questions, we must begin our searchings with a thoroughgoing understanding of present-day knowledge of human biology. Anything less deprives us of the necessary background of information against which any theorizing must be measured. Only by starting with certainties can sense be made of the uncertainties, and of our shared experience of transcending the mere requirements of assuring the survival of DNA. The entire long history of our species's acquisition of scientific knowledge is founded on the attempt to make sense of experiences and phenomena that, as long as they remained unexplained, were likely to be attributed to mystical forces. Endless observations have been made about human capabilities, adaptability, and response that continue to defy interpretation, perhaps because they have been thought to fall somehow beyond the purview of scientific study. But as Shakespeare tells us in *The Comedy of Errors*, "Every why hath a wherefore." Philosophy is hampered without science, as science is hampered without philosophy.

If every cell in my body has exactly the same DNA and is the descendant of a single fertilized ovum, how is it that I am made of so many different and easily distinguishable *types* of cells, amounting to some two hundred separate categories in all? And how is it that each category is specialized to perform certain distinctive functions? If a cell of the intestinal lining and a muscle cell, for example, have identical DNA, why do they look and act as though they are unrelated to each other? Why is a muscle cell able to contract (which a cell of the intestinal lining cannot) and a cell of the intestinal lining able to manufacture digestive enzymes (which a muscle cell cannot)?

The answer is actually not difficult to come by, even though it has taken many decades of research in hundreds of laboratories to find it. What a cell looks like or does depends on the proteins of which it is made. Different proteins build the constituents of a cell into different patterns, thus causing them to look different and do different things. Each type of cell uses its DNA in its own way to create products that are unique to the specialized job it needs to do.

All cells do have many proteins in common because they must do certain similar basic things to stay alive. We might, for want of a better term, call these *housekeeping* proteins. But in addition to housekeeping functions, each type of cell also has certain task-specific proteins that either are not present in other cells or are present in only small quantities. The main job of the red blood cell is to transport oxygen, and for this it carries many molecules of a huge protein called hemoglobin. For other of its lesser responsibilities, it has task-specific proteins unique to it. Of course, it also carries its full complement of housekeepers. In like manner, the intestinal cell has digestive enzymes. And so it goes.

The cell accomplishes its journey to uniqueness by using only that part of its DNA that will create products enabling it to perform its assigned functions. It does this by a process equivalent to switching genes on or switching them off. Of all the vast array of proteins that it is possible for the DNA in our twenty-three pairs of chromosomes to make, each cell has the ability to control which ones are actually produced, in what quantity, at what rate, and precisely at what instant. The mere fact that a gene exists for a given polypeptide chain does not necessarily mean that this particular polypeptide chain will actually be made. In fact, the great majority will *not* be made.

Another way to say this is that each cell has methods of controlling how genes express themselves. It does this by repressing some parts of the DNA and activating others. This ability is called selective gene expression, and it is carried out by enzymes, hormones, and molecules called, naturally enough, gene repressors and gene activators. It is selective gene expression that allows cells to control not only which gene products are made but also the amounts of those products, their rates of production, and just when in the life and function of a cell they appear.

The specifics of bringing all of this about have not yet been completely elucidated, and they are, in any event, the stuff of textbooks of molecular biology. For our purposes, it is worth pointing out that gene expression can be affected at any of the several steps between DNA and its ultimate products. The first point of control, and perhaps the most important, is at transcription from DNA to RNA: Mechanisms exist within the cell to determine which genes are actually transcribed into RNA, and how much RNA is produced. The next point at which changes can occur is that the RNA can be modified, either by chemical changes in the molecule or by actually having a bit cut out of it by specific enzymes. Still later mechanisms exist to choose which transcripts will be translated into proteins, or even to change the character of the protein that is produced. While some or all of these processes are having their various effects on gene expression, enzyme activity is also being controlled in such a way as to affect the rate and amount of protein manufacture.

Gene expression continues even after a cell has become highly specialized and is carrying out its particular job in the life of the total organism. An example of this is the way a mature cell responds to signal substances like hormones by activating or deactivating specific genes within itself. The fact that these processes are reversible allows a dynamism that adds immeasurably to our ability to react and respond to changes in our environment, whether the stimulus is at the level of the cell or the entire body.

The result of selective expression is that only about 7 percent of genetic

DNA sequences in any given cell are ever transcribed into RNA. Still, this allows for the manufacture of plenty of proteins to carry out all the huge number of activities constantly going on in each cell. Some idea of the magnitude of these activities comes into focus when a few numbers are contemplated: a typical human cell synthesizes some twenty thousand distinct types of protein molecules from its DNA, of which many are present in only small amounts. But some two thousand of these twenty thousand types of molecules are present in quantities of fifty thousand or more identical copies. This means that even if we count only those molecules present in amounts of more than 50,000 each, the total is still a very minimum of 100 million protein molecules in each cell. Such a staggering figure gives some idea of the swarming immensity of biochemical activism within us, all in the service of constant vigilance and response in order that life may continue. Add to this the fact that some of the molecules present in much smaller numbers carry out duties that are among the most important activities of the cell. It is a measure of our biological success that we go on day by day, living out our lives in total unawareness of the turbulence it takes just to keep us alive. We are unaware also that the multitude of processes intended only to sustain us carry with them the potentiality of being used to enrich us. Overseen by the human neocortex, the enrichment of human existence has been the unpredicted outcome of countless trillions of trillions of cellular transactions.

Because gene expression is under the control not only of enzymes and hormones but of an assortment of other variables within and around the cell, it is clear that the *milieu intérieur* influences these variables. Such influences are important not only as the fertilized ovum is growing into an embryo and then a fetus but also when it is a fully formed human being. They continue to be important throughout our lives. More than the genome alone, then, determines development, adult functioning, the interpretation of surrounding events and—ultimately—perceptions and behavior. The environment (some of the internal aspect of which is, in turn, dependent on the genome) also plays a substantial role. At all steps of the undertaking, the totality of what we are is subject to the way in which we adapt to what is going on around us, at every level from the cellular to the entire worldly circumstance in which we lead our lives.

A name has been given to the process by which gene expression allows a single fertilized ovum, or zygote, to give rise through the course of embryonic development to those two hundred types of cells, each distinctive in size, shape, and function. Because its result is the production of a wide variety of different kinds of cells specialized to different kinds of functions (and usually located in different kinds of tissues, organs, and systems), the process is called, not surprisingly, differentiation. It is gene expression that induces differentiation into such as muscle, blood, intestinal, thyroid, nerve, and liver cells, to name just a few. It is differentiation that molds precursor organic clay into a recognizable human being, or a rose.

Before any of this, there is the zygote. And before the zygote, there is fertilization. And for human beings at least, before fertilization there is a singular event called sexual intercourse.

To understand the extraordinary phenomenon that is the sexual act, it is first necessary to have some knowledge of the biological equipment we bring to that (quite literally) seminal enterprise. Our reproductive organs differ from the rest of our bodies in one specific respect: Except in regard to the action of certain of their hormones, they contribute nothing to our survival. As though to make up for such a glaring exceptionality, they contribute *everything* to our ability to reproduce ourselves.

Ladies first.

The entire female reproductive system exists to serve the needs of the ovary. The whole complex of uterus, tubes, vagina, and external genitals has as its sole function to ensure that the ovary's product, the ovum, is properly cared for. The hormones that take part in the process do have other auxiliary roles in addition to that, but they are subservient to their primary function, which is to aid and abet the single objective of the ovary.

The ovum wants only one thing: not to die unfertilized. All of those passions and poems, all of those rages and roses, all of that sexiness and subtlety with which humankind has adorned the sweet prelude to fertilization, are in the service of the ovum's need. Empires fall, ids explode, great symphonies are written, and behind all of it is a single instinct that demands satisfaction—the ovum must have its way. So purposeful is the ovum in its quest that it figuratively contorts and drives the mind of humanity in ways we have barely begun to understand. With disguises and subterfuges to entrap and allure, the little seductress makes the goal so desirable that it sometimes seems beyond attainment, and therefore more to be pursued. In professional attempts to make some sense of a moment that, behind all the multitudinous veils of distortion, is biologically meant to be among the simplest and most direct in the entire panorama of life, our psychiatric brethren have given all manner of descriptive names to the psychosexual deceptions of females and males alike.

Suppressed, repressed, regressed, cathexed, sublimated, symbolized, inhibited, fixated, displaced, transferred—these are some of the names given to the maneuvers of our minds as they attempt to avoid direct confrontation with the reproductive cell's essentially simple desire to meet its mate. It should be so easy, and among animals whose cerebral limitations prevent the manufacture of a psychological maelstrom like our own, it *is* easy. But to let it remain easy is not the human way. The human way is that the simplest stimuli are sufficient to send all manner of signals along all manner of pathways, and to make something vastly complex of what was biologically meant to be quite clear. Sexual reproduction is apparently too direct for us—we have taken on this enormous bundle of baggage we call sexuality. The first small stirrings of sensual thought inevitably lead to an uproar of possible responses, taking us finally to regions of mental reconnaissance unknown to any other creature, and demanding choices that determine the pattern of an individual's life—and through him and millions like him, the pattern of a culture.

The profusion of psychological possibilities available to members of the species *Homo sapiens* as they contemplate the sexual act is the consequence of all the neurological and biochemical firepower we have been granted by natural selection, well beyond our closest evolutionary kin. The existence of such an abundance is at once our glory and our albatross. That soaring spirit of ours, so much of which finds its fundamental source in sexuality, is companioned by consequences for which we pay dearly in coin measured by peace of mind. And yet, we would not change it, because losing the grim means losing the great. Without our sexuality, we are little more than a complex of strategies for staying alive, and an ovum's mindless resolve to get itself fertilized.

The ovum's blind quest, though it has a discoverable basis in the laws of chemistry and physical science, is one of the most powerful primordial forces in the creation of what we call human nature. Whether or not sublimated or misdirected sexuality is the driving energy behind absolutely everything, as some claim, it inescapably pervades so much of human thought as to be at the very least among the prime movers. We know that the urge to reproduce is a prime mover in all other animals—why not ourselves? Were it otherwise, our species would die out.

Accordingly, there may always be a chicken-egg controversy, but there should be no equivalent argumentation when the subject is humans and the ovum. The ovum came first and ordained that her commands be followed. Though she has hidden it behind the coquettish skirts of the garment we call sexuality, Ms. Ovum embodies the principle of preservation of the species. Even more than that, she embodies the principle of life.

The ovum must have its way, and the instinct of the entire organism is to see that it does. We seek a course toward reproducing our kind through the maze and morass of contradictory drives, and we find ourselves buffeted by the competing forces of instinct and conscience. We often lose our way and unexpectedly come upon something wonderful, such as music or faith. The psychiatrists call this phenomenon the sublimation of the sexual drive, but we see it as a kind of serendipity and consider ourselves fortunate to have discovered it. But sometimes we find instead something difficult to deal with, and that, too, must become incorporated into what we are. The complex and uncertain journey is not made one iota easier by being undertaken largely in the realm of the unconscious mind, or being directed toward what is ultimately, under its many-layered raiment of sexuality, the simple need that an ovum be fertilized.

At the journey's start and end is the ovary, which makes not only the ovum but also the female sex hormones, estrogen and progesterone. Anatomically, the whole structure of uterus, tubes, and ovaries, plus the broad expanse of folds of tissue holding them all together, look like nothing so much as a small bird seen from below, its wings widely spread. The bird's short, squat body is the uterus and its outstretched wings are the thin layers of supporting tissue, along the thickened upper edge of which lie the two oviducts, or fallopian (for Fallopius, the Italian anatomist who described them) tubes. Suspended on the underside of each spreading wing is the ovary, a ½-by-1-inch rubbery-firm egg-shaped structure lying laterally out near the open mouth of the tube.

As the ovaries are being formed in the embryo, they come to contain a total of almost 2 million diploid cells called oocytes, all with the potentiality to become ova. Each of the oocytes lies surrounded by a structure called the follicle, consisting of fluid within two layers of cells. A female begins meiosis I during her embryonic life but stops while she is still a fetus in her mother's womb, before the stage at which the homologues separate. Not until puberty does the process resume, by which time only some thirty thousand to forty thousand oocytes remain. A mere four hundred or so will eventually leave (at a rate of one a month for the next thirty-five or forty years), each from one or the other ovary, without any regard to an orderly sequence.

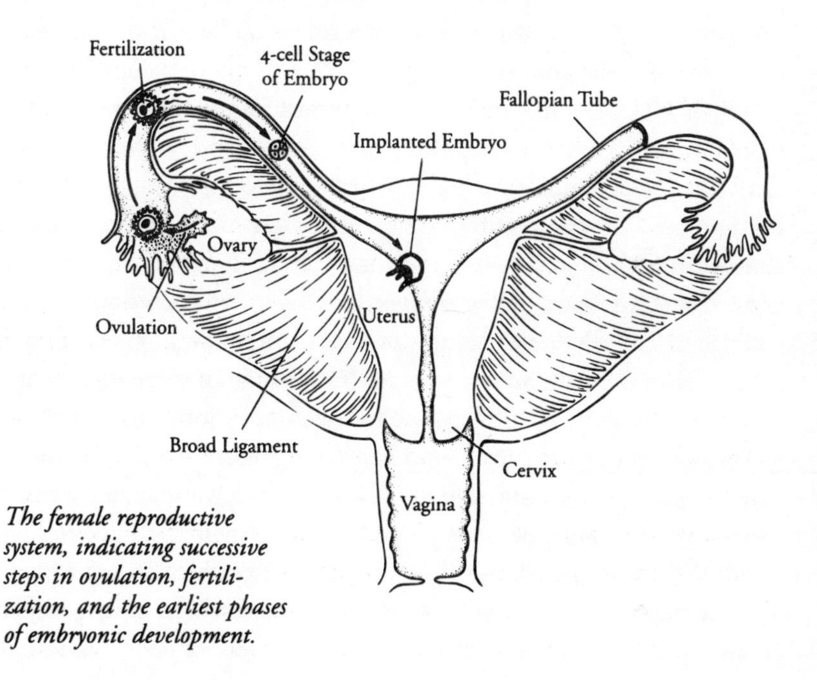

The female reproductive system, indicating successive steps in ovulation, fertilization, and the earliest phases of embryonic development.

The entire process begins in the brain. *Toujours le cerveau*—"always the brain." Conscious or unconscious; automatic, autonomic, or controlled by the will; asleep or awake—there is brain in everything. The autonomic nervous system (see chapter 4) is largely under the control of the hypothalamus, which integrates the system's responses to changes in the internal and external environments. The hypothalamus has been called "the brain of the autonomic nervous system." It and the medulla are involved in a wide variety of the responses of moment-to-moment existence, such as appetite, digestion, circulation, the balance of water in our bodies, and the general category of emotion, particularly sexual behavior, pleasure, rage, and fear. If you've ever wondered how body temperature is regulated, for example, think of the hypothalamus. Within it is a clump of monitoring cells that are sensitive to the temperature of the blood and keep it stable.

One of the functions of the hypothalamus is to act as a gland. It manufactures what are known as "releasing hormones," whose job is to cause the so-called master gland, the pituitary (also known as the hypophysis, lying just beneath the brain), to secrete hormones of its own, which in turn activate other glands. Several examples of this will be described in this book, the first of which plays a predominant role in the process about which we are here concerned.

When a girl reaches puberty, the hypothalamus begins the rhythmic secretion of a chemical called gonadotropin-releasing hormone, or GnRH. The etymology of the word *gonadotropin* is a fine example of the basic simplicity of most scientific terminology. Because *gonad* comes from the Greek *gone*, meaning "seed," it is the general term for either the testicle or ovary. *Tropin* is derived from *tropos*, "a turning," which has come to refer to any sort of change. So a gonadotropin is a substance that makes an ovary or testicle change in some way. Elevations in GnRH cause the pituitary gland to release two hormones called follicle-stimulating hormone (FSH) and luteinizing hormone (LH), which activate the girl's ovaries to begin adult functioning, unarguably a change of considerable magnitude.

Some time after this, the girl starts to menstruate. At the beginning of a menstrual cycle, FSH stimulates the follicle of one of the oocytes to enlarge and mature into a sphere of many layers of cells. Some of these cells produce a great deal of the female hormone estrogen and some make progesterone. By the fourteenth day of the cycle, there is enough estrogen in the bloodstream to stimulate the pituitary to increase markedly its secretion of luteinizing hormone (LH), which until this time was being manufactured in only threshold quantities. By now, the mature follicle and its contained oocyte have migrated to the surface of the ovary and are protruding from it in the form of a blisterlike projection almost half an inch in diameter. The ovum within has resumed its meiosis and completed the first division into what is called a secondary oocyte, which contains the vast majority of the cytoplasm, and a tiny useless cell (called a polar body) with very little of it, soon to shrink away and die.

The sudden outpouring of LH triggers ovulation, in which the follicle opens and squeezes the oocyte out into the abdominal cavity like toothpaste out of a tube, very close to the adjacent entrance of the fallopian tube. Following ovulation, what is left of the follicle remains on the surface of the ovary, but, under the continuing influence of LH, some of its cells transform themselves into a yellowish solid glandular structure called the corpus luteum (Latin for "yellow body"), which secretes primarily progesterone but also some estrogen. Should a pregnancy develop, the corpus luteum will continue to secrete hormones to support it; if, however, fertilization does not occur, the little structure slowly degenerates after about twelve days, and blood levels of progesterone and estrogen fall. But during these twelve postovulation days, the hypothalamus, influenced by the elevated progesterone and estrogen levels, decreases its secretion of GnRH, thereby signaling the pituitary to lower its output of FSH and LH. At the end of the twelve days, the progesterone and estrogen levels wane in the absence of a pregnancy, causing GnRH to raise FSH, signaling the start of a new cycle.

There is no certainty about why, in the absence of a pregnancy, the corpus luteum should degenerate after about twelve days, but it is thought to be the result of yet another chemical influence: The process seems to be one of self-destruction, due to the corpus luteum's own secretion of locally acting hormones called prostaglandins, which interfere with its function.

All of these steps in the menstrual cycle can be seen as being the result of a series of what are, in essence, feedback loops. A fall in the blood level of one hormone is compensated by increased activity in the next gland in the loop, whose hormone activates yet another gland or glands to elevate hormones sufficiently to slow down production of the one whose elevation caused the original fall. Specifically: As a woman's cycle ends, a low progesterone level affects not only the pituitary directly but also encourages the hypothalamus to make more GnRH. The result is an increased FSH, which stimulates the maturing follicle to raise estrogen levels enough to result in an outpouring of LH in much higher than threshold levels. This, in turn, stimulates ovulation, markedly elevating progesterone as the corpus luteum is formed. The progesterone and estrogen tell the hypothalamus to make less GnRH, thus adding to the message they send to the pituitary to make less FSH and LH. Toward the end of the cycle, the corpus luteum disintegrates, with the result that progesterone and estrogen levels again become low, which increases GnRH—and the merry-go-round spins off on another turn.

This reciprocity of one hormone responding to the fluctuating levels of another and determining the output of yet a third and more is one example of the multiplicity of feedback mechanisms that control levels of various proteins and other chemicals secreted into the bloodstream; these proteins and chemicals function as hormones and enzymes throughout the body. It sounds complicated and it is, especially when compounded by the introduction of local hormones like prostaglandins, which are not part of the loop but affect it nevertheless. Still, the principle behind all of these feedback loops is quite simple: As always, the body is constantly compensating as it strives to maintain its equilibrium at the same time each of its organs is striving to do the job for which it exists and for which it is uniquely qualified.

Estrogen and progesterone do far more than what has thus far been described. In the blossoming young woman, estrogen causes the appearance of the various so-called secondary sexual characteristics, such as breast development; the typical female distribution of fat directly beneath the skin, particularly in buttocks and thighs; an increase in the number of small blood vessels in the skin; the widening of the pelvis; and a distinctive fat deposition under the newly hirsute skin of the pubis, resulting in a slightly protuberant curve that has since antiquity been called the mons veneris, or mound of Venus. Venus having been the Roman goddess of love, the appropriateness of the term would appear to be as obvious as the seductive little hillock itself when it pushes against snug garments. The hormone also has the effect of enlarging the young woman's reproductive organs.

During the first week of the menstrual cycle, the increasing estrogen levels cause the inner lining of the uterus to proliferate and thicken, preparing it to receive a fertilized ovum. As noted earlier, relatively small amounts of progesterone are excreted during the twelve to fourteen days of the cycle, but as the corpus luteum is formed, a great deal more is produced, along with the large quantities of estrogen. The increased progesterone causes the thickened lining of the uterus to produce a fluid rich in nutrients. Blood vessels increase in number to support the needs of the new tissue and to transport the components of the nourishing broth, a kind of uterine equivalent of mother's milk, or perhaps mama's manna.

Should a fertilized ovum now make its appearance, the uterine lining is altogether ready to welcome it, so thick, soft, moist, and warm a haven as to justify every meaning of the word *womb*. Neither an obstetrician nor a psychiatrist could have designed a better place in which to prepare for the challenges of the outside world.

But should all the preparation prove to be unnecessary, the uterus wastes no time in futile regret. The decline of estrogen and progesterone levels in the bloodstream induces the constriction of the profuse new arteries in its readied lining, thereby cutting off the supply of oxygen and nutrients. The newly established tissue dies, capillaries bleed, and the whole proliferated mass sloughs away over the next three or four days, to be discharged as menstrual flow through the vagina.

Much nonsense and some good sense have been written about premenstrual syndrome, and I have no wish to add to it. But at the end of this long description of ovarian function and the hormonal fluctuations that accompany it—where such metaphors as merry-go-rounds, spinning, and loops have been invoked, and others like roller coaster and tempest would be equally appropriate—is it any wonder that so many women have mood swings during these days? The real wonder is that those swings are not much more severe than they are.

Compared to women, men get off lightly. Males need not occupy their minds with the practical issue (pun incidental and unintended) of monthly meteorics. They need contend only with puberty at the beginning of the reproductive years and a variable period at their end, a decline that some call the male climacteric. The male sail is otherwise smooth. In fact, the only physical concern the average boy or man ever has about his reproductive apparatus is to avoid being hit in the testicles—that and getting to use it as often as he would like, of course.

The testicles are kumquat-shaped organs about two by a little more than one inch in size and slip around easily within the pouchlike bag of skin and fibrous tissue called the scrotum. Immediately inside the scrotal skin and closely adherent to it lies a thin layer of involuntary muscle called the dartos, whose function is to help prevent testicular injury of any sort. Ever on the alert to stimuli of various kinds, whether associated with fright, pain, or pleasure, the dartos contracts to tighten the scrotum, thereby drawing the testicles up toward the abdomen, where they are somewhat less likely to be exposed. It is aided in this commendable activity by the cremaster, a thin layer of muscle surrounding the blood vessels entering and the sperm-carrying tube leaving each testicle. Contraction of the cremaster—like that of the dartos, always involuntary—pulls the testicles up toward the body, supplementing its companion muscle's action. Its behavior is so predictable that it can be used as part of a physician's physical examination to check reflexes. When the inner aspect of the upper thigh is stroked, the cremaster contracts and the testicle rises.

The dartos and cremaster are particularly responsive during adolescence and young manhood, sometimes giving rise to amusing and/or perplexing circumstances. As a fifteen-year-old, I one day found myself standing in the unaccustomed position of shortstop in a pickup game of baseball being played on an open expanse of flattened dirt near my home in the Bronx—an area called the Jerome Avenue lot. Not being a particularly gifted infielder even on the best day of my life and having been assigned an unfamiliar and quite terrifying position, I soon let my attention drift from the game's action because I became preoccupied with devising strategies to defend my body against any fast grounders that might come my way. As luck would have it, a low, hard-hit one came rocketing toward me in the very first inning, before I had developed all the details of my plan for self-preservation. So deep in thought was I that the speeding, skipping sphere was upon me before I became fully aware of it. I tried to position myself properly, but at the last instant, the stony-hard thing took a sudden bad hop on the irregular surface of the cinder-covered field, bounced off the tip of my outstretched glove, and scooted up the inside of my left leg. The two balls, mine and the horsehide, collided with a nauseating thud, which, though almost soundless in itself, made me roar like a bull elephant with a strangulated hernia.

Needless to say, my dartos and cremaster contracted with lightning speed. Somehow, I kept standing, probably unwilling to fall for fear any change in position might worsen my already-intolerable distress. After what seemed forever but must have been only a few minutes, the pain wore off, my dartos and cremaster relaxed, and my scrotum cautiously eased back into the fabric of my Jockey underwear. I switched to my accustomed position in right field, where I might be beaned by a fly ball but at least stood no reasonable chance of being emasculated. Ball-shy, in both the athletic and anatomical sense, I stayed away from the Jerome Avenue lot for the next two weeks.

It is impossible to study human anatomy, human function, or human behavior without being impressed at every level, even one as seemingly superficial as this scrotal story, with the way in which responsiveness and adaptation constantly reveal themselves as the key to sustaining and reproducing ourselves. What does a muscle fiber in the dartos know of the sperm it rescues, or of a kiss? And how is its reflex guarding of the testicle, savage even while it is protective, linked to an impassioned need to hold another person closely in the radiant moonlight of a gentle spring evening—an embrace meant to lead toward another embrace and finally a culmination that is also both savage and protective, even as its ultimate function is to perpetuate the species?

In the scientific microanalysis of such reductive phenomena as feedback loops and the biochemical principles that make possible a genome, it is easy to lose sight of the path that leads inexorably to the sexual act and the reinforcing relationship between it and love. From those two driving forces, both the result of natural selection, we have gone well beyond even those responses directed toward our survival and our species's continuity. To what was at first only molecular biology, purpose *has* been added: Humankind has created romance, beauty, order, society, and culture. Somehow, all of this has soaked itself into our awareness, even as we transcend our origins and nature's intent. In transcending, we have achieved something lofty, something beyond what might have been only our simple selves.

Part of the something is to be found in our poetry. Here follow a few lines written by a poet ignorant of so much as an eyeblink of genetic or evolutionary knowledge:

> *Each loves itself, but not itself alone,*
> *Each sex desires alike, till two are one,*
> *Nor ends the pleasure with the fierce embrace;*
> *They love themselves a third time in their race.*

Long before Darwin and long before the psychobiologists, Alexander Pope sensed that the drive for continuity is so deeply woven into the fabric of humanity that we give it form in the way we express love. We have learned to use love to nurture the outcome of the unity—a child—that is its fulfillment, and to find joy not only in the unity but in the posterity it assures.

To all but a few of the hundreds of trillions of sperms a man produces in his lifetime, the ovum's call is a siren song, promising nothing but early decay if unheeded or ignominious destruction on the figurative rocks and reefs of the female genital tract if followed. But to those few, it whispers that something of them will persist into the next generation and beyond. Configurations of nucleotides will live on in the distinctive hue of an offspring's hair, or the double-jointedness of his pinky. That is all it takes, and 300 million sperms at a shot are off and running, or, rather, lashing their tails mightily in an attempt to propel themselves and their twenty-three chromosomes apiece toward life and a bit of longevity. A small proportion never lash their tails at all, because they have none. When added to those that have two, or whose tails are coiled, or who have some abnormality of the head, the total figure is 20 percent that stand no chance of making any headway toward their destination.

Packed into each testicle is a total of some eight hundred to nine hundred tiny tubules, each of which is about thirty inches long. Lining the walls of this total of approximately two thousand feet of coiled conduit are the spermatogonia that will undergo meiosis and mature into sperm cells, otherwise called spermatozoa. In the small spaces between tubules lie clumps of cells that produce the male sex hormones.

A sperm has an oval head—containing its twenty-three chromosomes—a short cylindrical body, and a very long tail. Affixed atop its head like a cone-shaped hat is a structure called the acrosome, which contains enzymes that break up certain chemical barriers and thereby facilitate penetration of the ovum in the very unlikely event that the whole voyage is successfully negotiated and their services are required. The sperm's tail carries a great deal of ATP to provide energy for the lashing motions that will propel the entire structure up the female reproductive tract. Along the way from the tubules to their eventual release point at the tip of the penis, the sperms pass through several small accessory organs and tubular structures that add various enzyme and nutrient-containing fluids to maintain their good health. The largest of the organs is the prostate, a chestnut-shaped gland that lies at the base of the bladder, just before the point at which the sperm reach the entrance to the penis. The mixture of sperm and the secretions of these various structures is collectively called semen.

From the tubules in the testicle, the sperms pass into a single tightly coiled tube (called the epididymis, nine to twelve feet long) that lies on top of one of the organ's poles. Here, they can be temporarily stored and undergo certain physical and chemical changes to the acrosome (called capacitation) that enhance its ability to penetrate the ovum; further capacitation will occur when the sperms reach the tube of the female. The epididymis empties into a very long, thin whipcordlike conduit with muscle in its wall, called the vas deferens, or, more simply, the vas. This is the structure divided by a surgeon's scalpel in the male sterilization procedure called a vasectomy. It is readily identifiable by any man who while reading this page or at some later time is prepared to reach into his pants, pick up his scrotum between thumb and first two fingers, and feel for it.

Accompanied by the testicle's arteries and veins as well as some nerves, the vas travels eighteen inches from its origin, at first upward into the abdomen and then downward behind the urinary bladder, finally emptying into a very short tube called the ejaculatory duct. Actually, the ejaculatory duct is formed by the confluence of the terminal portion of the vas and a small cul-de-sac at its very end, the seminal vesicle, whose function is to add fluid ingredients to the semen. The two ejaculatory ducts enter the prostate, converge, and empty into the urethra, the same passageway that carries urine from the bladder out to its exit at the tip of the penis.

The semen's destination is the penis, a pendulous shaft hanging off the lowermost part of the trunk, just below the pubic bone. Of the many remarkable characteristics of this unique structure, perhaps the one least commented upon is that it contains no fat. This means that no matter how much weight he gains, a man cannot make his penis one millimeter larger. Wishful thinking doesn't help, either.

A more visible uniqueness (unless it has been surgically or ceremonially trimmed away by circumcision) is the folding over of skin called the prepuce or foreskin, which provides a hoodlike covering whose value has been debated since long before the swashbuckling David removed it from two hundred dead Philistines to provide Saul with what must certainly have been the most gruesome bride-price ever demanded. (Actually, the scheming king had asked his daughter's ambitious suitor to provide only half that number, assuming that the cocky young upstart would be killed in the effort. But David, like the overachieving biblical Batman he was, doubled the ante. In succeeding, he attained both of his objectives: He got the girl and intimidated the dickens out of his already cowed monarch. The story ends with this portentous statement: "And Saul was yet the more afraid of David.")

With or without a foreskin, the penis consists of three separate cylinders of spongy tissue, geometrically arranged in a form roughly approximating a triangle, in such a way that two lie alongside each other on top and the third lies below and between them. The upper cylinders, called the corpora cavernosa (because they are cavernous as well as being spongy), are somewhat bulkier than the lower, the corpus spongiosum. It is through the center of the corpus spongiosum that the urethra runs. The penis expands at its tip into the glans, an acorn-shaped structure whose exquisite sensitivity is due to the vast numbers of nerve endings in its velvety-soft skin.

Also of exquisite sensitivity and also containing spongy tissue is the analogous structure in women, the clitoris. It has two miniature corpora cavernosa and a small spongy tip that functions very much like the male glans. Less than an inch long and a fifth of an inch in diameter, the clitoris lies close to the point at which the two smaller lips, or labia minora, meet, toward the front of a woman's external genitalia. Immediately behind it is the opening of the urethra. These structures and the entrance to the vagina are protected by two fleshy front-to-back folds of skin-covered tissue, the labia majora (large lips). The pubic hair extends down to provide a thin cover for this entire anatomic area.

The vagina itself is a tube about three to four inches long, whose wall contains fibrous and muscular tissue. Immediately inside its opening and lying just internal to the labia minora, the entrance into the vagina of females who have never had sexual intercourse is partially occluded by a thin curtain of tissue perforated in its center. This fragile guardian of virginity is the fabled hymen, or "maidenhead," renowned since the first glimmerings of the storyteller's art as the venerated and indisputable evidence of a maiden's virtue. Minnesingers, poets, scribblers of certain novels, and all manner of their predecessors in ancient civilizations have enthralled their audiences with mesmerizing tales of bloodied sheets and the bloodying of entire tribes when the bed linen was found to be post-nuptially without evidence of the hymen's wholeness. It is in its own way the most paradoxical of structures: Its integrity in a woman has ever been used as a measure of the integrity of some man, or of the masculine values he protects.

Inspection of the hymen, in fact, is hardly the moral equivalent of DNA testing—its absence would be useless in a court of law, and even its presence is not necessarily dependable. It is hardly an infallible marker of virginity. The thin curtain has a variety of configurations, in some women presenting a formidable obstacle to Venusian delight, in others giving way with minimal pressure and hardly any staining of blood, and in still others being absent entirely. Not only that but—to quote a line that bemused several of my classmates and me when we came across it in the twenty-fifth edition of *Gray's Anatomy* more than forty years ago—"It may persist after copulation, so that its presence cannot be considered a sign of virginity." Those few dry (dare I say bloodless?) words represented the only editorial comment discernible in the massive tome's 1,478 pages. After much speculation about the reason for their inclusion, it was finally decided that the man who penned them must have been recently deceived in a love affair.

The vagina extends inward and upward to receive the lowermost inch of the uterus, a rounded collar-shaped structure called the cervix. Closely adherent to the back of the vagina is the front of the rectum, the only thing separating the two being a thin layer of fibrous material.

If one were to seek a structure in nature whose shape somewhat resembles that of the uterus, it would surely be an inverted pear. The very bottom of the inverted pear would correspond to the cervix, and its upper, widened portion would be the uterus's thick body. The muscular uterine wall is lined on the inside with a lush gland-filled mattress of membrane called the endometrium. It is the endometrium that responds to cyclic hormonal fluctuations by alternately building itself up and then tearing itself down, the sloughed tissue being discharged into the narrow uterine cavity and passing out as menstrual flow via the cervix and then down into the vagina.

The body of a bird is roughly pear-shaped. If the uterus can be seen as the body of a pear-shaped bird, then its outstretched wings would look very much like the two thin, broad sheets of tissue that extend from the organ's sides. As noted earlier, the fallopian tube lies along the upper border of the wing, extending outward from the lattermost point of the upper portion of the uterus on each side. Because the wing is broad, the two sheets of tissue are called the broad ligaments. Each ovary is attached to the underside of one of the broad ligaments, lying so close to the open far end of the tube that an ovum leaving its ruptured follicle can easily find its way there.

The purpose of this detailed description of male and female anatomy and physiology is to supply the setting for that sublime moment at which a couple couples and the man's sperm takes off on its hazardous journey toward the ovum of his dearly beloved. Although quite obviously not an absolute requirement, the entire process is facilitated when it takes place in an atmosphere of courtship, seduction, fantasy, caring—in short, a romantic aura that not only humankind but a vast variety of citizens of the animal kingdom enjoy. What makes it different for humans is the degree of commitment of which we are capable and the emotional (call them psychological, if you prefer) connotations of what we are doing. I refer here to the entire range of consequences of the existence of the sexual act—specifically sexuality. We have endowed the sexual act with a culture all its own, one that permeates every other aspect of the greater culture in which we live. It is in many ways a combination of the ultimate act and the ultimate fantasy that fashion our lives. As much as any other single factor, its fulfillment or sublimation is a motivating force in the molding of the human spirit.

For a man, the first physiological requirements for sexual arousal are fulfilled by fantasy and his five senses. He has thought something, or seen or felt or smelled something. Perhaps a seductive word has been said to him, or he has been put into a predisposed state of mind by the lyrics of a song or the rhythmic beat of sensual music. Not only the touch of a kiss but its taste may be enough to whet his desire. Most commonly, it is a combination of several or all of these factors, but when a man is psychologically prepared, it doesn't take much of a stimulus to make his blood begin to redistribute itself toward his pelvis and its hangings. Although usually in a more gradual fashion than the somewhat hasty stirrings of her mate's, a woman's ardor is inflamed by very much the same proddings, and, perhaps more than his, by those that arise in her sense of fantasy. To perceive a potential partner beginning to respond to such rising excitements is to respond oneself, with added fervor. There is a synergism in sexuality.

At the outset of a man's concupiscence, the arteries and arterioles to the penis dilate in response to messages reaching them via parasympathetic nerves, thus allowing increased amounts of blood to enter the spongelike spaces within the tubular corpora of its shaft. The organ enlarges in all dimensions as the three cylinders quickly fill up, raising the temperature and pressure within the progressively bulging tissue. The expanding corpora compress the veins that are trying to empty them of their rapidly increasing contents, making it even more difficult to prevent engorgement by the rising force of the dammed-up blood. These factors, abetted by the compressive action on the veins of the parasympathetically stimulated *erector penis* (also known as the ischiocavernosus muscle), make the internal pressure rise, stiffening the shaft even more, until it is standing firmly upward like a thick rod of flesh projecting at an angle of approximately twenty degrees from the center of the man's eager groin. Cocked up on the far end of the rod is the softly flanged protuberance of the smoothly rounded glans. Expectant and at the ready stands the turgid male generative organ in its state of full erection.

At the same time, the entire physical and emotional mechanism of arousal has built itself up toward the anticipated crescendo. There is a general tightening of muscles in various parts of the body, as well as an increase in blood flow to certain distinct areas such as the lips, earlobes, breasts, and the inner membrane lining of the nostrils, where a small amount of spongy erectile tissue is to be found. The dartos and cremaster have contracted, propelling the testicles up against the body exactly as mine were when hit by that baseball, but under far preferable circumstances.

By this point, a sense of urgency is becoming stronger by the second, and there begins to be some loss of mental self-control, particularly when the man is young and inexperienced. Caution may be thrown to the winds in the heedless necessity to gratify an overpowering drive of such immediacy that its consummation cannot be denied for another second.

If all is going as it should, the man now inserts the bulging tip of his congested glans between his partner's labia and slides the full length of his rigid penis into her vagina. The action, called intromission, is smoothed not only by her own internal wetness but also by a small amount of mucilaginous secretion that has leaked out of the end of his penis. He then begins the process that sex manuals, in their characteristic unimaginatively workmanlike style, call thrusting, in which he rocks his pelvis forward and backward in such a way that his glans over and over again traverses the entire length of the vagina, all the while rubbing smoothly against its lubricated walls. The slight friction of the outside of a man against the inside of a woman sends countless erotic messages from the abundance of nerve endings in the glans up to the sensory nerves of the penis, and from there not only to the brain but also to the spinal cord. From the cord, impulses are fired back down to the pelvis, completing what is called a reflex arc.

Although the reflex arc is a mindless circuit, it is reinforced by further messages from the intensely excited consciousness of the steadfastly thrusting man, whose pulse and breathing have by then begun racing to keep up with his elevated blood pressure and the mounting (in both senses of the word) frenzy in his consciousness. At the instant when the rate and forcefulness of excitatory stimuli traversing the reflex arc and coming down from the brain reach a critical threshold, the process suddenly bursts out into the almost simultaneous phenomena of emission, ejaculation, and orgasm—which are set off like the instantaneous explosion of perfectly timed fireworks.

Other than those to some of the voluntary muscles, all of the messages have been passing over fibers of the autonomic nervous system. At this instant of culmination, these fibers fire off a blast of sympathetic signals to the involuntary muscle in the wall of the vas deferens, forcing it to contract rhythmically, propelling sperm from the epididymis into the ejaculatory duct. The same group of fibers causes the prostate and seminal vesicles also to contract, with such precise timing that their contents and the sperm arrive in the duct simultaneously. By this means, everything is mixed together into the fluid variously known as semen, ejaculate, and come. Ninety-five percent of the ejaculate has been provided by the secretions of the prostate and seminal vesicles. The rest is sperm.

Synchronization is so perfect that with everything all at once in place this way, messages are at that instant fired along a tiny branch of the nerve to the entire area, the pudendal (here hangs a tale: the pudenda, as the external genitalia of both male and female are often called, derive that name from the Latin *pudere*, "to be ashamed"), to a thin but vastly consequential muscle called the bulbocavernosus. This feather-shaped structure originates just in front of the anus and extends forward to embrace the base of the penis. The nerve impulse throws the bulbocavernosus muscle into a series of powerful spasms of contraction that suddenly spurt the semen out through the urethra and into the receptive warmth of the waiting vagina. The forceful pulsatile nature of the semen's discharge explains why the process is called ejaculation, a term originating from an ancient word meaning "to throw."

But there is more, and it is a more without which I suspect very few of us would indulge in this inelegant and physically awkward form of ridiculous pelvic calisthenics. The more is called orgasm.

The male orgasm can occur without ejaculation and ejaculation can occur without orgasm, but generally they take place together. The word derives from the Greek *orgamein*, which means "to swell." Somewhere at a point lost in time, the idea of swelling to the bursting point must have been introduced—if an orgasm is nothing else, it is the bursting out of an intensity that can no longer be contained,

In essence, an orgasm is a release. Its psychological component floods forth in such a torrential outburst that the conscious mind confuses it with those of its aspects that are purely physical. At the peak of muscular contraction, messages of a rapturously pleasurable nature are sent up to the brain, producing what can perhaps best be described as a sudden outpouring of ecstasy that overpowers the senses at every level of awareness. At the very same time, the tightened muscles in the pelvis and elsewhere begin to relax. An emotional outflow is felt that has no parallel in human experience. It was no doubt with moments like this in mind that Nietzsche wrote, "The degree and kind of a man's sexuality reach up into the ultimate pinnacle of his spirit."

As ejaculation is ending, sympathetic nerves signal the penile blood vessels to constrict somewhat. At the same time, the ischiocavernosus and other tightened muscles ease up. As blood leaves the penis, the erection subsides. Orgasm fades gradually, leaving behind a quite powerful physical afterglow emanating from the pudendal region, lasting a variable number of minutes. A sense of relaxation ensues and a pleasant tiredness is felt. Sometimes, feelings of lassitude or even sadness overcome the spent lover. It is just then, when the perhaps as-yet-unconsummated drives of his partner are being expressed as a need for more lovemaking or even merely the continued cuddling and comfort of love's gentler manifestations, that the man often wants to lie back and withdraw into his own thoughts.

It takes at least ten minutes, and usually longer, before fantasy or feeling can start the series of sensual reflexes up again. Commonly, the man will have no further strong sexual longings for hours or days, and for some men, weeks. If a lover is wise, he will use the time of his afterglow well.

The coordination of sensual peaks in men and women is not commonly so perfectly timed as adults-only television and gothic fiction would have their devotees believe. Not to say that it doesn't happen, but expectations of simultaneous orgasm are in real life more likely to be frustrated than fulfilled. The first obstacle to concomitant coming has to do with the slower arousal of women. The second is anatomical. Having assured itself that the species will reproduce with or without orgasmic synchronicity, natural selection has never been required to favor anatomical or physiological variations that would permit such a devoutly wished consummation. Were it otherwise—had early *Homo sapiens* or Neanderthal females resisted intercourse unless they peaked exactly when their mates did—the world would now be populated with quick-climax women whose clitorises are inside the vagina. Either that, or our species would be assured its continuity only by forcible rape, and civilization would be very different—if it existed at all.

The origin of the disjunction lies in the fact that the main source of orgasmic potential of the vast majority of women is not the vagina but the clitoris, the rod-shaped little structure that exists only for that purpose. Being the analogue of the penis, it consists basically of two membrane-covered corpora cavernosa and an immensely rich network of sensory nerve endings. With sexual arousal, the corpora become congested and erect by exactly the same mechanisms as does the penis. Because the clitoris lies so far forward of the vaginal opening, it receives very little direct stimulation during the process of thrusting, although the consequent rhythmic traction on the labia does cause a degree of excitation. In certain positions (woman on top, resting on her knees, and leaning forward, for example), the man's pubic bone tends to rub across the clitoris somewhat, but not usually enough to induce orgasm. Effective and dependable clitoral stimulation must usually be direct, whether manually or by some other stratagem. The timing of this is difficult to coordinate well enough for both partners to attain climax together, especially since most women reach their peak more gradually than their mates.

With appropriate sensual prelude, the woman will have, at some point near the beginning of intercourse, spread her legs to enable the man to position himself between them. When her thighs part, the entrance to her vagina is exposed by the separation of the labia minora. If she has become sufficiently aroused by stimuli meaningful to her, the vaginal canal is by this point relaxed and its walls are separated, sometimes ballooning out, to receive her partner's penis. Even in the early stages of arousal, messages coming in via the autonomic nerves have increased pelvic blood flow, stiffened the clitoris, swelled the labia, and caused the secretion of a slippery, somewhat pungently scented liquid to lubricate the vagina.

The amount of the lubricant varies from woman to woman, the same degree of sexual excitement producing a plenitude in some and scanty volume in others. Not infrequently, especially in a young woman, the fluid is so profuse in amount that it escapes through the labia and wets the skin, hair, and any undergarments she may be wearing. Just as in the man, other areas of the woman's body experience an increase in blood flow at the same time; this is particularly manifest in the breasts, causing spongy erectile tissue in the nipples to become rigid. The heart and respiratory rates increase, and breathing may become heavy.

Meantime, muscular tension in the pelvis has increased to the point where involuntary spasms may be occurring in the vagina and surrounding area. By now, the penis has entered and its entire length is riding in and out, gliding firmly against erogenous areas of the vagina even as the clitoris is being jiggled and bounced by the rocking motion imparted to the labia. For some women, this is the point at which an orgasm will be set off, but most still have a need for more direct clitoral stimulation. At whatever point a woman's orgasm takes place, its physiology is usually in many ways like that of a man, but she experiences more contractions in her pelvis. Her partner may be able to feel rhythmic muscular spasms within the vagina. Recently, evidence has emerged indicating that women may exhibit a far wider range of orgasmic experience than men do, in both degree and kind. While there has been controversy about these findings, the one certainty is that variations from the most common patterns of response have been shown to be more frequent than previously thought.

What has been described here is no more than the bare outline of the anatomy and physiology of an act whose sole biological function is to permit fertilization of an ovum. Writing about it is like trying to compose a universally fitting description of a person's home. Because all members of our species share certain fundamental needs for shelter, our many kinds of home will have basic similarities. The assorted cultures and societal groups into which we are divided will each dictate variations on the general theme, and beyond even those there will be a plenitude of chosen and circumstance-controlled differences that are highly individualistic and personal, to the point of intimate privacy. My house is very different from the one next door. There is probably no human function more genetically predetermined and yet more susceptible to nurture and individual variation than the method by which we reproduce our kind.

The body has been compared, with good reason, to an ideal society whose trillions of citizens are divided into special-task groups, all striving to benefit the nation: The special skill of a muscle cell is to contract—it is joined with a local group of similarly skilled citizens to form the community organization called a muscle; a fat cell is able to store a particular kind of nutrient—it is joined with its fellows to form the community organization that is a fatty area; a glandular cell is equipped to manufacture and secrete a certain chemical substance needed for maintenance of some other tissue—it is joined with identical cells that make the same chemical substance, thus forming a community organization that is a gland like the thyroid, or becoming part of a structured sheet like the lining of the stomach; a nerve cell integrates signals from within and without us and transmits them via long extensions called axons—cell bodies are joined together in small community organizations called nerve centers and ganglia, and a huge one called the brain; their extensions are found linked up to form the community organizations that are the transmission trunks called nerves.

The body's work is indeed carried out by division of labor—we are made of individual tissues, each with its own job to do. There are four major categories of tissue, called epithelium (or lining tissue); connective tissue; muscle tissue; and nervous tissue. Subspecialization occurs within each of the major categories. One form of lining tissue, for example, is glandular epithelium, such as the inner lining of the intestinal tract, whose cells secrete mucus and digestive enzymes into the gut. Among the many subspecialized connective tissues are tendons, bone, and fat. There are three distinct categories of muscle, called voluntary (alternatively known as skeletal, striated, or striped), involuntary (alternatively known as smooth), and cardiac. The different types of nerve cells have a wide variety of tasks, of which much will be said later.

In one way or another, every one of these tissues is responsible for some highly specialized contribution to the total effort of maintaining the moment-to-moment activities that take place within the body. In addition to engaging in the myriad functions that permit its own survival, each type of tissue is also carrying out an assigned task that contributes to the welfare of the whole.

To pool their efforts for maximum efficiency, tissues of various kinds serve together in the structural units called organs. Each organ coordinates the contributions of all its constituent community organizations of specialized tissue in order to perform one or more of the major tasks for which it is known: The kidney consists of tissues of several sorts, working together to clear the blood of toxic materials and to maintain the stability of the body's fluid and chemical balance; the presence of the liver's various tissues means that it can detoxify certain by-products of the body's far-flung chemical reactions, as well as make bile to aid in the digestion of the fat we eat; some tissues of the stomach churn up the swallowed foods and push them along into the intestine by the wavelike motion called peristalsis, while others are secreting chemical substances to aid in their digestion; the distinctive tissues of the ovary combine in that organ's ability to generate reproductive cells and also to be the source of certain of the female hormones—and on, and on, and on.

It is as though several communities of different capabilities are joined together to form the cities that are organs. Several organs work together in a confederation called an organ system, dedicated to the coordination of some major division of the body's labor. Twelve organ systems are recognized: skeletal; muscular; integumentary, or skin; digestive; urinary; respiratory; circulatory; lymphatic; endocrine, or hormonal; nervous; immune (whose activities merge into those of the lymphatic system); and reproductive.

Some systems, such as the digestive and urinary, are formed into a physical continuity of their constituent parts, but others, such as the endocrine and immune, consist of widely scattered organs, tissues, or even groups of cells that do similar kinds of work. The endocrine system includes a spectrum of structures, from distinct organs such as the thyroid to widespread groupings of cells within a variety of organs dedicated primarily to other duties. The immune system is dispersed throughout the body in small units such as the lymph nodes and within such organs as the spleen, liver, and lungs.

Although each system is primarily responsible for a particular category or perhaps two of contribution to the stability of the whole, it also usually has secondary functions beyond those of its major tasks. The pancreas, for example, is part of the digestive system because its cells secrete certain digestive enzymes into ducts that lead to the intestine. Nevertheless, it also functions as an important component of the endocrine system because other of its cells manufacture the hormone insulin and secrete it into local capillaries. Endocrine tissues are those that help control body functions by means of hormones and other chemical signaling substances they produce, which enter the bloodstream directly from the cells. Because they do not require the intermediary of ducts, endocrine glands are also known as ductless glands.

It is their spatial and functional *organization* that is the most striking characteristic of the various categories of cells, tissues, organs, and systems, specifically the way their multitudes of activities are coordinated with one another. Had not some organizational plan evolved, there could have been no life beyond single-celled structures (of course, the life of even single-celled organisms depends on a vastly complicated integration of activities within them). But *plan* is a word that presents difficulties, because it is usually understood to imply intent, or even conscious will.

To imply intent is not my intent, and as for conscious will—it is precisely what I am arguing against. My intent is only to imply—to state unequivocally, in fact—that multicellular organisms have survival advantage over those that are unicellular, and increasingly complex organisms tend to increase their advantages as their complexity increases, for the simple reason that they are less subject to the whims of environment than are their more primitive cousins.

As the animal kingdom evolved, complexity grew until the backboned, or vertebrate, category came into being. It continued to evolve until mammals appeared some 200 million years ago. Mammals have been earth's dominant animals for 70 million years because their physiological complexity has made it possible for them to carry out tasks that permit survival in a variety of environments (dominant in all senses but number; some 850,000 species of insects have thus far been described, and there are probably hundreds of thousands more remaining to be discovered). When the order of mammals called primates made their appearance about 75 million years ago, it was specifically the complexity of their brains that enabled them to adapt so well to such a wide range of changing circumstances. Their highly specialized brain allowed highly specialized forms of behavior, leading to success in competing for available resources.

When higher species of primate evolved larger and ever-more-complex brains, the period of postnatal dependency of their offspring lengthened. The average brain size of the earliest forms of *Homo* was about 60 percent that of modern man (800 ml to 1,400 ml, or 27 ounces to 50 ounces). For the high primate's bigger skull to make its way out of the pelvic outlet of the mother, it had to leave the womb at a relatively earlier stage of the brain's development than did its predecessors—before becoming too large to get through. This meant that the infant was still far from being able to care for itself. With the appearance of progressively more complex mental abilities (involving larger brains and birth at an even less ready stage), it took longer and longer for the newborns' thinking patterns to mature and for enough learning to occur that they might be assimilated into the social systems that primates were building.

The appearance of the kind of abstract thinking of which modern *Homo sapiens* is capable necessitated further lengthening of the period of dependency. The elongated dependency provides more time for knowledge to be imparted and for the young to absorb the thought patterns that the species has already developed over millennia. In addition, the added interval means that the young will have more experiences during a time when impressionability is high and thinking processes are not yet fixed. It enables the growing members of the species to absorb and respond to the numerous events in which they participate during their prolonged immaturity. The result is that the experiences of the various phases of childhood and adolescence are soaked up by a brain still richly responsive to new and ever-varying stimuli acting on its receptive plasticity. In this way, the mind is gradually shaped. It is a manifestation of the organized behavior of the brain.

The brain is an organ whose cells and tissues are specialized for receptivity to stimuli, integration of signals, transmission, and responsiveness. Like all other organs, its specialized powers contribute irreplaceably to the animal's survival and the facilitation of its reproduction.

Just as each individual cell within us responds and adapts to received signals in such ways as to maintain dynamic stability and perform with optimal efficiency, the vast array of functions we call "the mind" must also respond in such ways as to maintain dynamic stability and make optimal use of its potentialities. The mind is, after all, a product of the tissues of the body, and it should therefore ultimately be subject to the same rules of maintenance and survival. Like the brain, which is its main anatomical and physiological basis, the mind no doubt functions in such ways as to attempt to provide a state of what may be called mental homeostasis. That it is less successful at accomplishing this than is the body from which it arises does not change the general principles that guide its performance.

Like all communities of cells throughout the body, nervous tissue adapts to circumstances it encounters. But because it is uniquely equipped, qualitatively and quantitatively, to exchange information with its fellows, it has much more potential to do its adapting by coordinating itself with the requirements of other parts of the body. It is the integration of parts of the brain with its other parts and its ability to act (at a tremendous rate of speed) on the basis of signals from every other tissue of the body that are the basis of the mind, of abstract thinking, and of much that has gone into the creation of the dazzling array of faculties that I have called the human spirit.

All of this is made possible by the way in which cells and their capacities are organized. As biologists are fond of pointing out, it is not so much a plant's or an animal's organic structure that is the key to life and species survival, but the organization and coordination of the parts of that structure into the needs of the whole. The organization and coordination that explain *our* singularity, the singularity of humankind, is the organization and coordination within our biologically advanced brain and its enhanced cerebral capacities, but also the order and control with which our species has learned to use them.

The coordinated functioning of the various parts of the body and their integration toward the unified purpose of stability or homeostasis—even as they are constantly undergoing the little changes required to maintain life and reproduce—are nothing short of breathtaking. At least since Aristotle wrote the first treatise on embryology, *De Generatione Animalium*, but probably long before that, some of the most profound thinkers of every era have turned their attention to the problem of understanding just how it is that a fertilized ovum becomes a fully functional human being, every tissue and organ within whom does exactly what it is supposed to at exactly the proper instant, all in the service of the dynamic stability that is the basis of survival, growth, and reproduction. Recent researches elucidating some of the elements of the process have only served to magnify the never-ending fascination.

It all begins, of course, with the penetration of the ovum by one of the 300 or more million sperm cells deposited in the vagina during emission. It takes only about eight minutes for the sperms to enter the cervix, make their way through the uterus (which they do blindly, lashing about so directionlessly that only a few find the right way), and swim up the tube to reach the ovum. Millions lose their way every minute and die; only about a thousand ever get to the tube.

For human conception to take place, the sperms must start on their journey during the five- or six-day period ending with ovulation. The chance of conception is one in ten when intercourse takes place five days prior and rises to one in three when it is on the same day. After ovulation, it is too late. This means that the single sperm that will ultimately be victorious is usually waiting in a crowd of its fellows when the ovum reaches the tube. Imagine it figuratively staring up at the new arrival, a structure some 85,000 times its own size. It is an absolute necessity that the winner be right on the spot, because an ovum lives only from twelve to twenty-four hours, while a sperm can survive up to six days.

Once proper contact is made, a combination of chemicals produced near the surface of the egg and by the sperm create conditions that allow one sperm's head to penetrate the ovum, leaving its tail behind to perish. At that instant, the ovum's cell membrane undergoes an as-yet-poorly understood biological transformation that renders it impenetrable to any other of the many sperm that have succeeded in reaching its surface.

Instantaneously, several other events now take place. The ovum completes meiosis II and the tiny polar body is separated from it; the nuclear membranes of both sperm and ovum disappear; and the father's and mother's chromosomes combine into the diploid number of forty-six. Following this, the zygote divides into two. Each of these cells divides again and again, with the resulting cells becoming smaller and smaller because they must all share in the cytoplasm. During this time, the cleaving zygote is being propelled down into the uterus by contractions of the tube and the beating of hairlike projections called cilia, which line its inside.

At about three days, the rounded clump of what is by then sixteen cells has reached the uterus. During the next three days, the number of cells continues to increase and the center of the ball hollows itself out. By day seven, the hollowed ball of cells has implanted itself in the lining of the uterus, which had been made lushly ready for it by the hormones secreted by the corpus luteum. The uterine lining soon envelops the group of cells and covers it completely.

Cells are now beginning to differentiate, in a fashion to be described a bit further on. All major organs are formed during weeks four through eight. By eight weeks, the tissue in the uterine lining has combined with outer cells of the embryo to form a spongy structure, filled with nourishing blood vessels and called the placenta, from the Greek *plakous*, "a flat cake." Considering its function, which is to nourish the embryo, it makes symbolic sense to call it a cake. When fully formed, the placenta is actually an organ of about eight inches in diameter and one inch in thickness, situated usually on the wall of the uterus that is toward the mother's back. It secretes large quantities of estrogen and progesterone as well as other hormones, and it provides for the exchange of oxygen, carbon dioxide, nutrients, and wastes between mother and developing embryo. Among its other hormones are gonadotropins, which help to keep the corpus luteum secreting progesterone so that the uterine muscle does not contract and expel the developing embryo. So effective is the placenta as a secretor of female hormones that after a relatively early point in gestation the mother's ovaries could be removed without preventing the baby from going to full term.

The embryo has by this point become surrounded by membranes that form a sac to protect it; the sac is filled with fluid. This so-called amniotic fluid serves as a cushion and a buffer for the developing child. During this time, the umbilical cord, which will grow to be an inch thick and four and a half feet long, has formed. Through its center run two arteries and a vein that carry blood between the embryo and the placenta. By the end of the eighth week, early forms of almost all of the internal organs are present, even though the embryo is only a bit more than an inch long and weighs less than a fifth of an ounce. Tiny arms and legs are recognizable and even fingers and toes can be seen. The embryo's head has begun to form, and ears and eyes are detectable. The central nervous system and muscles have acquired the ability to respond to delicate stimuli. Actually, the former zygote has ceased being an embryo and is from this point on called a fetus.

This perfectly timed sequence of events has been set and kept in motion because the appropriate genes have been turned on and off since conception. An impressive array of hormones has appeared, whose sources are not only the ovary, uterus, and placenta but also the hypothalamus, pancreas, thyroid, pituitary, and adrenal glands. Enzymes of all kinds have also arrived on the scene from both parent and child to regulate the various processes of development and sustenance of the pregnancy. In time, hormones originating in the fetus itself come to play an ever-expanding role in the successful conduct of intrauterine maturation and delivery. Right up to the end, new hormonal effects appear exactly when needed, instigated by the presence of the fetus. Among them are those produced by the mother (and the placenta, too), which facilitate delivery by relaxing the pelvic joints and ligaments and make the uterus contract with sufficient force to expel the baby. In all of these ways, pregnancy is an interaction between mother and child. Starting with conception, these two people undertake a journey that, in its various manifestations, will never end.

We should return to the beginning now in order to review the cellular events that have been taking place within the embryo, by which differentiation gives rise to the sequence of consequences just described.

From its inception, embryonic development is all swarming vitality. It is as though human biology has burst forth with the exuberance of creating a new life. Every detail is supervised and seen to with a combination of controlled precision and protectiveness that make these processes seem thoroughly automatic even while they are being induced, incited, and coaxed by a veritable concert of molecular substances and events. When such seeming miracles are contemplated, even observers with the apparent sangfroid of laboratory researchers—perhaps they more than others, in fact—cannot help but be moved to a sense of profound kinship with those awestruck ancients who, millennia before the insights of science began to elucidate them, bestowed the honorific of Mother on the vast complex of physical and chemical forces we call nature—or Nature.

As magical as the visible development of the embryo must have appeared to those who first observed it in animals, it seems almost matter-of-fact when compared to the underlying cellular activities that are its cause. Only in the past few years have some of them been elucidated, but enough is already known to give some general direction to studies that will eventually identify the chemical signals that tell a zygote how to become a child. The precise manner in which the vast multitude of cells that constitutes a zygote's progeny is directed to differentiate into distinctive tissues and organs is at present one of the most exciting of the many remaining frontiers of biological exploration.

There is a single underlying principle that explains cell differentiation: It occurs as a result of selective gene expression. All of the cells ultimately resulting from the zygote carry the same genes. It is the activation or repression of various of these genes during appropriate times in the course of development that produces cells of differing structures and functions as they are needed, and positions them in the proper part of the embryo. Most of the controls that determine genetic expression are the result of regulation of transcription.

All kinds of complex rearrangements of cell groupings take place throughout the various phases of embryonic life. Cells receive chemical signals telling them their position relative to one another and how they must align themselves spatially in order to form the fully differentiated tissues and organs. Biologists use the term *pattern formation* as a general heading under which to list the many mechanisms that lead to the specialization and positioning of tissues.

The very first element in pattern formation actually occurs even before fertilization has taken place. In what must surely be a symbolic act of maternal foresight, the oocyte—the early cell that will develop into an ovum—as it matures, distributes its various organelles, proteins, and RNA molecules into prescribed areas of its cytoplasm. When the zygote cleaves after fertilization, each new cell receives different parts of the cytoplasm, and it is therefore distinct from its one, then three, then seven, then fifteen, and so forth siblings. Not only does each of the new cells find itself in some specific and predetermined location in the early ball of cells but it is as though it has received instructions in its cytoplasm from the ovum, telling it what tissue it must eventually create from itself.

As groups of cells form, they interact with other groups. One way in which this happens is that a cell group may produce a hormone or other molecular substance that signals a nearby group to do something, such as move to a somewhat different location or synthesize a new protein. This kind of chemical or physical interaction, in which cells influence other cells, is termed *embryonic induction*—the cells are induced to some action. The inducing substances accomplish their mission by diffusing from one cell group to the other, moving through the *milieu intérieur*.

Each part of this serves the overall purpose of growth, but not everything is a straight line of building. Certain of the tissues are needed for only a brief period in embryonic development; although they are necessary to carry out a specific function at a specific time, they would get in the way of further development were they to persist. They are eliminated by a genetically controlled process called *programmed cell death*. The program, like every other part of embryonic life, is in the genome. A simple example of this is the way human hands and feet start out as paddle-shaped structures with raylike portions within. At the proper time, the tissue between the rays dies, leaving fingers and toes. In the very tiny percentage of cases in which this sequence fails to complete itself, the baby will be born with webbed digits, and the plastic surgeon must do the job left undone by the genome.

In all these ways, therefore, cells respond to various chemical signals around and within themselves, telling them to migrate in groups to different parts of the developing embryo. There, they come together with other groups to form structures positioned in very specific places. These movements of rearrangement take place at distinct points in the life of the embryo. The intricate choreography that produces a woman or a man is controlled by its genome, which carries instructions not only for the manufacture of proteins but also for the timing and sequence with which they are made. As the embryo is taking shape, genes are influenced to switch on and off by signals from within the cell, and also by signals from its surroundings. Depending on the needs of nearby fluids and tissues, molecules in a cell's environment signal it to express or not express the appropriate genes. By such sequences of events, the two hundred or more types of cells of which we are composed differentiate and are directed to their ultimate locations in the body.

Among the genes that all cells in an embryo's or an adult's body have in common is a set of at least thirty-eight of what may be thought of as master regulatory genes, whose function in earliest embryonic life is to lay out the general geographic plan of the body. Known as Hox genes, they lie in the same order on their chromosomes as the segments of the body for which they provide the blueprints. Thus, genes involved in the formation of the head lie well above those for the abdominal area. The Hox genes govern such factors as top-to-bottom axis and the way in which primordial groups of cells will align themselves and eventually develop into specialized tissues and organs, each in a predetermined location. As noted heretofore, even the cells that will make arms and legs are committed very early to their future, probably as soon as the very beginnings of the rearrangement phase.

The result of the embryo being laid out this way is that the cells arrange themselves into three distinct early layers. From the three layers, all of the body's tissues are destined to be generated. From the outside in, they are the ectoderm, the mesoderm, and the endoderm. Each of them will soon split into subpopulations of cells, marking the onset of the formation of more specific tissues and of organs.

The ectoderm will give rise to the nervous system and the outer layer of skin; the mesoderm gives rise to the skeletal, muscular, circulatory, lymphatic, excretory, and reproductive systems, as well as the spleen and the connective tissue layers of the digestive tract and skin; from the endoderm will come the inner lining of the respiratory, urinary, and digestive systems. For this to happen, various cell groupings in the three primitive layers will migrate, twist, turn, glide, fold, bend, lengthen, branch, fuse, split, thicken, thin, dilate, constrict, hollow out, form pockets, pinch off, adhere, separate—it is all like some Busby Berkeley movie sequence, but on a much grander scale. Hundreds of millions of dancers appear and they all participate, forming themselves into the shapes of the various tissues and organs. Every movement and change is choreographed by the genetic code as it governs the synthesis of protein molecules by individual groups of cells that tell nearby participating cells what to do. What is happening is that gene activity in certain cell groupings produces chemicals which affect gene activity in other cell groupings. In different forms, this kind of signaling will go on throughout the individual's life to fulfill any of a huge number of the organism's needs. But those in the embryo are among the earliest manifestations of the process whereby one area of cells influences another. As so well put by Charles Sherrington in his version of "The Wisdom of the Body": "The whole astonishing process achieving the making of a new individual is thus an organized adventure in specialization on the part of countless co-operating units." So early do these tissues begin their differentiation that at the beginning of the fourth week a distinct tube-shaped structure has been formed; this will eventually become the heart. Its presence is unmistakable, because it is already beating.

As noted earlier, so much organ formation has occurred by the end of the eighth week that the developing human being is no longer an embryo. Now called the fetus, it grows as its organs are going through the various phases of maturation. The pregnant woman carrying it is preparing to become a mother. High levels of estrogen and progesterone from the placenta and a rising level of a pituitary hormone called prolactin cause the breasts to enlarge considerably, often to twice their ordinary size. Much of their usual content of fat is replaced by glandular tissue. Progesterone produced in the placenta has the effect of inhibiting milk production, so the breasts produce none until the placenta is out of the mother and the newborn is ready for a meal.

Upon childbirth, the mother's milk will not automatically flow with ease through the breast's ducts and out of the nipple. Instead, it must be forced out by contraction of certain modified muscle cells that surround the glands of the breast. Nature, acting here, too, the role of a nurturing mother, has provided a signaling mechanism to allow this to happen. It is a signaling mechanism of the most remarkably maternal sort. When the infant suckles, nerve endings in the mother's nipple fire sensory impulses up to the hypothalamus, telling it to order the pituitary to release a hormone called oxytocin. Oxytocin stimulates contraction of the muscular cells around the milk glands. This is the underlying mechanism for the process called the suckling reflex.

These same sensory impulses from the nipple also stimulate the hypothalamus to tell the pituitary to secrete a continuing supply of prolactin, which persists as long as milk is taken from the breast, whether by the child or by some artificial means. If milk is allowed to remain unremoved, prolactin secretion is inhibited and within a week the breasts lose their ability to make milk.

There is something heartwarmingly right about a strictly biological phenomenon that fulfills the physical needs of both mother and child while it also fulfills their emotional needs. Although both aspects are attributable in large part to instinct, there would appear to be more than just that. The suckling reflex is only one example of those provisions of nature that serve a maternal force that I believe our species has made into something beyond mere inborn drive. Mothering begins with instinctual behaviors that, for the vast majority of women, at some point come to cross the boundary supposedly separating what is favorable only in the pragmatic eyes of natural selection from what is favorable in the eyes of the mind and spirit, as well. It appears to be an example of what might be called "the Loewi principle" in restated form: The beauty of a nursing mother can never be explained by a little oxytocin around the milk glands.

Biology contributes its share of the totality by giving far more than just nerve impulses, hormones, and molecular messengers. It also provides an anatomical esthetic that, although its effect is much enhanced by emotion, stands alone in its twinned splendor as one of the glories of structure. By this, I mean the human breast, which has fascinated not only poets and lovers but scientific investigators, too, for varied (and sometimes similar) reasons. Sir Astley Cooper of Guy's Hospital in London was England's leading surgeon during the second and third decades of the nineteenth century, and among his contributions were detailed studies of the normal and diseased mammary gland. In his published work can be found the following appreciation of the breast of a nursing mother:

> *The natural obliquity of the mammilla, or nipple, forwards and outwards, with the slight turn of the nipple upwards, is one of the most beautiful provisions of nature, both for the mother and the*

child. To the mother, because the child rests upon her arm and lap in the most convenient position for suckling; for if the nipple and breast had projected directly forwards, the child must have been supported before her by the mother's hands in the most inconvenient and fatiguing position, instead of its reclining upon her side and arm. But it is wisely provided by nature, that when the child reposes upon its mother's arm, it has its mouth directly applied to the nipple, which is turned outwards to receive it, whilst the lower part of the breast forms a cushion upon which the cheek of the infant tranquilly reposes. Thus it is we have always to admire the simplicity, the beauty, and the utility, of these deviations of form in the construction of the body, which the imagination of man would lead him, a priori, to believe most symmetrical, natural, and convenient.

The human emotions prompted by a newborn have been felt by billions of men in every remoteness of this planet. And yet, they seem unprecedented each time they appear. To describe them in somewhat more detail, I will tell of the birth of one of my children.

Sarah and I were married in 1977, when I was forty-six and she was twenty-nine. Long before that, she and I and my two adolescent children had comfortably and almost without realizing it formed ourselves into a close-knit family. Although my first marriage had not been a happy one, it had two happy consequences. I will resist the temptation of a father's pride and say only that Toria, then fifteen, and Drew, then twelve, had inspired the unquenchable motivation that, in spite of my own sometimes overwhelming doubts, brought me through the very worst period of a life already too darkly dappled with struggle and sorrow. The several years before my divorce had left me confused, depressed, and tremblingly uncertain. I lived during those days as though enveloped in an impenetrable mist of despair, in whose oppressive bleakness I had become separated from the buoyant self-assurance that had since earliest childhood been my unfailing guide. Unable to rediscover it and unable to navigate without it, I had lost my way. What I had really lost was myself.

The kids—the very existence of the kids—was for me the one reassuring truth that sustained me. No matter how despondent I felt, day after dreary day, I fixed my thoughts on Toria and Drew as though on the radiant promise of some far-off luminosity beaconing and beckoning me back to them. As long as they were somewhere there on the other side of the mist's opacity, I would will myself to find the place where they waited for me. And I did.

As though to make up for all that had preceded it, the period of emergence was a time of incalculable enthusiasm and of renewal, effervesced by the unanticipated entrance into our three lives of the lovely presence that is Sarah. It was a glorious period for all of us, culminating three years later on that May afternoon when she and I were married.

Within a year, we had left the seaside cottage I had been renting and bought ourselves one of those suburban homes of the well-tended sort that used to appear regularly in movies of the thirties: garden, verdant front lawn, and a substantial mortgage—the whole package on a quarter of an acre. "The squares shall inherit the earth!" I proclaimed, and we went on to create the squareness we have celebrated since that time. It is because of the mists and dark woods in which I have in the past been lost that I bear a heightened sense of the value of simple things.

The only cloud on my otherwise sky-blue horizon in those early days was Sarah's wish that she and I have a child together. I realized it was not the sort of cloud that foretells a storm, but I resisted the consequences of its hovering presence nevertheless. In the final days of my fifth decade, the mere thought of playing daddy to a newborn made the collar of my shirt feel suddenly a size too small. Actually, I really hadn't wanted to get married in the first place, presumptuously assuming that Sarah would be content with continuing in perpetuity the domestic arrangements I had found so conducive to my well-being. But she had finally gotten her way, and with quite happy results. She was now determined to get her way again. Needless to say, she succeeded.

I was not surprised by the enthusiasm of the kids when Sarah became pregnant, because they had long been electioneering for a little sister or brother. But I was astonished to discover that I, the most reluctant of procreators, was enjoying the whole gestational process as much as they were. The excitement mounted as this embryonic Nuland became more overtly obvious, mitigated for me only by Sarah's insistence that I attend Lamaze classes. It is one thing for an expectant father to do this kind of communal thing when in his twenties or thirties, but quite another at fifty, which was my age when the incipient little person was approaching arrival.

At first, I took it all in good humor, even though I was nearly two decades older than the closest father-to-be in the group and was the only one to have been through pregnancy before. But I am not a very gregarious, chummy kind of fellow. My age as well as my predisposition not to do the male bonding thing distanced me from an easy conviviality. The ultimate alienation took place at the second Lamaze meeting when the chirpy young nurse who directed the course asked each of us to tell what we did out there in the real world. She called this "sharing yourself with your peer group," and its purpose was to cement our fellowship. It was not to be. From then on, I was kept at a respectful and slightly uncomfortable distance.

The worst session was not the share-yourself one, however, but another, in which the hypervoluble instructor brought back some of her former student couples to tell us of their childbirth experiences. She must have chosen them using the same criteria developed for selecting TV game-show participants. One after the other, the gushing twosomes gleefully regaled us with tales of the rapturous excitement to be discovered in sharing the joys of heavy labor and a fresh placenta, or perhaps that is all I could focus on. My mind flailed about in the turbulence of cascading clichés, all bubbling and burbling until I thought I would drown in the overflow of verbal corn syrup.

But still, I persisted. I learned all the breathing techniques and every other of what would be my assigned tasks during delivery, determined to become the perfect coach and partner. There had been no such thing as a Lamaze class when Toria and Drew were born, or at least none their mother or I knew anything about. But the planet had taken more than a few very vigorous turns since then, and one of them concerned the way in which childbirth is conducted. During my initial experiences with delivery, an infant's first view of the world was through the opening in a sterile green surgical drape, as its head slid antiseptically between mother's shaven, germ-free labia and into the highly trained arms of a scrubbed, gowned, and gloved Fellow of the American College of Obstetrics and Gynecology. The nail-chewing daddy of that era was meanwhile pacing a furrow through the rug in the waiting room, anxiously hoping for news of his wife and child. Being a member of the hospital's surgical staff, I was permitted entry into the delivery room directly after the birth of each of my two older children, but very few fathers were granted that privilege. The others were left to fend for themselves, with only their cigarettes for entertainment and a few similarly afflicted brethren for company. Childbirth in those days was a totally medicalized procedure. There was precious little opportunity for sentiment or marvel, at least until after the baby was in its mother's arms and the all-powerful obstetrician had deigned to grant his assent.

The women's movement changed all that. Their basic approach to the problem was direct and simple: Pregnancy was to be treated as a natural process that needs a certain amount of help by people experienced with it; obstetricians should guide it only to the extent of preventing complications and getting mother and child out of trouble should nature pull one of its infrequent tricks; in the absence of indications for cesarean section, the vast majority of deliveries do not need medical intervention; delivery, if it is to be vaginal, by a properly trained midwife with skilled obstetrician backup is an option that should be available to all women.

The clinical field of obstetrics has made enormous contributions to the solution of the problems of both mother and child. But it is precisely as a result of those contributions that we are currently at a point where physicians' surveillance and availability, rather than their active participation, are sufficient to protect the safety of vaginal deliveries except in unusual circumstances. There are historical and sociological, rather than medical, explanations for our modern treatment of childbirth as a process so fraught with the imminent potentiality for disaster that every one of its steps must be taken with the moment-to-moment invasiveness of a doctor's technological skills. At a routine vaginal delivery, to have them at the ready is one thing; to use them as a matter of course is quite another.

So much that is wondrous is going on in every cell and tissue of our bodies and there is so little of it in which we can consciously take part, or even observe. Why should we further deprive ourselves by taking one of the few universally acknowledged visible marvels of the body in which we can fully participate and turning it into a medical specialist's virtuoso performance?

Like its end, we have surrendered too much of life's beginning to the doctors. The women's movement asked us to remember what our culture had gradually forgotten—childbirth by vaginal delivery is a natural process. It is the doctor's job to keep it that way, and to be ready to implement a course correction should nature veer. In the great majority of instances, the trappings of an operating room are as seldom needed during our entry into the world as is the futility of last-ditch resuscitation efforts at our exit.

Fortunately, the community of obstetricians wisely yielded to claims whose validity is now being proven thousands of times each day in delivery suites and homes through much of the world. And I—I, whose professional career has been lived at the most advanced outposts of technological medicine—was paradoxically the beneficiary of the enlightened retreat from precisely the sorts of positions on which I had always stood my ground. And so was my wife and so were the two children we would bring into the world by taking advantage of the new insight that was nothing more than the old wisdom.

Listen to Sarah describing what it is like to be pregnant. Nearly half the world has experienced it, or will; the other half has been told of it in detail, or will. But it remains, even more so than fatherhood, ever new.

> *There was something mysterious about it, and very private. In the first trimester, I didn't want to tell anyone except our immediate family—the kids, my parents—because if there's a miscarriage, you don't want people draping you in crepe. So you have this feeling very much to yourself. It reminds me of Mary as she was thinking about the baby Jesus right after his birth. She "kept all these things and pondered them in her heart," but for me the pondering began to happen as soon as I knew I was pregnant. It was that same sort of feeling that all first-time mothers probably have: This is something to keep, and to ponder in my heart.*
>
> *You know your body will change, but it doesn't for the longest time. You're tired, your breasts get bigger, and you start getting thick around the middle, but the world doesn't really know you're pregnant for about four or five months, and all that time this private world is yours. This little person is developing inside you.*
>
> *I remember being fascinated by looking in your books to see what was happening to this embryo, thinking about how the cells were dividing. It was just amazing to think that this was going on inside my body. It was all so private and extraordinary. And there was wonder in it.*
>
> *It was mine; it was something I could keep, and ponder. All*

that time, organs were being formed and a human being was being created inside me. And no one knew but us—and mostly, no one knew but me.

The pregnancy was very easy. I just kept doing everything I always do, only my body was changing. In the last month, I began to feel like it wasn't my own—it belonged to this creature inside me. Nothing made any sense. My body had changed so extraordinarily that it seemed as though another universe were sticking out from my middle. You know the old jokes—I couldn't reach my feet; I had trouble sitting down; I had trouble getting up; I had trouble sleeping; I had trouble digesting food. Toward the end, I was really feeling miserable.

Why? Because there was no room. There was no room for me. *This child had taken over my body. I was constantly uncomfortable.*

The day before I went into labor, a friend saw me from a distance and said, "You look like a ripe peach." It was this thing we always call the estrogen flush. My skin color had changed. My skin felt tight—everywhere. I felt like every part of me was a balloon. My hands and feet were a balloon, my face, too. I did have a tiny bit of swelling, but it wasn't that.

I remember just before this little fellow was born, I went to sleep with about six pillows to prop me up, and I was thinking, God, I can't wait for this child to come out of me so I can get a good night's sleep—and then realizing what a jerk I was being. Once he was born, I'd never get that good night's sleep, because I'd be up every three hours to nurse him. But that almost didn't make a difference, because I had this intense feeling of just how I absolutely couldn't wait to get rid of this load I was carrying. It's like some basic animal instinct: You can't wait to get this thing out of your body, to reclaim what is you. At that point in your pregnancy, you feel like you aren't anything but a vessel. And that's exactly what you are—a vessel.

At about 2:00 a.m. on the morning of September 15, 1981, Sarah awoke with what she thought was the urge to urinate. She got up and immediately became exasperated with herself—she had wet the bed. "Oh great!" she thought. "This adds insult to injury—just what I need." She had indeed wet the bed, but not at all the way she assumed. What was coming out of her was a slow leak of amniotic fluid.

She was experiencing what is colloquially called "breaking the bag of waters." In many women, rupture of these membranes is the first evidence that labor has begun. In others, labor is heralded by the so-called "bloody show," which is the expulsion of the plug of mucus that forms in the cervix of pregnant women, along with small amounts of blood from local vessels.

By this time, Sarah was beginning to feel some mild lower abdominal cramps. These were not much like what she had been told to expect, but were, as she put it, "without a beginning, middle, or end." She woke me up and described them as being like the menstrual cramps she had experienced as a young girl. She has never forgotten my drowsy response: "Fine. Now go back to sleep." By then, she was much too excited to take my advice, so after a short while she nudged me awake again. This time, I tried to distract her by telling her to draw up a list of all the people I would have to call after the baby was born. Sarah has a great fondness for making long meticulous lists of various kinds, and my only thought was to set things up so that her mind was sufficiently occupied to prevent her from interrupting my sleep again, until it was really time to do something. A long familiarity with abrupt middle-of-the-night awakenings has inured me to their effects, and I can return to the arms of Morpheus in an instant, which is exactly what I did.

Sarah hadn't done any packing in preparation for going to the hospital, so after completing her list, she decided to get some of her things together. She had a small book containing instructions for what to take, but she couldn't find it at first. After rooting around for a while in our third-floor bedroom, she went downstairs and continued her search. Unable to locate the book, she took up a small bag and began filling it with items she thought she would need, and even some she thought she wouldn't.

> *It was as though I was nesting—getting the nest ready for whoever this person was. I'm convinced that's what it is; it's the nesting instinct—you have to get things ready in order for the person to be born.*

I awoke at about 5:00 a.m. and saw that Sarah was definitely not going to go back to sleep. Her discomfort was beginning to have some structure to it, as though it was gradually transforming itself into labor. When she suggested calling our obstetrician, Marshall Holley, I tried to hold her off with, "No, I think he needs his sleep more than he needs to hear from us right now. And anyway, it's about time I had a haircut." Sarah had begun cutting my hair within a few weeks of our first meeting, and I have never set foot in a barbershop since. We went down into the kitchen and the tonsorial proceedings began. I was determined to jolly Sarah along while I still could, hoping not to call Marshall until the more civilized hour of seven. When we finally phoned him, he suggested we meet him in his office at nine.

By the time we arrived at the office, labor was moving along at a good pace. During the few minutes Sarah was on the examining table, the opening of her cervix widened almost an inch, from three centimeters to five. Widening, or dilatation, of the cervix is defined as the first of the three stages of labor, the others being the actual childbirth and the expulsion of the placenta. At ten centimeters, or about four inches, the tight stretching of the cervix and upper vagina is stimulating the maximal outpouring of the hormone that will signal the uterus to push the baby out. But with Sarah only halfway to the goal, there was obviously still plenty of time.

Although the child would enter the world on that day, the hours still to go made Marshall offer us the choice of being admitted to the hospital just then or going home to wait. Sarah preferred the second option, especially after Marshall's midwife associate, Linda Lisk, said she would come over to check in on her.

I took my wife home, got her settled, and went right back to the hospital, because I had an operation scheduled for 10:00 a.m. With what I thought was absolutely certainty, I promised to be back at noon.

Having been told by Linda that labor proceeds at a more regular pace when a woman is active, Sarah proceeded to get things ready for that evening's dinner, taking a chicken out of the refrigerator and putting it into the oven. The contractions were developing a definite rhythm as she puttered around the kitchen. They had become so strong that she had to stop what she was doing each time one reached its peak, until it subsided.

Sarah's back was now constantly aching, but at the top of a crescendo's contraction the pain in that location was so severe that "I just wanted to die." Linda was phoning every twenty minutes, and at this point she suggested that Sarah place a pile of pillows on the floor and let herself sink facedown onto them. The gentle pressure is theoretically meant to stimulate the baby, lying within her in a head-down position, to make the turn into the birth canal. No sooner had she done this than the pains began to speed up, increase markedly in forcefulness, and focus more in her abdomen.

Although it was proceeding at a somewhat faster rate than usual, Sarah was undergoing a typical evolution into hard labor. The uterine contractions were forcing her baby to descend gradually into the birth canal and position itself for delivery. From the instant of conception, to childbirth, and then on to nursing—that any of this could have taken place is due to one of those seemingly miraculous sequences of nature by which a complex multiphasic process is shepherded through a series of enormously intricate events made possible by an orchestration of enzymes and hormones each of which becomes activated precisely at the juncture where its presence is required in order for the next step to take place. It is no accident that I have written all of this in one breathless sentence. Even the presence of a dash and a few commas near the beginning gets in the way of the seamless continuity I am trying to emulate.

What is it, exactly, that initiates the events of labor? When does a woman's body reach a point at which it will no longer tolerate being "a vessel," when it must "get this thing out" and reclaim itself? The answer seems in large part to be found, once again, among the hormones.

Oxytocin, the pituitary hormone that squeezes the milk out, also stimulates powerful contractions of the smooth muscle in the uterine wall. There is something about the gradually increasing levels of estrogen during the course of the pregnancy that makes the uterus more and more sensitive to oxytocin's effect. Near the end of the full forty-week term, the estrogen levels are at their height and so is the uterine responsiveness to oxytocin. It is thought that the increasing stretch on uterine and vaginal tissues at the very end of pregnancy finally reaches a critical level, at which nerve impulses are sent up to the hypothalamus, which in turn signals the pituitary to release oxytocin. The stretch resulting from the pressure of the baby's head contributes mightily to the entire oxytocin effect, but it also sets up a direct neurological reflex arc through the spinal cord that adds to the effectiveness of the contractions.

This may be the first time a person uses his head in playing the game of life. But in childbirth, as in all other things, the head is sometimes not enough—a bit of hormone helps. So the little squirt kicks in with a little squirt: Oxytocin released in the pituitary of the fetus contributes to contraction of the uterus. Not only that, but other hormones produced in the adrenal and pituitary glands of the fetus have already increased the number and activity of oxytocin receptors in the uterus.

There is something quite marvelous about this eruption of events that sets the stage for labor, initiates it, and carries it through to its successful conclusion and aftermath. In a sense, the child itself lets its mother know exactly what is needed from her, whether some strong uterine contractions to allow it to leave the womb, or a bit of a squeeze on the milk glands to provide its first worldly food. These are striking examples of the mutually rewarding interaction between mother and child that, in one form or another, will continue, whether physically or emotionally, throughout the lives of both of them. It gives perhaps an even deeper meaning to John Donne's famous aphoristic line that "No man is an island, entire of itself."

Back to Sarah, last seen in hard labor and alone in the house. With the oxytocin doing its uterine job so effectively,

> *The contractions started coming fast and furious. At that point, I called your office to find out where you were, because it was 12:30 and you weren't there. It was clear that this child was coming soon, and I was getting into just a little bit of a panic. Your secretary said you were on your way. Well, about a minute later, you galloped through the front door like the Canadian Mounted Police. As you rushed in, you shouted, "Okay, Sal, I'm here—let's go!" At that point, I was positive I had started what they call transition, that advanced active phase. The pains were one on top of the other. You got really close to me, wanting to help, but all I wanted to do was just push you away because I couldn't get enough air. I felt that all of my animal instincts were at play. I knew that I had to get somewhere to have this child—I knew that my husband was loving and kind and wanted to help me but that nobody could help me at this point. So I had to push you away and say, "That's fine. It's fine for you to say, 'Let's get going,' but I have to do it in my own time."*
>
> *So we made our way to the car and I managed to get in. It was*

drizzling and rainy, and that didn't help matters one bit. To make things worse, a crew was working on Whitney Avenue [the main road into New Haven from our home in a suburb], and we had to go through a back way in a roundabout route to get past the construction. You were trying to chat me up, but I didn't want to talk. I just wanted you to get me to the hospital. We finally passed the digging and pulled out onto Whitney. Just then, some guy cut you off and you started swearing, like you always do when that happens. That was just a little too much.

I remember driving through all the back streets and finally reaching Whitney Avenue. The other car cut me off just as I was picking up speed, and I began my swearing with the opening words I customarily use when there is someone to listen: "Did you see what that son of a bitch just did? I can't believe the damn—" Sarah now lost patience with me. "Shep, give it a rest, will you—just get me to the hospital."

With that, everything became very concentrated, and from then on I drove with a single-minded determination that carried us through several red lights and a few stop signs and then up to the admitting entrance of Yale–New Haven Hospital. Here, too, there was construction. The patient drop-off place had been temporarily changed. When I found it, I rushed into the building to grab a wheelchair, but Sarah couldn't wait. She somehow managed to hoist herself out of the car and laboriously (hah!) trudged her way through the back entrance, where I found her shuffling awkwardly around, looking for the admitting office. That, too, had been moved, down a series of corridors.

I settled my bursting wife into the wheelchair and propelled her toward the new office, speeding like a hungry farmhand at dinnertime. The hustle of our entrance had no effect on the lethargic admitting clerk, who slowly began to process a hospital bracelet. Sarah emphatically let her know that there was no time for such formalities, unless she wanted to assist in a delivery right there on her neat little desk. That, followed by the unmistakable sound of an incipient mother's groan, energized the clerk sufficiently to make her scribble "Nuland" on a bit of paper, which she hurriedly affixed to a plastic band and slapped on Sarah's wrist. We were then rocketed up to the delivery floor.

The contractions had begun to come every two minutes or less, and they were lasting about sixty seconds, rising to an excruciating crescendo before easing off. Between the contractions, a steady and quite severe pain never abated.

It wasn't clear whether Marshall Holley would get to the hospital in time, but Linda was waiting for us. She helped Sarah put on a hospital gown, got her into the birthing bed, and examined her. The top of the baby's head was pushing through Sarah's vulva, which encircled it like a crown. For obvious reasons, this stage of delivery is called crowning.

Almost immediately, it was time to push. Marshall had not arrived, but his associate, Dr. Effie Chang, came in and examined Sarah to confirm Linda's findings. We had become very close to Linda over the preceding months, and Effie willingly assented to our wish that this skilled midwife do the delivery. A few moments later, Marshall arrived and simply stood by with the proprietary air of avuncular geniality that is his way, smiling as he watched wordlessly while Linda was gently guiding Sarah with soft words and her reassuring touch. Her aim, it was quite clear, was to be as unintrusive as possible but at the same time not to allow Sarah to get in the way of what her body had to do.

Sarah would later remember that there seemed an angelic quality about Linda all through the delivery. Her blond hair was long and done in curls, and a circle of ringlets around her forehead was strongly suggestive of a golden halo. If not an angel, she was at the very least the good fairy of childbirth. Her voice was calm and almost hushed. As I watched her and occasionally glanced across the room at Marshall, I was grateful for the sense of peace and naturalness they conferred on a situation that might have easily been chaotic.

The birthing bed on which Sarah was about to have our child looked no different than the comfortable, spacious bed in our room at home, with the exception that its lower section could be slid out if necessary to allow the accoucheur to get closer to the field of action. It was without stirrups, and nowhere in that room were to be seen any of the impedimenta of an OR. A sterile just-in-case delivery kit was stored in a wrapped package on a table in the corner, but otherwise the place was as homey as the room in which the baby was conceived.

A nurse took up a position at the left of the bed and was soon using her hands as a kind of human stirrup to keep Sarah's knee bent and her thigh as far up and back as she could. I moved to the right side and did the same, all the while trying to remember what I had learned in Lamaze class about the correct way to coach and to encourage all of the proper breathing techniques. I recalled just enough to be what I thought was helpful, but it wasn't Lamaze that was filling my heart at that moment. It was the simple fact of being there, steadying my wife's still-stockinged foot in position with one hand and clasping myself into continuity with her by holding her hand with the other. There was no way I could so much as imagine the physical aspect of what she was experiencing, but the essence that passed between us was the unity of our life together. For me, those moments were its fullest expression.

I was there because I wanted to be part of my child's birth and I wanted to support Sarah as much as I could, but I had not anticipated what was beginning to happen to me. A fullness rose up within my chest. I couldn't move my gaze from that beloved face, and even its grimaces had the beatific appearance of a sacredness we had created. I saw her through the tearful haze that was the only outward evidence of the infinity of emotion that was welling up out of some great depth of me, seeming to find its only outlet by overflowing from my eyes. If it could have, it would have burst through my ribs.

But Sarah had more immediate things on her mind:

> *It was such a short period of time that we were in that room. Almost immediately, it was time for me to push. There was no time for anything—why, I was still wearing my socks. I never even had the fetal monitor.*
>
> *The urge to push was so powerful that Linda had to tell me when not to do it. My whole body wanted to bear down, but she made sure I did it at exactly the right time. It was here that the Lamaze training was so helpful to me. When I wasn't supposed to push, the Lamaze breathing techniques gave me something to focus on, and that was much better than just telling myself not to. If you don't know how to do anything but tell yourself, it doesn't work—the body goes ahead anyway and does what it wants to do. But focusing on the breathing technique and all that goes with it lets your mind manipulate your body, instead of the other way around. It's mind over matter, and that's very important.*
>
> *I was making sounds like I never heard come out of me before.*

It was guttural and low, like a growling animal—all instinct. That was the most connected I've ever felt with the rest of the biological world.

I think I pushed maybe three times, and Will was born. It was just amazing, to have this person come out of me. What did it feel like? Any woman will tell you the same thing. Perfectly frankly, it feels like a huge bowel movement—isn't that an awful way to put it? But that whole part of the anatomy is so joined up. All the nerves and muscles seem bound up into one whole, so you have the sensation of eliminating the largest thing that has ever left your body. What you feel down there is very much like a burning, rather than raw pain. You can feel the tissues stretching. That's when they do the episiotomy, and it's a real relief.

An episiotomy is an incision made into the tissues around the vagina in order to facilitate delivery. Especially in women having a first child, there may be some danger of a tear appearing in this area as the baby's head is forcing its way out. The episiotomy is basically a controlled tear whose function is to prevent a ragged, badly crushed wound, or even a laceration into the rectum. It is usually done by infiltrating the tissues with local anesthesia and then using scissors to make a short incision at about the four to five o'clock position of the vaginal opening. It is stitched up immediately after delivery.

When Will was born, Linda immediately put him on my abdomen. Then Marshall gave you scissors and a couple of clamps to cut the cord. I looked at my baby, and he had this very direct gaze, looking right at me. I was a little startled by that, because it wasn't supposed to happen. He was wonderful—it was such a wonderful feeling. I wanted to nurse him right away, because that's what the books say to do, but Linda said, "Don't do it just yet. He's heard your voice for nine months—now he just wants to look at you." What a glorious moment—he'd been hearing your voice and my voice for nine months and all of a sudden here we were and here he was.

It's a peculiar thing for a woman. Clearly, I didn't get pregnant by myself. I wanted very much for you to be a part of this, focusing me and patting me—but you know something? I was so wrapped

> *up in me and in this process that I had an image of you to my right, and of your holding me, but I wasn't really aware of anything anyone said except for Linda. Somehow, my focus was on me and on her, and on trying to do exactly what she wanted me to. Even Marshall wasn't part of it. And then after Will was born, the picture opened up and you were very much a part of it. I became so aware of you, and of the two of us together, with our baby, and kissing and hugging and crying. The crying began as soon as I saw him. When he was put on my belly, I remember so well looking at him and saying those words we've both remembered through all these years: "He's magic."*

What Sarah had forgotten, probably because she had never really heard it, was what I said just after Will was out. It was the last thing I had ever expected to think, much less actually blurt out: "Sal, you know that corny stuff they told us in Lamaze class about how phenomenal all this would feel? Well, guess what? It's not so corny after all!" We are told there are no atheists in foxholes—let it be known that there are no cynics at the birth of their own children.

That night, Sarah would understand the magnitude of what she was coming to feel for her little boy.

> *One of the most profound memories I have is of something that happened in the middle of that first night. I was sound asleep when the nurse came in. She was a remarkably sweet woman, and she said to me, "Mrs. Nuland, your child wants you." She turned the light on, helped me up, and got me something to drink. Then she went out and brought in this Person. He had been freshly bathed, and it was the first time I saw him that way. His hair was reddish blond and he was all peaches and cream.*
>
> *She put him in my arms. Everything about him was clean and fresh and beautiful. I felt this unbelievable surge of mother love. It was a kind of love I had never known before, and I had not anticipated having this profound feeling. It was visceral.*

And so—no man is an island, and neither is any woman, particularly at the time of childbirth. It is from "this person who needs me" that the whole complex of human relationships is initiated, one that will bring this child into the community first of the family and finally of civilization itself. But perhaps the patterns begin even earlier—of empathy and mutual dependence that have fostered the development of the cultures that *Homo sapiens* has built. Perhaps they begin, in fact, as early as conception, the instant at which the relationship between a forming human being and another has its origin. It is then that the expression of specific genes initiates the sequence of consecutive processes by which two living things influence each other's development. Though we share these earliest bilateral dependencies with most animals, they go on to physical and emotional complexities in our species that give them significance beyond those in all others. Among the reasons for those rewarding complexities is our distinctive method of giving birth.

The process of childbirth itself can be seen as an extension of the principle of pattern formation, the process by which a sequence of events is laid out by signals coming from the environment of the affected cells. The extreme difficulty of delivery of the human pregnancy is unique in the animal kingdom. A woman's pelvis is the only one so shaped that the fetus must enter the birth canal facing sideways and rotate ninety degrees, finally coming into the world facedown. The necessity for this is thought to be due to evolutionary changes in the pelvis that narrowed it and otherwise changed its shape to enable our legs to be centered under us in such a way as to allow two-legged walking. This resulted in a passageway much less easily navigable than in other primates, and with a larger head trying to get through.

The narrowed pelvis and the larger head make delivery so difficult that the mother needs assistance in order to accomplish it. Even chimpanzees, our closest animal kin, usually hide at the time of birth and deliver themselves alone, but women have to be helped. A great deal has been made of this by some physical anthropologists, and with good reason. The constricted anatomy means that having aid at the time of delivery enhances the individual's chances of reproductive success. By this line of reasoning, it would appear that needing such help is a product of—and having it is a factor in—natural selection. It has been claimed that the physical problems caused by the difficulty of childbirth make empathy and close communication valuable assets, for without them, very few deliveries would succeed. If this is true, it follows that humans are *biologically* constituted in a way that preconditions us to factors tending to bring us together into groups of mutually supporting individuals. Not only is it the mothering instinct toward "this person who needs me," therefore, but the entire set of circumstances in which it is biologically ordained that birth must take place that are among the instigating elements in creating human relationships, societies, and culture. What makes such a thesis particularly attractive is its consonance with all of the events of the forty weeks preceding the actual delivery, during which the progeny of a single fertilized cell differentiate, form tissues and organs, mature, and are born, all because of a biological mutuality between mother and child.

These are compelling ideas, made all the more persuasive by their consistency with the general principles that govern the interactions within cellular systems and with those that guide the development of the embryo and also guide the responses of the mother's tissues and psyche. Only with caution should a thesis be extrapolated from cells to societies, but sometimes the speculations that follow from witnessing observable natural events are too suggestive to ignore. Granted their as-yet-unproven nature, there appear to be biological origins to the reasons why no man is an island. The direct line of chemical and physical principles linking molecules to cells, to tissues, to organs, to you and me—and to the community of humankind—increasingly reveal themselves as the basis of those products of our anatomy and physiology that we call mind and spirit.

The heart of creatures is the foundation of life, the Prince of all, the Sun of their microcosm, on which all vegetation does depend, from whence all vigor and strength does flow.

William Harvey, M.D., 1628
Letter to Charles I of England

9 | THE HEART OF THE MATTER

Thus in a single sentence does William Harvey, the discoverer of the circulation of the blood, sum up the glory of the heart in language both poetic and biologically accurate. He writes in the grand tradition of those enthralled by the omnipotence and the enigma alike of nature—which creates an edifice of flesh and sets it free, to be governed by concealed self-generating internal mechanisms of majestic authority. The lore and literature of all peoples is filled with such testaments to the heart's mystical sovereignty.

In dim caves, primitive men saw that on the one hand, the life of this thing was its own, and no force of will seemed able to alter by a single thump the rhythm or pace of its interminable bounding force; on the other, it sped up and sometimes skipped and stumbled at times of fear or rage, while the tranquillity of a summer evening softened and slowed its intensity as though its mood responded to the balm of an untroubled mind. But which, in fact, preceded which? Did the turbulence or calm of one's emotional seas arise from the patterns of the heart itself, or did passion originate elsewhere, to draw the heart into doing its bidding?

Hippocrates would teach one thing, Aristotle another. Aristotle was sure of his ground. To him, the heart was not only the seat of the soul but of the emotions, as well: "The motions of pain and pleasure, and generally of all sensations plainly have their source in the heart, and find in it their ultimate termination." To Aristotle, the heart was the central capital from which our lives are governed.

The followers of Hippocrates, however—and Hippocrates's contemporary Plato, too—believed that the seat of the emotions was the brain; all other parts followed after. What they shared with Aristotle was the thesis that life depends on an innate quality situated in the heart, called animal heat. To Aristotle and others, the pulse and cardiac contractions were the result of the heart's innate heat, which warmed the blood and therewith also warmed all parts of the body. Much of Aristotle's thinking in this regard was formed by his detailed study of the developing embryo, whose first animate sign of life is the heartbeat. With such visible evidence available to him, it is understandable that he would interpret his observations as indicating that the heart is the source of all life.

The belief in some form of innate heat—with the lungs there to cool the blood should its temperature become too high—persisted until the seventeenth-century explosion in scientific thinking. With the introduction of inductive reasoning and the beginnings of the experimental method, researchers began to think in terms of seeking a demonstrable chemical source for the body's heat. It would be a long time before they found it, but enough fanciful ideas about the heart's role in life had meanwhile become so deeply implanted in popular thinking that some of them have never quite left it, at least in the creative minds of poets and lovers—and we should be grateful for that.

In 1628, William Harvey wrote a small book of seventy-two pages whose publication would mark the most significant turning point in the entire history of Western medical thinking. In his *Exercitatio Anatomica de Motu Cordis et Sanguinis in Animalibus* (Anatomical Exercises Concerning the Motion of the Heart and Blood in Animals), he demonstrated that the blood is *not* constantly being replenished in the liver as it is used up, and it is *not* sent out from there via the veins, and the veins do *not* drench the tissues with it like some ceaselessly ebbing and flowing irrigation system. This had been by far the most widely accepted of several competing theories of blood flow, but *De Motu Cordis* (as it is usually called) would in time ring the death knell for all of them. Harvey used a combination of quantitative and qualitative methods to make a series of easily confirmable observations and measurements that led him to the conclusion that he proclaimed in one ringing sentence of the single paragraph that is the entire contents of his chapter 14:

> *It must therefore be concluded that the blood in the animal body moves around in a circle continuously, and that the action or function of the heart is to accomplish this by pumping.*

And then, to this compact package that would prove to be the greatest gift any researcher has ever bestowed on medical science, he added the final scarlet-ribboned bow: "This is the only reason for the motion and beat of the heart." With this, he no doubt meant to lay to rest the misguided confidence of his readers in those mystical attributes of the heart so dearly cherished by kings and commoners alike. Perhaps his message was not lost on Charles, though the royal patient must have taken consolation in knowing that his doctor still believed so strongly in cardiac supremacy that he compared the heart's indispensability to that of the king himself. (This high opinion of the monarch seems not to have been shared by all of his subjects, because Charles was executed as a public enemy in 1649, to be succeeded as "the Prince of all" by Oliver Cromwell.)

Although the implications of Harvey's contribution were not at first universally appreciated (and in some quarters not even accepted), the publication of *De Motu Cordis* opened the way to a profusion of studies aimed at elucidating aspects of cardiac function and the details of the various of its elements, such as blood pressure, the pulse, and the activity of cardiac muscle. Finally, in the late eighteenth century, physicians began to use the fruits of findings based on Harvey's principles to help them understand the diseased hearts of their patients. By the first decade of the nineteenth century, the technique of percussion had been developed, by which it was possible to estimate the size of the heart by tapping on the chest. In 1817, the stethoscope was invented. Postmortem examinations were becoming commonplace at that time. By a combination of physical examination of the living and autopsies of the dead, a great deal was learned about the heart during the course of the nineteenth century, but the advent of X ray in 1895 was the first of the technological innovations that have brought us to our present state of detailed knowledge of so many aspects of cardiac anatomy and physiology.

The heart is most easily and comprehensively understood by thinking of it as being composed of two pumps lying side by side in the center of the chest like a pair of Siamese twins attached along their entire length. Like most Siamese twins, they are just enough different from each other that they are easily told apart. Although they share a common electrical conduction system and must carry out their assignments simultaneously, they have different jobs to do. They are equipped to do those jobs because they are made almost completely of muscle, called the myocardium.

Each of the two pumps is composed of an upper collecting chamber called the atrium and a lower ejecting chamber called the ventricle. The atria do not need to be strong because their contraction squeezes blood no further than into the ventricles lying below them. The ventricles, on the other hand, have walls of encircling muscle as thick as half an inch or more for the left and a third to a half of that for the right.

Being stronger and more muscular, the pump on the left can build up a much higher pressure on the blood within it. The left ventricle's job is to squeeze down so powerfully that it can force its contents out of its chamber and into all the arteries of the body, no matter how distant. The reason it must be so strong is that it needs to push blood much farther, and against a great deal more resistance (think of the tens of thousands—yes, tens of thousands—of miles of elastic arteries, arterioles, and capillaries out there in the periphery) than does its partner. Under normal conditions, doing this demands a head of pressure equal to that needed to hold up a column of mercury 120 millimeters high, about 5 inches. That height of mercury is equivalent to about five and a half feet of water.

The right ventricle has only to propel blood through the nearby low-resistance circuit of the lungs. Accordingly, its pressure usually does not have to exceed thirty-five millimeters, about an inch and a half of mercury. These differences explain why centuries ago it became customary to distinguish between the so-called greater and lesser circulations. The greater is also called the systemic circulation, and the lesser is the pulmonary, from the Latin *pulmo*, meaning "lung."

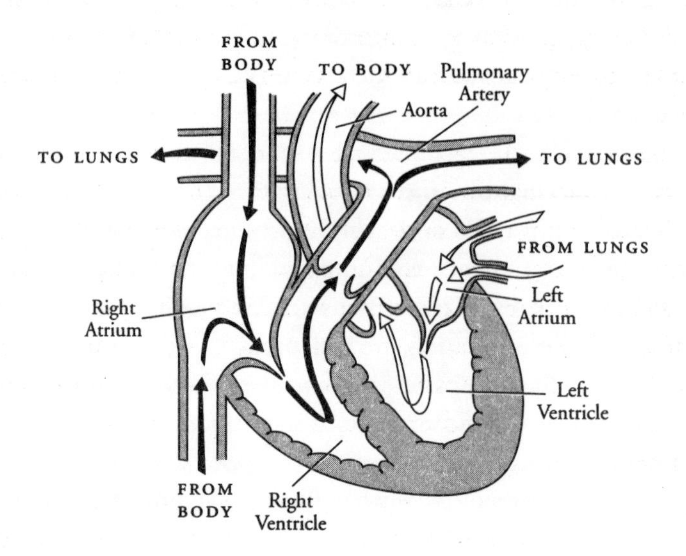

The heart, with directions of flow of unoxygenated and oxygenated blood indicated by black and white arrows, respectively.

The shared wall between the two sides of the heart is called the septum, which is thin between the two atria and much thicker between the ventricles. Although the two septa are in continuity, each has its own name because they separate different chambers—they are called the interatrial and the interventricular septum. Separating each atrium from the ventricle below it is a one-way valve, consisting of either two or three delicate leaflets, or cusps, which fit snugly together when the valve is closed but fall away widely from one another when it is open in order to allow blood to pass easily through. Because the valve on the right has three cusps, it is called the tricuspid; because the two cusps of the valve on the left come together in a shape reminiscent of an upside-down bishop's miter, it is called the mitral. The exit channel from each ventricle is guarded by a three-cusped valve, called the pulmonary on the right and the aortic on the left. All four cardiac valves permit only forward flow. They snap tightly shut as soon as the pressure rises in the chamber or vessel into which they lead, thereby preventing backflow.

Blood returning from the lower part of the body arrives into a large cisternlike vein, called the inferior vena cava, before emptying into the right atrium. Similarly, blood from the head and upper part of the body enters via the superior vena cava. Before describing the events in a complete cardiac cycle—which is another term for a single heartbeat—a pause is in order, to allow a panoramic overview of just what it is that is about to reveal itself.

On the face of it, the cardiac cycle is no more than yet another of those wonders of coordination and timing upon whose flawlessness and predictability all human life depends. "No more than"—by this point in reviewing the amazements of human biology, the promise of reading about one more example of natural selection's 3.5-billion-year history of ascending accomplishment is greeted perhaps with continuing interest but hardly with surprise. One begins to expect these prodigies of our biology and ourselves and to develop a smugness about our machine's smoothness.

And yet things obviously do go wrong. Not only that, but they are going wrong all the time. Whether from the constant molecular bombardment that is part of cellular life or because of the constant environmental bombardment that is part of daily existence on our planet, DNA is constantly being altered, cells are losing orderly control over their growth patterns, the products of ordinary digestion are gumming up the walls of our arteries, tissues are being poisoned by the chemical agents we ingest in the name of pleasure—the threats to ongoing life are perpetual and omnipresent. That we survive so many of them says as much about nature's process of selection as does the very existence of the organs that are so constantly under attack. A major key to our persevering continuity is the ability of our body's systems to right countless small wrongs that every instant threaten to destroy the workings that evolution has taken eons to create. We adapt; we change our molecular responses; we repair ourselves—we parry every blow, and all in the service of keeping the gyroscope of life spinning smoothly. It is the changing that enables the constancy. Even the constancy is constantly changing in order to maintain the dynamic equilibrium of cellular life that allows the dynamic equilibrium of our lives in the world. The changing constancy of living things underlies the homeostasis of the multicellular, thinking organism that is a human being.

By no organ system of the body is any of this better exemplified than by the one we call circulatory. A small part of the story of the circulatory system's ability to adjust has already been described in telling the saga of Marge Hansen. Now we come to the heart of the matter.

In an average lifetime of some seventy-five years, the cardiac cycle must repeat itself at least 2.5 billion consecutive times and pump out a million barrels of blood. While you are sitting quietly at rest and reading this book, your heart is constantly doing twice as much work as your leg muscles would were you running at top speed. All of the terpsichorean brilliance we have seen demonstrated in molecular interaction and cellular differentiation will avail nothing in the absence of the sustained dependability of a tireless heart. The sustained dependability of a tireless heart relies, in turn, on the performance of the trillions of chemical reactions occurring in its aggregate of cells during every instant of its function.

For most of us, what holds the greatest interest is the cardiac activity we are physically aware of. Whether consciously or without thought, we are ever on the alert for the signals our organs transmit to us; we often respond to those signals not only instinctively but with deliberation and with understanding of the probable consequences of the decisions we make. And this, too, distinguishes humankind from the beasts. *Homo sapiens* has an awareness of the workings of his body unknown to any other animal. We feel the beating of our hearts, we hear the growling of our intestines, we see and smell the quality of our excreta, as do so many other animals—but only we have a useful sense of what they mean. On conscious and unconscious levels, we are responding and adapting as can no other animal to internal signals, both the strong and the subtle, telling us to alter something in our behavior. We are aware of our tissues and perhaps our very cells in ways possible only to an organism with a mind. We persist in life, therefore, by responding and adapting to stimuli not only coming from outside but coming from inside, as well. Only by answering the inner voices can we fortify ourselves to answer those that come from the external environment. When we say that we follow the dictates of our hearts, we are talking about far more than conscience or romance.

The heart does the physical aspect of its dictating so clearly that much of it is audible, providing that the cardiac voice is enhanced by conduction through a stethoscope or some electronic source of amplification. Its sounds are produced by its muscular contractions, the snapping shut of valves, and the flow of blood through its chambers—in a word, by its pumping.

When it was stated a few pages earlier that the twin pumps carry out their assignments simultaneously, what was meant is that both atria contract together and so do the ventricles. The synchronization is such that the atria are relaxed when the ventricles are contracting and vice versa. When a chamber is contracting, it is said to be undergoing *systole*, which is a Greek word going back to Galen in the second century A.D., meaning "a pulling together" or a "contraction"; the relaxation phase of the cardiac cycle is called *diastole*, from the Greek verb *diastello*, meaning "to separate," because the relaxation is a pause between two systoles. Both *systole* and *diastole* are pronounced with a long *e* at the end, and the accent on the syllable containing the *s* (*sys* and *as*, respectively)—*sys*tole and di*as*tole.

The cardiac cycle begins as blood flows passively into the right atrium from the venae cavae and into the left atrium from the pulmonary veins. About 70 percent of the blood reaching the ventricles from the atria flows directly in, and then the atria contract (atrial systole) to propel the rest of it (none of the four chambers ever empties completely; there is a small amount of residual blood left after each contraction). Following this, the atria relax (atrial diastole) and the ventricles contract. The ventricles' systole raises the pressure inside their chambers and forces the tricuspid and mitral valves shut, effectively preventing blood from regurgitating back up into the atria. No electricity or hormones are required—cardiac valves shut for the simple mechanical reason that elevated pressure pushes their leaflets upward into the closed position. This is precisely the way flap valves work on a water pump.

Systole forces the ventricular contents out into the pulmonary artery and aorta, raising the pressure in those large vessels and in turn forcing the pulmonary and aortic valves shut to prevent backward flow. When the cycle is completed, the entire heart is in diastole for a brief interval, and then the whole sequence begins again.

The closing of each set of valves is forceful enough to make a slight thudding sound, which is easily heard when the listener's ear is placed on the subject's chest. The first of the thuds issues from the tricuspid and mitral closures and the second from the pulmonary and aortic. They can be simulated by saying, "Lub-*dup*, lub-*dup*, lub-*dup*" at a rate just a bit faster than once per second. Doing that produces as accurate an imitation as the human voice is capable of, and one that is very close to the real thing. Medical students have been parroting hearts this way for generations.

Certain diseases may affect a valve's ability to function, usually because its leaflets become scarred or because calcium and other substances are deposited on them, or both. When this happens, blood will regurgitate back into the preceding chamber when the leaflets cannot close (this condition is called valvular insufficiency), or be impeded in its progress when they cannot open fully (valvular stenosis). Occasionally, a child is born with either a tight or a floppy valve, which results in difficulties similar to those observed when the problem is caused by an acquired disease. Insufficiency and stenosis produce distinctive sounds called murmurs, occurring at specific times in the cardiac cycle. By taking note of the quality of the murmur and its timing, a physician can often diagnose the disease that is causing the problem. Should it not reveal itself to the stethoscope, the abnormality will be identified by one of the vast number of diagnostic technologies available nowadays.

As will be seen, the heartbeat has an independent synchronicity originating within its own tissues, but it does, in addition, also listen to signals coming from afar. The heart receives messages conveying the needs of other parts of the body, and it responds to them by speeding up or slowing down, and by strengthening its contractions when called upon to do so. The messages telling the myocardium to do one thing or another are conveyed to it by the fibers of the autonomic nervous system, or directly through the hormones adrenaline and noradrenaline, secreted by the central portion of the adrenal gland, called the adrenal medulla. (This should not, of course, be confused with the medulla of the brain. Being the Latin word for "the marrow," the word *medulla* is used by anatomists to convey the sense of the center, or core, of a structure.) In this, it is similar to numerous other organs of the body, whose function is responsive to the autonomic system and to certain hormones.

In response to the adrenaline and noradrenaline liberated by sympathetic signals, the heart's rate and forcefulness of contraction go up; in response to the acetylcholine liberated by parasympathetic signals, the heart slows, but there is no effect on contractile force. This is because the ventricles, although richly endowed with sympathetic fibers, have virtually none from the parasympathetic system. Such an anatomical arrangement is an example of the dominance of one of the autonomic systems over the other, as described in chapter 4.

The autonomic nerves affect cardiac rate and the force of contractility by their action on the heart's internal pacemaker (called the sinoatrial node, which will be discussed shortly) and on the myocardium. Electrical activity in the node and myocardium is influenced by the concentrations of calcium and potassium in their cells, which in turn are determined by how many of the molecules of these two elements can pass through the cell membrane. The amount of binding of noradrenaline or acetylcholine to the appropriate receptors in those membranes changes permeability in such ways as to facilitate or inhibit the passage of calcium or potassium, thereby affecting electrical activity as required by the body's needs. When I was frightened by the screeching tires of the suddenly braked car I encountered in chapter 4, sympathetic activity and the release of hormones from my adrenal medulla initiated by messages coming down from my hypothalamus sent a flood of calcium into the cells of my heart, setting off reactions that induced them to fire off and speed up its rate and forcefulness.

Like other of the body's structures, the heart is an intriguing mixture of dependency and splendid solo automaticity. Although hormones and an autonomic nerve supply influence its rate and forcefulness, its rhythm is independent of the outside influence of either the nerves or the chemicals. Its faultless regularity and the very fact that it beats at all are the result of a self-generated stimulus, triggered anew by the SA node for each pulsation.

The sinoatrial, or SA, node is situated in the back wall of the top of the right atrium, very close to the entrance of the superior vena cava. It is a tiny ellipse of tissue that serves as the pacemaker, driving the coordinated beat of the heart. Its messages are carried by an arborizing bundle of fibers to a relay station between the atrium and ventricles, called the atrioventricular, or AV, node. From there, they are transmitted to the wall of the ventricles by another arborizing network of fibers called the bundle of His, named for the Swiss anatomist who discovered it in 1893. The SA node that generates the signals, the AV node that relays them, and all the fibers that carry them are composed of specialized muscle cells that have become adapted by eons of natural selection to do their work. They are known collectively as the cardiac conduction system.

The purpose of all the circuitry and squeezing is to carry the blood along the pathways first described by William Harvey as being "in a circle continuously." Finally, we are ready to pick up that circle by following a drop of blood as it enters the right atrium from one or the other of the cavae.

Whether in the portion of the atrial contents that flows down into the right ventricle by gravity or the lesser amount that is forced into it by atrial contraction, the drop soon finds itself being propelled by ventricular systole into the pulmonary artery and then to either its right or left division. It moves easily through the low resistance of gradually narrowing pulmonary branches until reaching the capillary adjacent to one of the 300 million air sacs, or alveoli (singular: alveolus), in our two lungs.

It is in the alveolus that the real business of respiration takes place. Here, only two thin layers separate the blood from the air that will provide the oxygen to refresh it following its return from the long, exhausting journey just taken to the far-off peripheries of the body. The two layers are the very thin single-celled capillary wall and the very thin single-celled alveolar wall. Their very thinness allows minimal interference with the diffusion of oxygen into the blood and the diffusion of carbon dioxide out of it and into the alveolus.

Once it has passed through the two layers and entered the capillary, almost all of the oxygen—98 percent of it—combines with the iron-containing protein in the red blood cell, called hemoglobin (a perfectly huge molecule, containing some ten thousand atoms), to form the compound oxyhemoglobin. The remaining 2 percent is dissolved in the blood plasma. It is in oxyhemoglobin that the oxygen will be carried to the body's tissues. At the same time, the carbon dioxide that the blood has brought along from the distant cells as a waste product of their metabolism diffuses through the capillary and alveolar wall and enters the air that will be exhaled.

Though I have presented it in terms of its overt outcome, the fact that all of this exchange can take place so smoothly is the result of another of those wide-ranging arrays of highly intricate but exquisitely coordinated processes that characterize the workings of all living things but veritably adorn those of the most complex—us. Respiration is enabled by our body's orchestration of a polyphonic harmony of certain principles of the physics and chemistry of gases and solutions, the mechanics of breathing, the structure of the air passages, the distinctive characteristics of lung tissue, the molecular behavior of hemoglobin, and a set of specialized reflexes. Its depth and frequency respond to subtle changes in the blood's acidity and its content of carbon dioxide and oxygen, always in the direction of maintaining a state of readiness. We yawn; we gasp; we breathe faster or slower—all automatically and all to address some specific need of the blood and the cells it supplies. The entire sequence and timing of the melodious concertante of respiration is conducted, as are all of our bodily compositions, by neurological and hormonal controls and by the anatomic and physiological qualities of the participating tissues. They work together to provide constant surveillance and assure that the process will occur with such spontaneous regularity that even an act as conscious as breathing will take place without deliberate thought.

On leaving the alveolar capillary to enter the tiniest tributaries leading to the pulmonary vein, the refreshed blood is a bright scarlet, the color of oxyhemoglobin. It passes into the left atrium and then goes on to the ventricle. Ventricular systole drives it out into the aorta and thence toward the body's peripheral capillaries, where yet other kinds of exchanges take place.

Except in bone, cartilage, cornea, testicles, and the inner ear, there is not a normal living cell in the body that is more than twenty microns away from a capillary—a distance one-third the diameter of a human hair. The capillary interacts with the cells in its neighborhood through an exquisitely thin rim of *milieu intérieur*, the interstitial fluid bathing some or all of its surface. Exchanges between blood and these fluids take place across the one-cell thickness (*thinness* would be a better word) of the capillary wall.

Flow through capillaries is much slower than it is in the arteries, for several reasons. The total cross-sectional area of the gigantic capillary bed in any bit of tissue is so much larger than that of the artery feeding it that this factor alone accounts for much of the decreased velocity. And then there is the tightness of the fit of the blood cells moving single file through these tiny vessels, which at six to seven microns (1/4000 of an inch), are just about the same size as the red cells traversing them. The blood cells, being somewhat pliable, actually bend just a bit in order to get through. Now add the resistance of the arterioles that guard the entrances to capillary beds, changing with local conditions and responsive also to signals from afar. Clearly, there are plenty of factors to explain the leisurely pace of the circulation within these tiny vessels, so slow, in fact, that it has been estimated to be approximately one-thousandth of the torrentially rapid flow in the aorta. It is imperative that the blood move at such a crawl, because otherwise there would not be sufficient time for the exchange of gases and nutrients to take place. Even with all the meandering, blood spends only about a second in any capillary before passing into the tiniest veins, the venules. It must do all of its delivery and pickup in that brief interval.

Once arrived in the capillary, the oxygen separates itself from oxyhemoglobin like a passenger getting off a railroad car, then passes through the thin layer of interstitial fluid to enter individual cells along with the nutrients that have also been brought by the circulating blood. In the process, the hemoglobin molecule loses its bright red color and darkens somewhat. Carbon dioxide from the cells diffuses into the capillary. By dissolving in the plasma (9 percent of the carbon dioxide does this), combining with hemoglobin into a compound called carbaminohemoglobin (27 percent), and in the form of bicarbonate (64 percent) produced by its chemical interaction with water, it travels back toward the lungs. The chemical changes reverse themselves in the pulmonary capillaries, and the carbon dioxide is freed to enter the alveolus, where it is blown out with the next exhalation. The blood is purified of other waste products of cellular life as it passes through the liver and kidneys.

No wonder William Harvey said of the heart that it is "a sort of internal animal" that "hath blood, life, sense, and motion." He went further: "The heart is as it were a Prince in the Commonwealth, in whose person is the first and highest government every where; from which as from the original and foundation, all power in the animal is deriv'd, and doth depend."

Even princes are susceptible to disease, and sometimes to failure. When the heart needs help, it is available in the form of a variety of medical treatments, some of them involving pharmaceutical agents and some of them involving surgery. When naught avails, a heart, like a prince, must be replaced. Almost always, replacing a prince is not particularly difficult, especially because so many candidates wait in the wings. In the matter of hearts, things are different, perhaps because Tennyson was right in more ways than he knew when he wrote that "Kind hearts are more than coronets." Not only are there too few available replacements; not every kingdom (read person) will benefit. But should an appropriate donor organ be found, the process of implanting it into a needy patient's chest is remarkably standardized and straightforward. Cardiac transplantation, while not quite a routine surgical procedure, has become so commonplace that it has long since lost its status as a cause for amazement. Here follows the story of a transplanted heart and the man in whose chest it has now peacefully pumped for seven years.

The idea that animals could exist whose parts come from two or more different creatures is as least as old as the *Iliad*, and almost certainly a great deal older. Homer writes of an animal called the Chimaira, "a thing of immortal make, not human, lion-fronted and snake behind, a goat in the middle, and snorting out the breath of the terrible flame of bright fire." So preposterous were such fantastical beasts thought by less imaginative citizens than writers of epics that the word *chimera* entered modern English to mean—and here I quote *Webster's Unabridged Dictionary*—"an impossible or idle fancy." Well, my *Webster's* is almost fifteen years old, and even when it was compiled, it seems to have been out-of-date, at least as far as chimerae are concerned. Among biologists, the word has come to be used to designate any animal made of cells from more than one genetic origin, or as the latest *Oxford English Dictionary* now puts it, "an organism whose cells are not all derived from the same zygote." The man of whom I write, and every other person who has survived tissue transplantation, is a chimera.

Legends of successful transplantations are well known in the Western tradition. The two fourth-century brothers Cosmas and Damian were canonized for being credited with achieving a number of miraculous cures, but the most famous of those claimed for them was the grafting of the healthy leg of an Ethiopian donor onto the body of a bell-tower custodian who had undergone amputation. Such myths of transplantation are retold down through the ages because they appeal to that part of the human mind which clings to the hope that life can be made eternal. If only it were possible to replace worn-out or sick body parts with healthy new ones, there would be no need to die; at the very least, waning vigor might be restored.

But not all transplantation stories are unconfirmable legends. As early as the seventh century B.C., Hindu surgeons were reconstructing injured noses by using the patient's own skin. In no sense, of course, were such operations chimerical, but it is known that in late-eighteenth-century London, the Scottish surgeon John Hunter somehow managed to transplant a human tooth into the comb of a cock, and testicles from one chicken into another. These were isolated successes, and, though tantalizing in their implications, they led nowhere.

Things have changed, however: We are now living in an age of scientific astonishments. Every day, in medical centers throughout the world, people of all ages leave operating rooms carrying within them hearts or lungs or kidneys or pancreases that began life in the bodies of others. In July of 1989, for example, a man I will call Anthony Cretella underwent a heart transplant. There was nothing about his operation, his postoperative course, or his recovery that his cardiologists and surgeons did not consider routine. And that may be the greatest astonishment of all—that an operation so complex and technologically demanding, an operation that as recently as the middle of this century seemed too far off to hope for, went so rapidly from being a fantasy to being standard surgical practice.

I am a general surgeon. During my years of training, which ended in the early 1960s, all of us in the program learned to perform the various cardiac operations that could then be done (even at that time, the list was not short). Although it was clear to me that I had neither the personality nor the temperament of a cardiac surgeon, I have never lost a certain awe at the intricacies of the heart and at the skill of the doctors who treat it. And, like almost everyone else, I have been astonished at what has been accomplished in the field of transplant science in such a very short time.

One day, I decided to see for myself. Although I had observed parts of the transplant process, I had never followed an individual patient through the entire course of his operation and recovery. I asked Yale–New Haven Hospital's transplant-research fellow, Dr. George Letsou, to call me when his team was readying itself to perform its next cardiac transplant.

The muscular wall of the heart is nourished not by the blood within its four chambers but by a group of vessels that are embedded in its surface and encircle the top of the heart like a crown. For this reason, they are called coronary arteries. Major branches of the coronary arteries descend toward the heart's tip, diverging into smaller vessels along the way, to irrigate the myocardium with nourishing bright red blood from the aorta. When the coronaries become narrowed by spasm or by thickened plaques (the condition called arteriosclerosis), or both, the heart responds the way an overworked calf muscle does when it cramps. The resulting pain is called angina pectoris—from the Latin verb *angere*, meaning "to choke," and the noun *pectus*, meaning "chest"—and, as anyone who has experienced angina pectoris can testify, never has a symptom been given a more aptly descriptive name. What the sufferer feels is a severe squeezing or crushing in the front of his chest, as though a fist or a constricting band were trying to extinguish his life. The sensation often radiates up into the neck and down one or both arms—most commonly the left. Fortunately, the agony of most angina pectoris lasts only a short time, frequently less than a minute.

Early accounts of angina, dating from centuries before its cause was known, describe patients who are ghostly pale and terrified by a sense of immediately impending death, but then become quite calm and normal in appearance as soon as the episode passes. It was long ago recognized, however, that once attacks begin, only rarely does a patient become free of them. The great English physician William Heberden wrote in 1768, "The termination of the angina pectoris is remarkable. For, if no accidents intervene, but the disease go on to its height, the patients all suddenly fall down, and perish almost immediately."

They perish because they sustain a major myocardial infarction, or heart attack. A section of the myocardium supplied by a narrowed branch of one of the main coronary arteries is deprived of blood long enough so that it cannot recover, and it infarcts; that is, it dies of blood deprivation. If the infarction is large enough, it garbles the heart's intrinsic rhythm and converts it into a mass of uncoordinated wriggles and squirms, called ventricular fibrillation. Death follows within minutes because the fibrillating ventricle is incapable of the effective ejection of blood—there is no systole.

One of the most detailed premodern descriptions of angina pectoris was left by John Hunter, who succumbed to the disease himself in 1793. Hunter was a short, redheaded fireplug of a man, renowned for the fury of his contentious nature. He never learned to avoid confrontations, even though his blood-starved myocardium cried out its anginal warning each time he allowed his anger to get the better of him. He would often position himself in front of a mirror during an attack so that he could record his appearance as well as his symptoms. Hunter virtually dared his coronary arteries to stake their final claim. He was fond of saying, "My life is in the hands of any rascal who chooses to annoy and tease me." It was in the boardroom of St. George's Hospital in London during a heated staff meeting that Hunter's rendezvous with massive infarction occurred. He rushed out of the room in a torrent of anger and chest pain. Moments later, he collapsed lifeless into the arms of a colleague standing in the corridor.

The first infarction of most patients is relatively slight. The heart muscle recovers, and the injured area heals into a scar. As episodes reoccur, more and more muscle is replaced by scar, and eventually the heart becomes too weak to pump with sufficient strength. This is what physicians call heart failure. When that occurs, blood backs up into the lungs, liver, and other organs because the force of the heartbeat is insufficient to keep it circulating properly.

Anthony Cretella's heart had not yet become weak enough to fail, but he had the most crippling case of angina pectoris that some of his physicians had ever seen. The pain had reached the point where it was unpredictable. Cardiologists call such a condition unstable angina, because the attacks occur without the stimulus of physical or emotional stress. Even at complete rest in the coronary intensive care unit of Yale–New Haven Hospital, Cretella was having frequent episodes, requiring narcotics and also large doses of nitroglycerine intravenously and by mouth to help his coronary arteries resist spasm. In desperation, his doctors had consulted members of the anesthesia department's pain service, but even they had nothing to offer.

What led up to all this was thirteen years of known heart disease and probably many more, during which obstructing plaques had been building up. Anthony Cretella, who was fifty-one at the time of his transplant, had his first myocardial infarction in 1978, following at least two years of angina. Shortly after the attack, his hometown physicians, in Mystic, Connecticut, sent him to New Haven to consult Dr. Lawrence Cohen, a professor of cardiology at the Yale School of Medicine.

At that time, Cretella's cholesterol reading was 260, well above the upper limit of normal, which is approximately 200. Even levels up to 239 are considered borderline high, but 260 is unquestionably a bad number. He was a smoker, and he had done nothing about twenty or twenty-five extra pounds he carried on his stocky five-foot-eight-inch frame. Considering that Cretella was only thirty-nine, his physicians were apprehensive about his future. They were also concerned about an additional problem, and they wanted to be sure that Dr. Cohen was made aware of it: In their referral letter, they described Cretella as "a restless Type A emotional person who can never sit still."

Cohen's evaluation was more encouraging than he himself expected it to be. His studies of Cretella's heart showed that the arteriosclerosis was thus far limited to the right coronary artery; the left was still completely normal. The involved portion of the muscle was not large, and the angina had disappeared once the process of scarring was completed. Cohen felt that weight reduction and giving up cigarettes might prevent the plaque formation from progressing. In addition, he counseled Cretella about the necessity of adopting a more relaxed behavior pattern.

That might have been adequate for most patients, but it did not change a thing for Anthony Cretella: He preferred to believe that the prime cause of his heart disease was a factor for which he bore no personal responsibility—his family history. In the blunt-spoken manner that characterizes not only his talk but his entire approach to the world, he one day, a dozen years later, told me his theory. He spoke in short, economical sentences. His voice had no inflections—it never rose and it never fell. Like his face, it betrayed no expression, as though he was afraid that even such a small effort would consume his remaining store of energy. Cretella's skin had the pallor of chronic illness, and of exhaustion. The contrast with his hair, which is black, and his thick brush of a mustache gave his skin a gleaming whiteness, making his expressionless face look like an alabaster mask.

> *I was born with it. They never could do anything—it's my family history. When I was in my early thirties, I felt like I was already getting old. But nobody would listen to me. I had lots of jobs—I worked on a truck and I worked all different things—and I always felt good. And then I got a desk job, and that's when I started feeling like I'm getting old. That's when the arteriosclerosis started taking over. I believe that it was in my system all those years. As soon as I sat down, that blood thickened up. Maybe six, seven years before the heart attack, I argued that something was wrong with me. You see, I should have known, because my mother was sick with her heart and her arteries back then. And I did pursue it, but they didn't find anything. It was like that all the time. Of course, I refuse pain. I just ignore it. I used to get pain all the time, and they thought it was my hiatus hernia.*

Most of what he said is verified by the record. Cretella had been given a cardiac stress test when he complained of chest pain two years before his first coronary, and the results were negative. His pain was not typical of a hiatus hernia, which tends to be more burning in nature, but to his doctors, the stress test seemed convincing.

"I used to pass out—break out in a cold sweat and pass out," he went on. "My wife used to go nuts. Four, five, six years."

I agreed that even then it had probably been his heart, and said that in retrospect his pain appeared to be little bits of angina.

"*Big* bits," he said. "Finally, I had the heart attack. I wanted to say, 'You're a bunch of assholes, all of you!' "

The result of Cretella's recalcitrance and his refusal to change his eating and smoking habits was another heart attack four years later. In March of 1982, Cretella returned to New Haven to have his first cardiac operation, a coronary artery bypass graft, or CABG—universally called "a cabbage." In this procedure, short segments are excised from a leg vein and attached to the aorta and to points beyond the obstructed areas in each of the narrowed coronary arteries. It is tantamount to rerouting traffic around a highway pileup. All three of Cretella's main coronary arteries required grafting.

The results of the CABG were good, and Cretella remained lax about changing his habits. He seems to have thought that he had his problem licked. His only post-CABG concession was to cut his smoking from a pack a day to half a pack. He had been through a series of unsatisfactory jobs and an unhappy marriage, which ended in divorce in 1969. Even though he was pleased with his current job, and even though he had had a strong second marriage since 1975, he was unable to achieve a satisfactory level of equanimity. He was tense, easily upset, and not inclined to optimism. Nevertheless, the CABG lasted seven years, during which time Cretella was without symptoms and led a life that in all its essentials was quite normal.

And then, late in April of 1989, the angina pectoris returned. Cretella's local cardiologist admitted him to New London's Lawrence and Memorial Hospital on May 5 for a diagnostic workup. On that day, Cretella smoked his last cigarette.

The outcome of the studies was sobering. Two of the bypass grafts were completely occluded, and the third was only 5 percent open. The few of his smaller peripheral coronary vessels that were still open were described by the cardiologist as "diminutive." No vessel was wide enough beyond an obstruction to permit another CABG. On May 10, Cretella was transferred to Yale–New Haven Hospital.

The angina was rapidly growing worse. Since there was no possibility of an operation that might restore blood flow to the cardiac muscle, there remained only one hope. Dr. Cohen told Cretella that without the transplant of a normal heart, with its healthy coronary arteries, he had less than a 20 percent chance of surviving as long as a year. Reluctant and terrified, Cretella agreed to be screened for placement on the list of patients awaiting a donor organ.

Cohen had thought hard before offering Cretella a transplant. Donor organs are in short supply, and a transplanted heart often undergoes a process cardiologists call accelerated arteriosclerosis, which is one of the major causes of failure over the long term. The coronary arteries of the grafted organ are much more susceptible to plaque formation than are those of a normal person's heart. Transplant patients who continue to smoke and to maintain elevated cholesterol levels are at high risk of a swift recurrence of coronary occlusion, and Cohen's choice of patients for transplantation depends largely on his assessment of their willingness to comply with the rules.

Cohen weighed the conflicting factors.

> *There were two quite different threads that ran through my thoughts about Cretella. The first was that he seemed hell-bent on destroying himself. He was perhaps best characterized as a wise guy who had a very hard exterior, trying to be blasé. What really got to me and frustrated me in my contacts with him was the idea that here's a guy of forty-nine or fifty years of age who had been through triple bypass surgery, has a bad family history—and he's still smoking. He was also taking his cardiac medication very irregularly. The other thread was the absolute devotion to him that was demonstrated by his wife. She was one of the most genuinely concerned individuals I have ever seen. I subsequently saw much of her, and I could tell that there was nothing fake about this.*

Clearly, another factor in the equation was the character of Dr. Cohen. Unlike many high-ranking university physicians with a heavy load of academic duties, Cohen has a large consulting practice. The reason for his popularity as a consultant has as much to do with his personality as with his clinical skills. There is a gentle kindness in him, which pervades his relationships not only with patients but with colleagues and with the young doctors whom he trains. His voice is soft; he speaks very slowly and deliberately, in the manner of someone who is accustomed to being asked his opinion, but he is not a ponderous man. He has a sagacious sense of humor and gray-blue eyes that seem full of sympathetic understanding. Although he represents the peak of superspecialization, Cohen is an old-fashioned doctor. What interests him most is people, and people know it immediately. In spite of Cretella's history of irresponsible behavior, Cohen did not wish to deny him the kind of emotional support needed to carry him through an operation and its aftermath.

Cohen had already established that Cretella fulfilled the basic criteria for acceptance into the transplant program: He was less than fifty-five years of age; he had end-stage heart disease, with a life expectancy of less than twelve months; he had no other significant diseases and no infection; and his lungs were healthy. Cretella now underwent a psychosocial evaluation to make sure that everyone, not only Cohen, could be convinced that Cretella and his family (he has a son and two daughters) were equipped to handle the new challenges that would be presented by the transplant. His attitude toward his illness, his ability to comply with highly specific living patterns, the strength of his family support system, and his potential for a personally satisfying future—all these were assessed by a team of social workers, nurses, and physicians.

It was also necessary to evaluate Cretella's medical insurance. To everyone's immense relief, it was discovered that his company had provided a policy that covered every cent of the cost. When such coverage is not available, social workers and others have to scramble to find different sources of payment. No hospital can absorb the price of a heart transplant, which averages more than $125,000, and very few patients can afford the permanent out-of-pocket cost of drugs, which was at the time at least $5,000 a year. Currently, that figure is more than $6,000.

Discussions were held with Cretella and his wife, Susan, concerning the transplantation and its consequences—to make sure that they understood the entire sequence of events, including possible problems and complications. Faced with the virtual certainty of death, Anthony Cretella made a commitment to change his ways. Later, he would be his usual self when he was describing to me his feelings on May 18, 1989, the day he was told that he had qualified for the final candidate list.

> *I was scared shitless; I felt like I wasn't in the real world. But I decided to do it. It wasn't that I wanted a new heart; it's just that I was hoping to live. I'd do anything to live. I guess some people would say, you know, "I'm so sick, I want to die," but some people say, "I want to live—do anything you can for me." So that's what I said.*

Once it was recognized by surgical researchers that the heart's regularity is maintained not through the outside influence of nerves but by a self-generated stimulus arising in the SA node, a supposed major obstacle to its transplantation from one person to another was shown to be nonexistent: After a heart is stitched into place, it is capable of starting a spontaneous beat on its own.

But before cardiac transplantation could become a reality, two serious problems had to be solved, one of which has not reached a completely satisfactory solution even today. The first difficulty was strictly technical—how to graft a donor's heart simply and safely into the chest of the recipient. Research teams working in various countries eventually developed the necessary methods, however, and the actual surgical process by which one person's heart is connected up to the major blood vessels of another has remained virtually unchanged since Dr. Norman Shumway and Dr. Richard Lower first described it in 1960. The technical maneuvers involved in cardiac transplantation, although complex, are now far from being the most challenging in the repertoire of modern surgery.

The more difficult challenge was, and is, convincing the recipient's body that it must not destroy the new heart as an unwelcome invader. As early as the sixteenth century, the Bolognese surgeon Gaspare Tagliocozzi described a phenomenon he called "the force and power of individuality." Tagliocozzi lived at a time when amputation of the nose was performed as punishment for a variety of crimes, and in trying to reconstruct the faces of the victims of such punishment, he discovered that while skin taken from the patient himself would heal properly into place, skin taken from someone else would not. He theorized that each individual is endowed with some mystical way of knowing which tissue was foreign, and of killing it.

The source of the uniqueness of every person's flesh remained a mystery for almost three hundred years—until the basic characteristics of the body's immune system were discovered early in this century. Some of the most sophisticated biomedical research efforts of our time have gone into explaining immunity, and yet their results can be distilled into a principle so simple that Tagliocozzi, had he known of the existence of cells, could have predicted it: The cells and the body fluids of the host recognize specific elements within the grafted tissue as originating in some other individual; they then create specialized killer substances whose function it is to destroy that particular tissue. In the long history of humankind, this variety of xenophobia has been a good thing, because it has enabled us to fight off infection and rid ourselves of toxic foreign proteins that might wreak havoc with our bodies. On the other hand, it can create problems of its own, among them allergies and immune responses, such as transfusion reaction and some drug or food hypersensitivities, that can cause death.

Modern technology brings modern predicaments. Blood transfusion, the most common type of tissue grafting, became safe only when scientists recognized that each of us belongs to one of four major groups of immunity producers. Although no matchup is perfect, the immune response is not strong enough to cause trouble if recipients can be paired with donors within their own groups.

Blood is a tissue, but a heart or liver is an organ, and because an organ is composed of tissues of several types, the problems of proper matchup and compatibility become more complicated; in fact, an immune system's rejection of a donated organ is certain unless the donor is the recipient's identical twin. The greatest challenge to modern transplant surgery has been to mute the recipient's immune response sufficiently so that the donor's organ will not be destroyed.

One way of doing this is to decrease the host's ability to respond to the foreignness of transplanted tissue; another is to make the donated tissue less foreign by inhibiting those factors within it (its so-called transplantation antigens) that make it so menacing. Thus far, biomedical scientists have not succeeded with this second approach, and they have concentrated their efforts on methods that decrease the force of the host's immune response. The difficulty is that inhibiting the host's immunity to foreign protein also inhibits his resistance to infection. The transplant surgeon works in a narrow, poorly marked area between the two. If he does not sufficiently suppress his patient's immune system, the organ he has transplanted will be rejected; if he is too zealous in his suppression, he will be faced with overwhelming infection. There may be general guidelines, but there are no infallible rules in the game: Every patient is different.

In 1970, after many years of searching and of experimenting with less-than-satisfactory drugs, a researcher vacationing from his job with the Swiss pharmaceutical house Sandoz Ltd. quite literally dug up an agent that would eventually be called cyclosporine. On a camping trip in Norway, the researcher, aware that ordinary soil has proved an antibiotic treasure trove, was on the lookout for substances that might prove useful in the hunt for new drugs. A sample of Norwegian earth that he brought back to his laboratory contained a new compound that several of his colleagues found interesting enough to experiment with. The compound was studied intermittently for four years, until it was discovered that it had the ability to suppress the immune system.

Between 1974 and 1980, intensive investigations of the compound were made. In 1980, Dr. Shumway, of Stanford, and Dr. Thomas Starzl, of the University of Colorado, the world's leading transplanters of the heart and the liver, respectively, were granted sufficient cyclosporine for experimental use. Their studies resulted in FDA approval of the drug in 1983. Along with drugs that are synthetic versions of hormones secreted by the peripheral portion of the adrenal gland (the adrenal cortex), called steroids and used in the form of intravenous Solu-Medrol, or oral prednisone, cyclosporine has become the mainstay of immunosuppression. Recently a new drug, Nizoral, has been introduced which reduces the need for cyclosporine by 80 percent and cuts the yearly drug bill by several thousand dollars, since it costs less than $350 a year.

Two other immunosuppressive agents also play a role. One of them, an older drug called azathioprine (its brand name is Imuran), is less effective than cyclosporine and produces more side effects; the other, a protein substance called antilymphocytic globulin, which is given in a brief postoperative course, acts directly against the host's killer cells without injuring other tisssues. In the seven years since Anthony Cretella underwent his transplant operation, study of different drug regimens has resulted in significant changes. At least 60 percent of heart-transplant patients have been found to do well on a program of only two drugs, cyclosporine and azathioprine. Antilymphocytic globulin is reserved for special situations involving severe rejection of the transplanted organ. The decreased use of steroids has not only avoided their side effects but has also reduced the incidence of accelerated arteriosclerosis, to which they are an important contributing factor.

Since 1983, progress in transplantation has been rapid, and, with cyclosporine so effective in suppressing the host's immunity, the main criterion for heart transplantation has become size. As a member of the surgical team told me at the time of Cretella's operation, "Basically, anybody's heart goes to anybody it will fit."

A successful transplant is a triumph of organization. Although the chief surgeon is director of the enterprise, the actual putting together of all the elements is done by a person called the transplant coordinator. At the time of Cretella's surgery, Yale was fortunate in having Gail Eddy for the job. Eddy, who had been in her eleventh year as an intensive care unit nurse when she was tapped for the transplant team, always carries a beeper in her purse; except for vacations and professional travels, she has been on call seven days a week since having taken on her duties in 1987. When asked what she does for the program, she would say, simply, "Everything but the surgery."

Eddy sets up the process of patient screening and organizes the workup; she integrates the family into the overall planning; she is in constant touch with the New England Organ Bank, located in Boston, about her recipient list; she supervises the activities of the doctors and nurses in the outpatient clinic for pre- and postoperative management; she correlates data about organs, medications, and fluid requirements for each candidate; she is the first person notified by the organ bank when a heart becomes available; she makes sure that all necesssary equipment accompanies the transport team obtaining the donor heart and she monitors everything along the way; she is the one who tells the anxious waiting family the good news of a successful transplant; and she is the one who helps them through their grief should the transplant fail.

Gail Eddy is not sure that she believes in the condition that has been given the name "burnout," said to affect so many people in her profession. After thirteen years of ceaseless responsibility at the time I met her, she said that she had never felt its agonies. She attributed this to rewards that more than compensate for the frazzle. During one of our conversations, I asked her what it felt like to carry such constant burdens, and she replied, "Well, it sure is life at the max, isn't it?"

While Eddy was evaluating Cretella, his symptoms inexplicably decreased somewhat, with the result that he was moved down a few spaces on the list of urgent candidates in the organ bank's computer. On May 24, he was allowed to go home to wait.

But the sabbatical was brief. On June 11, Cretella had a particularly prolonged episode of angina and was admitted to Lawrence and Memorial Hospital with a new, though small, infarction. After three days of stability, he was transferred by ambulance to Yale–New Haven, fifty miles away, and there he was taken directly to the coronary intensive care unit and put on an intravenous drip of nitroglycerine.

While Cretella was at Lawrence and Memorial, he had begun to have some second thoughts about the transplant. He was relieved when his parish priest came to visit and told him he would not be violating church doctrine.

> *It made me feel a little better about it. I'm not a religious person, but I do believe. When transplants first flowered, I was one of these people that didn't believe in them. I mean, fixing things is fine. Replacing things—well, I didn't think it was right. I don't know why. And when I thought about the transplant, I had a problem with it. I think if it was ten years ago and they asked me, I would have said no. This time, I had no choice. How many people refuse it? None, right? There's your answer. I don't think people know how to accept to die.*

When Cretella returned to Yale–New Haven, he was near the top of the list of transplant candidates. As his angina worsened, his apprehension increased. But by then, he had decided that, regardless of his fears, he would do everything that was required of him. The nurses' notes in the coronary intensive care unit describe his anxiety and his periodic depression, but they also describe a man who "is dealing well with his hospitalization, has a great attitude and is very hopeful."

> *I actually changed my personality during that period of time. I'm not a person that's real outgoing, but I know how to do it. We made a joke out of it. The nurses were fantastic. They'd come in; they'd stay with me; we'd do stupid things. I stole my chart once in a while and I read the report: "If he's not joking, something's*

wrong." The joking was a step above—a step above what I am—so I could do what I had to do.

While all this was going on, Susan Cretella was trying to keep her mind on her job—she is a secretary—and on the logistics of commuting an hour each way to visit her husband almost every evening. The week that Cretella was told he had been put on the final donor list, Susan learned that her mother, who lived nearby, had been found to have brain cancer, requiring high-dose chemotherapy. The Yale–New Haven social workers put Susan in touch with the American Cancer Society so that such burdens as her mother's transportation to the oncologist could be taken off her shoulders, but the strain of worry began to take increasing toll. Still, she somehow managed to remain a quiet, calming influence on her husband.

On the Thursday evening of Cretella's third week in the hospital, the cardiac surgery resident, Dr. Christian Gilbert, went to the coronary intensive care unit to tell him that a potential donor heart was available. Cretella tried not to feel either too hopeful or too panicky; he had heard the same news a week earlier, only to be told after a few hours that the heart had not proved sound enough to use. This time, though, everything was just right. Not only was the donor blood type compatible but the new heart was exactly the right size to fit into the space that would be left when his own heart was removed.

Susan Cretella had just arrived home from New Haven when she got a phone call from her husband telling her the good news. She called his son and his two daughters, and his cousin, who had chauffeured her earlier that day, and they all drove back to the hospital as fast as they could. When they entered the unit, they came upon a scene of celebration. Cretella described it for me.

> *It was like a big party. The nurses and everybody were almost cheering. The whole place was apeshit, the whole hospital. I started getting chest pain, so they loaded me up with Demerol, and I was flying. It was a big party.*

The New England Organ Bank, the central agency through which the Yale team focuses its transplantation efforts, had called Gail Eddy to report the availability of a potential donor heart at a large hospital in a Massachusetts city about 140 miles from New Haven. The heart was that of a young man who lay brain-dead as a result of an event that had involved no injury to his body, and he was being kept alive on a respirator. Although he had no chance of recovery, all his chest and abdominal organs were perfectly healthy.

Shortly after midnight, a team under the direction of Dr. Graeme Hammond left Tweed–New Haven airport in a twin-engine prop jet chartered by Yale–New Haven Hospital. George Letsou, the Yale transplant fellow, went along to act as Hammond's assistant. As always, Gail Eddy was with them to supervise every step. They touched down in the Massachusetts city around 12:30 and were sped by ambulance to the hospital. There they met with one of the frustrations that sometime accompany this most complicated of medical undertakings. In those early-morning hours, only two operating rooms were functioning, and both were occupied with emergencies. The donor could not be brought in until one of them became vacant.

Meanwhile, the anesthesia team at Yale–New Haven Hospital, not having anticipated a delay, had taken Anthony Cretella up to the OR suite, transferred him to the cardiac room, and sedated him in preparation for the general anesthetic. The whole family accompanied him to the OR, and as he was wheeled away from them, through the electronic swinging doors and into the sterile area, he gave them a broad smile and lifted his clenched fists in a victory salute.

When I arrived at 2:00 a.m., the anesthesiologist, Dr. Kevin Morrison, was sitting in the darkened room with him, speaking in a soft, reassuring voice, trying to allay the fears that had by then returned. Cretella drifted off into a light doze, and I retreated to the surgeons' lounge and took a nap alongside a snoring intern. A little after five, I was awakened by a nurse touching my shoulder. Gail Eddy had called to say that the donor had been taken into the Massachusetts hospital's operating room and that Hammond and Letsou were now preparing to start the harvesting procedure.

I always hesitate before using the word *harvest* to describe removal of a donor's organs for transplantation, and yet there is no better term for a process that is so like the gathering of a crop. A transplant team converges on the donor's hospital, coordinates its timing and surgical sequence down to minutes, and carries off the healthy fruits of life—sometimes liver, kidneys, pancreas, heart, and lungs—which are then implanted into the bodies of people who would die without them. Cultivated with care and nourished with antibiotics and saline solutions, the organs usually thrive, and the patients who receive them usually recover well enough to live rewarding lives.

Even today, it is all so new that our society has not yet gotten used to it as an everyday occurrence, which may be the reason that donor organs are so hard to come by. To many people, transplantation still seems futuristic. Thousands of people die waiting for a donor. Some twenty thousand potential donors are allowed to die each year with their organs unused and therefore wasted, either because no one was willing to approach a family grieving the sudden loss of a young loved one or because potential donors—and most of us are potential donors, even if only of skin—had not made a bequest when they were able to do so. Several European countries have approached the problem with what are called presumed-consent policies, by which physicians are empowered to presume that, unless the family objects, a brain-dead accident victim has consented to the use of his organs. Such policies have not yet been seriously considered in this country, and until they are, or until there is far greater awareness of the serious shortage of donors, organ availability will remain in its present hit-or-miss status, and patients will continue to die just because the rest of us don't understand their need.

Because I was with Anthony Cretella, I was not present at the harvest in Massachusetts, but I have witnessed the procedure several times, and I have heard the details of that early morning's work from Hammond and Letsou. The young man's family had donated only his heart; ordinarily, permission is granted for several organs, and the heart is removed last. Being able to proceed without further delay simplified the Yale team's job, and, after Gail Eddy ascertained that the donor had been given the correct intravenous fluids and medications and that his heart was functioning perfectly, the team went right to work.

Hammond and Letsou split the chest down the center of the breastbone, opened the fibrous sac in which the heart is enclosed (the pericardium), and cut away all the ligaments and tissue attached to it and to the great vessels—the venae cavae, the aorta, and the pulmonary artery. When they were ready to remove the heart, Eddy phoned Yale–New Haven so that the transplant team there could begin to operate on

Cretella. The donor had been given doses of steroids and antibiotics, and now his bloodstream was injected with the anticoagulant heparin to prevent clotting. The heartbeat was stopped with potassium solution.

Gail Eddy noted the time as clamps were put on the great vessels. It was 5:34 a.m. The two surgeons now began to work very rapidly, freeing the heart and the vessels with long, swift cuts of their dissecting instruments. Twenty minutes later, Eddy phoned New Haven again—she and the surgeons were leaving the hospital. They carried the heart in the innermost of four concentric plastic containers, each filled with ice-cold salt solution. The containers, in turn, were secured within an Igloo picnic cooler. It fell to George Letsou to cradle the Igloo on his lap as the ambulance sped to the airport. The waiting plane took off immediately and was soon landing in New Haven. Another dash to the hospital, and at 6:52, Eddy phoned the OR to say that they had arrived. The hospital-to-hospital sprint had taken exactly fifty-eight minutes.

During that time, Anthony Cretella was being prepared to receive his new heart. He had been dosed with the immunosuppressants cyclosporine and azothiaprine and would shortly receive a dose of steroids. He was loaded with carefully chosen antibiotics. As the surgical team was scrubbing, Kevin Morrison injected a rapid-acting anesthetic agent into Cretella's intravenous line and inserted an endotracheal breathing tube into his windpipe. When all the intravenous and arterial monitoring lines had been secured, Christian Gilbert painted the entire front of Cretella's body with an iodine antiseptic and placed sterile drapes over the patient.

A few minutes before the induction of anesthesia, Dr. John Baldwin, the chief surgeon of the transplant team, had come into the room. Baldwin does not smile easily—at least not during working hours. He is a husky blond with a boyish appearance, and he was at that time forty-one years old. Despite his youthful look, he is all business, and he conducts himself in the quiet yet uncompromising manner of a man certain of his importance.

On the night of Anthony Cretella's operation, Baldwin was in the middle of his second year in New Haven. He had come from Stanford University, where he was one of Norman Shumway's brightest stars in the world's foremost program of cardiac transplantation. His arrival had been heralded for months, and with good reason. Over the previous decade, Yale's cardiac surgery section had declined to the point where it lagged far behind that of the school's other generally excellent clinical departments.

Two successive deans and a hardworking search committee had failed to persuade any appropriate candidate to take on the challenge of rebuilding. One after another, outstanding cardiac surgeons would come to New Haven, survey the chairman's job, and pronounce it undoable.

To bring Baldwin to New Haven, the dean of the medical school had promised him a great deal. Assured of unquestioned authority, and promoted to full professor in a matter of months, Baldwin had not hesitated to undertake a series of brusque campaigns in an attempt to establish his suzerainty not only over the territory understood to be his rightful surgical domain but also over territory traditionally within the jurisdiction of radiologists, cardiologists, and other groups. A summa cum laude graduate of Harvard College, a Rhodes scholar, and one of the major contributors to the development of combined heart-lung transplantation, Baldwin has a rare combination of excellent surgical skill and first-class research ability. And he exudes supremely unrufflable self-confidence. It is impossible to exchange more than few sentences with him without recognizing his air of aloof assurance. Even praise seems redundant to a man so certain of his own merit.

When all was ready, the members of the team took their places, and Gilbert, under Baldwin's direction, made an incision through the scar left by Cretella's CABG. Next, he split the sternum up its middle with an oscillating saw. The dissection inside the chest was not easy. The previous surgery had left the heart encased in a mass of woody scar tissue, and it was tedious work to separate vital structures from their adhesions. The difficulty was enhanced by a persistent ooze of blood from the scarred layers. Unlike many surgeons, Baldwin does not speak much while operating. What he says is restricted to instructions and comments on the work as it proceeds: "Just cut down onto it—don't screw around so much." . . . "He's juicy, this guy."

The dissection of the scarred tissues proceeded with meticulous care. At 6:55, exactly eighty minutes after the incision, Cretella was given a dose of heparin to prevent his blood from clotting. In the early days of heart surgery, the difficulty of inhibiting the clotting mechanism and then reversing the inhibition was a gigantic obstacle. It is now done with ease, using an injection of a single substance in a previously calculated amount, and a reversing agent, protamine, when the anticoagulation is no longer required.

The purpose of stopping Cretella's clotting mechanism was to prevent his blood from coagulating in the pump oxygenator—the machine that would take over the job of his heart and lungs during the actual transplantation. The pump oxygenator is a marvel of technology that, like the heart itself, seems deceptively simple in concept. Through an incision in the right atrium, large tubes are threaded into each of the venae cavae and are then connected to the machine, which diverts the blood from the heart and drains it into a reservoir; there it is pumped into an oxygen-filled chamber, where it spreads out onto the surface of a wide area of extremely thin polypropylene membrane. This arrangement simulates the way a normal lung functions, thus allowing the oxygen to pass easily through the membrane into the blood. The freshly oxygenated blood is then pumped back to the patient and enters his body through a metal tube placed in his aorta.

Simple as the principle is, it took more than two decades of single-minded effort before Dr. John Gibbon of Philadelphia was able to develop the heart-lung machine with which he carried out the first successful open-heart operation in 1953. When I ran the pump oxygenator in Yale–New Haven's first such procedure in 1956, the entire team and I had spent hundreds of hours in the laboratory preparing ourselves; today, young residents entering their training in cardiac surgery take the machine for granted.

After the heparin was injected, the tubes were placed, clamps were adjusted, and the bypass was begun. At 7:19, Baldwin gave the order to turn down the thermostat on the pump oxygenator in order to drop Cretella's body temperature to 28° C, or about 82° F. (At decreased temperatures, metabolism is lowered and blood-deprived tissues survive longer.) The aorta was clamped and then divided just above the heart, as was the main pulmonary artery. Cretella's heart was allowed its last few anemic beats, and then Baldwin and Gilbert cut it away from his body, leaving a rim composed of the back of its atria.

The wonder of a beating heart is minor compared to the incomprehensibility of no heart at all. I stood at the head of the operating table alongside the anesthesiologist, staring down at the huge empty space in Cretella's chest. Though heartless, he lived, and even breathed—at least in a sense. Seven feet away, soaking in its nutrient bath on a small steel table, the new heart waited. Hammond brought it over to the operating table, where Baldwin trimmed its aorta, pulmonary artery, and atria until they were exactly as he wished them. Without a hint of ceremony, the heart, which the team was now referring to as "the graft," was lowered into Cretella's chest and Gilbert began to stitch it into its new home.

The two surgeons did not convey any sense of haste, or even of speed. They said very little to each other, and virtually nothing to the nurses and the anesthesiologist beyond sparse instructions. Baldwin directed the action in short, flat sentences, free of either encouragement or criticism. Gilbert knew how to do his job; a few words from his chief, and a very good stitch became a perfect one. There was an occasional warning of potential danger ahead, and a few tersely worded admonitions. I sensed the rapport of teamwork but no warmth. And there was none of the jocularity that surgeons so often use to lighten the tension when the stakes riding on each movement are so high. The scene reminded me of nothing so much as one of those old movies in which two safecrackers are working with exquisite care and no overt signs of hurry, but always against the ticking of the clock.

What I witnessed was the insertion of a painstakingly and accurately placed series of polypropylene sutures that united the back wall of each donor atrium to the remaining rim of Cretella's own. The wall of the atrium is surprisingly easy to stitch; it has a yielding, elastic quality, which allows donor tissue and recipient tissue to bunch up against each other just enough to provide a snug, leakproof fit. The heart now lay in place, with a circular stump of hoselike orange-yellow aorta jutting out from its top, and the donor's coronary arteries leading down into its myocardium from that short length of conduit. Gilbert circumferentially sewed this new outlet to the open end of Cretella's own waiting aorta, taking large and strong bites of tissue so that the suture line would hold against the full force of the powerful ventricle ejecting blood into it at a pressure of at least 120 millimeters of mercury. When the suture line had been secured, the clamp across the aorta was removed, and blood freshly oxygenated by the machine poured into the coronary arteries to irrigate the muscle of Cretella's replacement heart. From the moment of clamping in the other city, the donor heart had been deprived of blood for two hours and forty-one minutes.

The last suture line is the one that unites the pulmonary arteries of the donor and of the recipient; it was now expeditiously completed. The heart belonged to Cretella. Nevertheless, it remained a little unsure of its new surroundings, and although it was filled with blood and the bypass was reduced to only partial diversion, it seemed hesitant to begin is own automatic rhythm. Instead, it quivered, as though shy. It began the uncoordinated, every-which-way squirming of fibrillation. Baldwin tried to shock it into good behavior with a few jolts of electricity fired through paddle-shaped electrodes applied to the ventricles, but the myocardium remained resistant. He and Gilbert then stitched wires to the heart's surface and connected them to an external source of current—a pacemaker, which would force the beat until the SA node and myocardium felt a little more at home, probably not until some days after the operation.

The heart's reluctance was worrisome to me. Everyone who has witnessed enough transplants has seen a few in which, despite the pacemaker and the cardiac medications, the graft never does pick up the beat. But I may have been the only person in that room who was beginning to wonder whether Anthony Cretella had come so far only to lose his life to a diffident graft. Baldwin knew better, and so did his team. He waited patiently, adding now this myocardial stimulant and now that, and the beat gradually improved. After about forty-five minutes, it had strengthened to the point where it took on a vigorous, healthy tone, asserting its independence of the now-redundant pacemaker wires. A few tiny leaks in the various suture lines were repaired, the heart-lung machine was stopped, and the bypass tubes were withdrawn. At 9:40, a little more than two hours after his old worn-out heart had been removed, Anthony Cretella was on his own.

Protamine was given to reverse the heparin; a few large drainage tubes were brought out through the chest wall to suck away any blood that might later accumulate from oozing; and the closure began. The breastbone was brought together with thick stainless-steel wires and the skin was stapled. By 11:30, Cretella was in a specially prepared isolation room of the cardiac surgery intensive care unit.

Were I asked what impressed me most about what I had just witnessed, the answer would come easily; it was the calmness of the thing. There was none of the buzzing, hyperactive flutter I have often observed when an operating team is self-conscious about the magnitude of its undertaking. There was no sense of the excitement, risk, or impending danger that can sometimes seem to hover over such events. The pervading mood was low-key optimism and unquestioned expectation of accustomed success. Not only Baldwin and Gilbert knew their jobs thoroughly; so did everyone else, whether doctor, nurse, or technician. Baldwin had somehow instilled into the entire corps his attitude of "This is what we do, and we're good at it." His exalted self-confidence, which might elsewhere seem arrogance, found its proper place in that operating room and took on the quality of command. For those two hours, Baldwin's presence at Yale was worth whatever it had cost, and perhaps more.

New life or no new life, Anthony Cretella refused to believe the evidence of his improvement. For the next three weeks, he needed plenty of nudging. He experienced bouts of pessimism that yielded easily to depression; being sequestered in the isolation room aggravated every black thought that came into his mind. Neither the cheerful efficiency of his nurses nor Gilbert's frequent reiteration that he was progressing well seemed to make a difference.

Gilbert, who was a veteran of many heart transplants, saw in his patient an extreme example of nervous disbelief.

> *He was about one of the most uptight guys I've met in a long time. He was never really sure he was doing okay. He kept asking, "Am I doing all right?" And he never really believed us. He thought about every heartbeat and every hiccup.*

Cretella saw it differently.

> *Gilbert was the best of them and I got the most out of him, but I had to push him. They'd come in with the whole team and they'd do all their stuff, and then they'd say, "Oh everything's going fine." You'd be asking them questions and they'd be trying to get out the door. They kept saying the same thing: "Don't worry about it; we'll fix it." Every time I asked a question, they said, "You have nothing wrong with you that we can't handle." That's what they told me when my pulse was 170 to 200 for eighteen hours; "Don't worry about it." How do you not worry about it? They're a different breed of people.*

Dr. Cohen had receded into the background. With Gail Eddy and the surgeons directing the overall planning, his clinical role had become peripheral. "At that point, Cretella knew my visits were not necessary to his care," he said. "He knew that I came to see him really as a friend."

A friend was exactly what Cretella needed. It was Cohen to whom he could speak, and speaking to him made a difference. "Dr. Cohen would come down in the evenings and talk to me, to pump me up and encourage me," Cretella told me.

I asked Cohen about his patient's perception of the way he had been treated by the surgeons, and he said some interesting things.

> *I think there* is *something different about cardiac surgeons, and particularly about cardiac transplant surgeons. There is something about the mentality of a cardiac surgeon—he does play God, and I think he does feel that at some level he can fix everything. The downside of that, of course, is that if things don't turn out well, it's someone else's fault—the cardiac surgeon is not to blame. It's the patient's fault, some member of the team's fault, some other department's fault. It's rare that I've heard a cardiac surgeon say, "I fucked up." It was very important to Cretella for him to maintain control—he's that sort of person. In a way, what he's saying in his complaints about the surgeons is that they don't let him have any control. Some patients are very happy just to place themselves in the total care of their surgeon, and they really are encouraged when the surgeon says, "Don't worry, I can take care of it." But for the type of patient Cretella is, who needs to feel he is in control, that kind of response is infuriating.*

A routine recovery is not necessarily an uncomplicated one. In spite of the immunosuppressive drugs, almost every patient experiences some degree of rejection of the donor organ in the early postoperative period. The diagnosis of rejection is made by direct biopsy of the heart muscle, and a biopsy is routinely done several times during the hospitalization and at longer intervals after discharge. Any evidence of trouble can be reported by the pathologist within a few hours of the biopsy, and when such evidence is found, the trouble is treated with a brief period of increased steroids, and the patient usually responds with improvement.

I went up to the operating room to witness Cretella's first biopsy, on the seventh day after transplant. God may have rested on the seventh day, but transplant surgeons, notwithstanding their convictions about their own divinity, never seem to take a day off. Gilbert locally anesthetized the skin on the right side of the neck and inserted in the jugular vein a series of wires and catheters, the last of which was large enough to allow an instrument called a bioptome to be threaded down through it into the superior vena cava and then directly into the cavity of the right ventricle. The bioptome is a flexible wirelike probe with a pair of pincer-shaped jaws at its tip. Watching on a fluoroscope, Gilbert slowly advanced the bioptome toward the inside wall of Cretella's heart, and when the tip was pressed into the myocardium, he squeezed the bioptome's handle. The pincers bit tightly down on a tiny wad of muscle, and he withdrew it. A waiting technician took the specimen directly to the pathology laboratory. The first biopsy showed no sign of rejection. When the procedure was repeated seven days later, there were some mild changes, so a course of increased prednisone was begun. The next biopsy, done after another six days, indicated that the condition was resolving, but the one after that demonstrated a slight worsening, and three more days of high-dose prednisone was ordered. Finally, after a biopsy on August 11, it was clear that the problem had been solved. The heart looked good, and Cretella was ready to go home.

He had left the intensive care unit at the end of his second week; the mild rejection and a few days of slight temperature elevation were routine. But Cretella continued to be depressed. What finally relieved his joyless frame of mind was that he could no longer ignore the mounting evidence of his recovery. He finally became convinced that, like 80 percent of all patients with transplanted hearts, he would be alive and well a year after his operation. In fact, having survived the surgery and the post-transplant hospitalization, he was in a category well above the 80 percent group. Patients who do not resume smoking and who control their cholesterol stand a 70 percent or more likelihood of being in good condition five years after operation, but that figure drops to 50 percent if they wander from the straight and narrow, and it may dip as low as 20 percent at seven years.

Cretella was discharged from the hospital on August 12, and ten days later I drove to Mystic to visit him. The Cretellas live on the second floor of a three-story tenement across from the plant where he works. A hand-painted sign on the door read WELCOME—SUE AND TONY, and their apartment was cheerful and pleasantly furnished. Sitting in their comfortable, scrupulously clean kitchen, I could understand the importance of Susan Cretella's influence on her husband. While I was there, a radio in the next room was softly playing popular tunes of the fifties and sixties—the salad days of both Cretellas—and at one point I could hear Brenda Lee singing, "All alone with just the beat of my heart."

We talked about tobacco and we talked about liquor. (Cretella is a hard social drinker, and there have been times in his life when he has been more than that.) I was not encouraged by his answers to my questions—especially when I asked if he thought he could stand by his resolve to stay away from cigarettes for the rest of his life.

> *I hope so. But I'd never say no about anything. I don't believe in that. I would never say, "I'll do this or that for the rest of my life"—not tomorrow or any part of my life. Booze, I've always been able to control. I've stopped for as long as three months at a time. Cigarettes are something different—no control. And I've tried. I mean, I quit for six months after I had the bypass. I can't tell you how many times I quit for two or three weeks. I'm not saying just now I can quit for keeps. I have no idea what I'll be like tomorrow, a week from now, or a year from now. You smoke and drink because you feel shitty all the time. I was even tempted to take drugs, but I never did. I always felt I was hooked on cigarettes. I can get pissed off with myself for drinking and quit, but I don't know if I can do it with cigarettes.*

Toward the end of my visit, our conversation turned to a topic I had been hesitant to bring up. What does it feel like to live with another person's heart in your chest? It proved to be something Cretella was trying very hard not to think about.

> *I don't know yet—I really don't know yet. When I catch myself thinking about it, I try to forget about it. You know—I think, What is it? A female? A male? Black? Orange? White?*

I asked him what he would want it to be.

> *I don't know that yet, either. I can't answer any questions like that at all. I even get upset talking to you about it. When I talk about it, I get paranoid. I think mainly it's because I don't know what's going to happen tomorrow, and the reason for that is that I can be*

> *sitting here feeling fine and all of a sudden something clicks and I get nervous and everything just starts going. Something in my body changes, as if somebody pushed a button. I talked to another transplant patient—he's in his fifth year—and he says it still happens to him.*

Susan Cretella said that her husband occasionally seems to go into a trance, sometimes for hours at a time. He seems to be thinking about nothing, she said, but his mind is really trying to escape those thoughts about whose heart he is carrying. Of course, I couldn't give him details, but I did attempt to reassure him, as several of his doctors had done, by telling him of the vibrant good health of the donor.

But the health of the donor was not what concerned him; it was something far more complex. I told him, "The donor was the kind of person whose heart you would be proud to have in you. I'm talking about character, about the kind of person it was."

Even that was not exactly what Cretella had in mind.

> *When you say the kind of person, what do you mean? Race, or color, or everything? One of the things—I didn't know if it would worry me—was whether it was black or not. I think that would bother me. It came up in my mind one day. Because, you know, they tell you it doesn't make any difference what kind of heart you get. And I'm sitting there thinking, I don't believe that; I honestly don't believe it. You know, a female heart or a Chinese heart or a black heart, a boy's heart or a man's heart. These people have different stuff in their body. I don't care how many tests they've taken. Everybody tells me what a perfect heart I got—well, you know.*

Cretella's deep brown eyes looked troubled when he finished speaking. The momentary alteration in his appearance was surprising, because it was so unlike the usual changelessness of his expression. He told me he was concerned that I might construe his comments to have what he called a "discriminatory" tone. But I doubt whether feelings like his arise from simple bigotry. I think they have to do with that inherent rejection of otherness—the ancient xenophobia we have learned from every cell that has ever existed on earth. It is Gaspare Tagliocozzi's "force and power of individuality" that makes Anthony Cretella torment himself about "the different stuff" that has been implanted in his chest. It tormented me, too, because I worried about the possibility that this psychic form of xenophobia might in some way contribute to eventual rejection of his heart.

As I drove back to New Haven, I thought about the heart itself and the skills of the surgeons who operate on it. I thought also about xenophobia and the vestiges of earliest life carried in our cells through billions of years of evolution. I thought about my growing conviction—the conviction that has so often occupied my mind since I first began to study the new discoveries of the cell's biology—that there exists an "awareness," connecting the events within the cell to the events of our daily lives and perhaps our thoughts.

But more than anything else, I thought about my renewed concern for Anthony Cretella's future. And I thought also about something that Dr. Cohen had told me. "People who have had cardiac transplants sometimes think they can extend the fact that they now have a young heart into many other spheres of their lives," he said. " 'So now I'm going to feel like I'm twenty-five; now I can run like I did at twenty-five; now I'll be as sexually active as I was at twenty-five.' It's as if their heart was their total life and body. The fact that their liver might be fifty or their brain might be fifty or their legs might be fifty is irrelevant—they have the heart of a young person." Something of Aristotelian belief still pervades us, and it makes us see the heart as an organ beyond mere function—as more than the Sun of the microcosm. There is an enchantment about it. To the descendants of Eve and of Adam, no discovery of cellular biology will ever decrease by a single line of poetry the mystical faith that in some unknown way this leaping thing within our chests is the central capital from which our lives are governed. By this conceit, a new heart means renewed youth.

Such misconceptions can be as fatal as cigarettes, cholesterol, or rejection. For Cretella to achieve his full measure of longevity, he had to fulfill Cohen's confidence in his ability to transform himself. And—like the little train who could—he did. He found a power in himself to sustain his ability to be "a step above what I am," and he overcame the unruly sprite of a wise guy who had "seemed hell-bent on destroying himself." He never returned to cigarettes, and he ceased the disquieting ruminations of xenophobia. The "step above" became Anthony Cretella as he now is. Aristotle would surely say that his perseverance originated with the organ transplanted into him on that July morning in 1989. So at least in that sense, he is young in heart, and filled with the determination—even if not the liver or brain—of a twenty-five-year-old.

On the morning of Cretella's discharge, I had visited his room for the last time, to bid him good-bye and good health. He stood at his bathroom mirror shaving; he was actually cheerful. We chatted for a few minutes, and I prepared to leave. And then there occurred one of those small events that have demonstrated to me that, at least in the minds of its craftsmen, cardiac transplantation had become a procedure which, if not exactly commonplace, was already well beyond its pioneer phase. The door of the room swung open, and standing there was Dr. Kenneth Franco, one of Baldwin's associates. Cretella took a few steps toward him, but Franco held up his palm like a policeman stopping traffic—to indicate that it wasn't necessary for Cretella to dry his lather-soaked fingers for a handshake, or to advance another inch. "Well, anyway," Cretella said, "thanks for everything you did." He might as well have been expressing appreciation for a proffered Kleenex. Franco's reply came in the terse, lackluster tone that characterized the Baldwin team's approach. "No problem," he said. The door closed and he was gone.

Certain memories from recurrent episodes of childhood fuse into a single image in my mind. When the same incident has been experienced hundreds of times, each time much like the last, it is commonly recalled as though it happened only once. Especially when the episodes began so early in my life—began before I was born, in fact—that they seem always to have been there, no individual moment of them stands alone as an isolated experience to be remembered in distinctive detail.

And so it is that when I try to recollect my mother—or my grandmother or aunt—koshering meat in a ritual some two thousand years old, what I see are not individual occurrences but only a still life on the canvas of my mind. I see not the woman, but the inanimate objects she has put in place. What appears before my eyes is a thick, flat square of wood perhaps ten inches on each side, so positioned on the edge of the kitchen washbasin that it is in no danger of slipping out of place even though tilted at an angle of about ten degrees. Probably under its edge lay something to support it so securely there, and yet I have no remembrance of what that something might have been. But I do clearly recall the dozen or so grains of coarse salt invariably scattered along the margins of the wood.

Occupying most of the board is a large wet slab of fresh, uncooked meat, from whose lower edge a barely perceptible ooze of watery reddish fluid is leaking. Catching the neon glare of the kitchen light, the very fine fat globules on the wet layer's surface glisten like tiny specks of mica floating toward the down-tipped end of the board. In an exquisitely languorous drip, drip, drip, the fluid falls into the sink and spreads in a narrow rivulet toward the drain. In this way is the ancient injunction fulfilled: "Only be sure that thou eat not the blood: for the blood *is* the life; and thou mayest not eat the life with the flesh" (Deuteronomy 12:23).

So seriously is this obligation taken that the meat had been sliced from a healthy animal slaughtered in the abattoir with a single to-and-fro sweep of a scalpel-sharp knife, wielded by a rabbinically trained specialist in such things. By the rapid cut through the carotid artery, the ritual slaughterer was fulfilling his double responsibility of preventing the animal's suffering and of draining the maximum amount of blood out of its carcass.

Then, on the day the meat was to be eaten, all visible arteries were torn from it and the flesh was soaked in water for a half hour, then drained on that board imprinted into my memory. Finally, it was sprinkled generously with the coarse grains that have become known in modern supermarkets as kosher salt (when what is meant is *koshering* salt), because that way more blood is drawn out. After an hour in the resultant flavor-destroying briny soak, the meat was washed again and placed once more on the sloping board that I see through the clear lens of the years. This is how Orthodox Jews prepare the flesh they will eat. I never had any idea of how delicious a slice of beef could taste until I was twenty-four years old.

Thus are the mystical attributes of blood honored in our time, with the same zeal they have inspired since early peoples observed that blood flowing copiously out of the body of a grievously wounded man usually took his life along with it. With the direct logic that observation tells all, it was clear to our ancestors that "the blood *is* the life."

The legendary power of blood is a thematic thread running through the ages and cultures of the tribes of the earth, as though to unite all peoples in one of those singular symbolisms that express our shared humanity. "Then Jesus said unto them, Verily, verily, I say unto you, Except ye eat the flesh of the Son of man, and drink his blood, ye have no life in you" (John 6:53).

In the metaphor and symbolic displacement that allows the breaching of the taboos of advanced societies, blood becomes wine in the Christian tradition, and flesh becomes wafer—and they thus give life. For although more brutal men in more brutal places have drunk the freshly shed blood of their enemies or of their sacrificed fellows, horror at such an act emerges from some deep unknown place in the psyche of more civilized peoples, and prohibits it. The Talmud quotes Rabbi Shimon ben Yehuda ha-Nasi: "Man's soul has a loathing for blood" (Makkoth 23b).

When there are such powerful prohibitions and equally powerful metaphoric displacements, they must surely be needed to overcome powerful attractions. Whether the prohibitions come from within—"man's soul"—or are codified by law, they exist because an ineluctable, albeit deeply repressed, fascination must be brought under control. Blood has seemed not only the essence but also the continuity of what we are. The bloodstream is the stream of one's heritage. It is the connectedness of the generations. Think only of the variegated ways in which our language refers to a man's blood in order to express judgments about him or about his lineage. More than any other tissue, it is the one through which we give voice to our moral assessment of our fellows. It is in their blood that we seek the clues to their lives.

Blood seems always to have been associated in men's minds with the heart. As early as fifteen to twenty thousand years before William Harvey described the circulation, some unknown Paleolithic man ochered a red heart into the picture of a mammoth he had drawn on the wall of the cave of Pindal, near Altamira, in northern Spain. Between then and the classical period, concepts of life and soul were sometimes unified in the quasi-scientific formulations of the Hippocratic physicians, who linked heart and blood in their notion that the one is the source of the innate heat that warms the other. It was two thousand years before such ideas were debunked by research and experiment. But all the centuries of curiosity and studies, particularly those of the most sophisticated scientific sort, have only served to confirm what Goethe had Mephistopheles tell Faust: *"Blut ist ein ganz besondrer Saft"*—"Blood is a very special juice."

The special character of blood extends far beyond the mystical associations our culture has attached to it. In fact, even the mysticism has a basis in biological reality, for blood really does give life to the tissues, in the form of nutrients and oxygen. It brings not only life but news and instructions, as well. Together with the nervous system, it is the source of all communication between the far-flung parts of the body's empire. Blood refreshes and sustains the *milieu intérieur* and carries the hormones that tell tissues and communities of cells what they must do in order to contribute to the welfare of the whole organism. If the body is like an ideal society composed of special-task groups, it is the nerve impulses and the blood's delivery of hormones that keep all participating members informed, letting them know how to help in the joint effort of life. Until about a hundred years ago, a term was in common use to sum up this aggregate of all the events in this cooperative venture: *the animal economy.* The advent of the concept of metabolism in the nineteenth century seems to have displaced the older expression, but the two really mean quite different things.

Metabolism is defined as the totality of that huge group of molecular interactions within an organism that build up (this part of metabolism is called anabolism) or break down with the release of energy (catabolism) those substances required for the performance of that organism's necessary functions. It is a highly specific term. Each one of the multiplicity of the chemical components of human metabolism can be studied individually and measured by the methodological principles of laboratory science. The concept of metabolism grew out of the realization of researchers that living things are made up of chemicals and their reactions in specific environments. It is a term that expresses the ultimate in a mechanistic approach to our existence, and it purposefully reduces life to a textbook of molecules and measurements that can be studied. Without such a reductionist basis, we would not have a valid foundation from which to understand mind and spirit. But it seems not to tell the whole story.

The language of science, like all languages, never leaves off creating and discarding assorted bits of its terminology. Not only that, but existing words and terms are forever taking on new meanings and losing their old; an expression in common usage by one generation may mean something quite different when expropriated by the next. Unlike biological evolution, the evolution of language does not adhere to the principle of natural selection, and so the fittest terminologies are not necessarily those that survive. The scientific and every other tongue have lost valuable expressions.

Among them is "the animal economy," a term that biologists and physicians found very useful for several hundred years, until its apparent disappearance some time near the turn of the twentieth century. Here the word *economy* was being used in its original and most basic sense, almost synonymously with *housekeeping* or *management*—it referred (as it does in a society) to the entirety of all the producing and consuming activities that go on in the body, as well as the laws that govern them. Having never encountered the term (except in a historical sense) in the writings of a modern author, I was surprised to find it still included in my up-to-date medical dictionary, defined as "the system of operation of the bodily processes in organic bodies; also the body as an organized whole."

The concept of the animal economy refers to the sum total of all the activities that preserve living things, whether on the molecular level or on the level of willed behavior, the conglomerate of the multitudes of responses and adaptations that sustain the constancy of organisms. Its product is the stability of that balance to which a cell or a man always returns as it or he responds to ever-changing circumstances within and without. The animal economy is the expression of all the internal forces that preserve life at every level and in all its meanings—and blood is its vehicle. Blood maintains the changing constancy of the *milieu intérieur* and permits life along with all that is understood by that word.

No tissue or cell can survive without blood. Loss of blood supply to local areas causes strokes, heart attacks, gangrene of extremities, kidney failure, and an assortment of other lethal conditions. The legal definition of death in most countries of the world is death of the brain, which, except in cases of trauma, is almost always due to blood deprivation.

The primary reason that deprived tissues die is their lack of oxygen. The blood carries oxygen, carbon dioxide, nutrients, water, waste products, hormones, and certain cells that are crucial to the animal economy; it stabilizes the concentration of acid substances as the body produces them, the so-called pH; it maintains body temperature by carrying heat away from areas of high activity, such as skeletal muscle, and redistributing it or bringing it to the skin for dissipation; its white cells actively fight infection.

Human blood can best be described physically as consisting of a suspension of several types of cells in a protein-rich fluid called plasma, which makes up 55 to 60 percent of its volume. Although there are some sixty kinds of proteins within the plasma, functioning in a variety of tasks, such as fighting infection, clotting, and transporting fatty materials, the most important of them are the three types called albumins, globulins, and fibrinogen. Because of the cells and the proteins, blood is viscous and sticky—we are not a thin-blooded species. As your mother used to admonish each time you abandoned your kid sister to go off with your friends, "Someday you'll be sorry. Blood is thicker than water, you know!" In fact, it is three or four times as thick.

Seven to eight percent of our body weight is blood, amounting to a total of about twelve pints for a person of 180 pounds or eight for one of 120 pounds. At any given time, only about 10 percent of it is in capillaries, while 15 percent is in the arteries. Veins, being large and easily distensible, hold about 70 percent of the total.

Because of the nature of the capillary wall, water in the plasma passes readily through it, but many of the larger molecules cannot. This easy passage of water back and forth is an important characteristic of the circulatory system, because it allows rapid fluid shifts in and out of the capillaries, as needed to maintain the stability of the blood and the extracellular fluid. To a great exent, the amount of shifting is determined by the concentration of molecules dissolved in the liquid. In passing through a membrane that allows only water to go through (called a semipermeable membrane), the water moves in such a way as to equalize the concentration of molecules on the membrane's two sides: it passes from the more dilute side to the more concentrated to achieve a balance. This passage of a fluid through a semipermeable membrane in order to balance one side with the other is what is meant by the word *osmosis*.

More than 90 percent of the plasma consists of water. The protein molecules in the plasma (otherwise known as plasma proteins) play an important role in keeping water from simply leaking out through the capillary wall and into the interstitial fluid, where the concentration of chemical substances is so high. (Any propensity to leak is abetted by the pressure at the arterial end of the capillary acting to push the water out.) Being present in such high concentration in the plasma, the protein molecules balance these tendencies for the capillary to lose water. Their effect is equivalent to what might occur were they to exert an actual attracting force or negative pressure to hold it in. Because it prevents osmosis, this "force" keeping water in the capillary is called osmotic pressure. Osmotic pressure is a general term used in any situation in which fluid is held on one side of a membrane by the concentration of dissolved molecules.

Of the many kinds of proteins in the plasma, the most important in the maintenance of osmotic pressure is albumin, which represents as much as 60 percent of the total number of protein molecules in the blood. Albumin levels depend to a large extent on the state of nutrition. In the process of digestion, proteins in the diet are broken down into amino acids in the gastrointestinal tract. The amino acids are then transported throughout the body by the circulation, to be taken up by cells and rebuilt into many kinds of proteins, as needed by the body. Much of this manufacture for general body uses takes place in the liver, which releases the protein into the plasma. Albumin thus synthesized has as its main function that of keeping water in the bloodstream, but it also binds itself to other substances to help transport them through the plasma. Blood, in fact, has some sixty different kinds of proteins, functioning not only for transport and osmotic pressure but for such duties as clotting and doing combat against infection.

Albumin levels fall markedly in malnourished people, with the result that osmotic pressure drops and water can no longer be efficiently held in the capillaries. Some of it leaks into the interstitial fluid and causes the tissues to swell. This explains the swollen extremities of victims of starvation or cancer, and the abdominal distension that is sometimes clearly visible, a condition known as dropsy in an earlier era. The swelling of tissues is called edema, and it is most common in dependent parts of the body, since the force of gravity aggravates the process. The heartrending images we see of young children with swollen arms and legs and hugely protuberant abdomens in famine-stricken Third World countries are the result of appallingly low levels of plasma albumin. Fluid leaks not only into their tissues but into the abdominal cavity in large amounts. Their weakened abdominal muscles have insufficient strength to confine it, with the consequence that enormous amounts of liquid are contained in the bellies of these dreadfully afflicted little people. Unless the process can be reversed by improved nutrition, every one of these children will die.

After albumin, the most prominent proteins in plasma are the globulins. There are three kinds of globulins: alpha, beta, and gamma. Alpha and beta globulins transport fatty materials and vitamins that dissolve in fat (A, D, E, and K). More will be said shortly about the role of gamma globulins in immunity.

In addition to the proteins, plasma also contains sugar in the form of glucose; fatty materials called lipids (including cholesterol, triglycerides, and fatty acids); vitamins; hormones; amino acids; salts of such chemicals as sodium, potassium, and calcium; small amounts of iron, copper, and zinc; some dissolved oxygen, carbon dioxide, and nitrogen; and waste products of metabolism, such as urea, uric acid, and creatinine.

The color of blood is the color of hemoglobin—brighter when oxygenated, somewhat darker when not. It appears blue when we see it in the vessels close to the surface of the skin because they are veins, whose thin walls have a blue-violet color. The lighter-complected the skin, the more likely the blood to be visible. The term *blue-blooded*, in fact, seems to have originated in Spain during the Moorish occupation, when members of the old Castilian families proclaimed their racial purity by pointing out the blueness of the blood visible through their white skin—they called it *sangre azul*. In a classic example of selective observation, they preferred not to notice that when cut, Castilians and Moors shed blood of the same color. As Catholics, they seem also to have ignored Paul's statement to the Athenians, when he made clear to them that God "hath made of one blood all nations of men for to dwell on all the face of the earth" (Acts 17:26).

Red blood cells are called erythrocytes, from two Greek words meaning "red" and "cell." Like all the formed elements in the blood, they originate in the bone marrow from progenitors called stem cells. As a stem cell proceeds to differentiate into an erythrocyte, it loses its nucleus, a seeming deficiency, but one that has the advantage of allowing more room within the cell for hemoglobin molecules. Actually, erythrocytes don't need the nucleus. By the time they are fully formed, they no longer require protein-synthesizing instructions, since they have already manufactured sufficient enzymes and other proteins to live out their life span of about 120 days. Without a nucleus, they cannot reproduce, depending for their replacement on the stem cells in the bone marrow. For these reasons, erythrocytes are said to be terminally differentiated.

When fully formed, a red cell is biconcave. This is a more formal way of saying that it is shaped like a doughnut with a thin layer of dough filling its center, instead of a hole. Such a shape provides the advantage of bringing the hemoglobin close to the cell membrane for more efficient release of oxygen. Also, its shape, as well as its elasticity and flexibility, allows it to be easily bent or otherwise distorted so as to permit passage through the narrowness of the capillaries. As soon as the tight quarters are departed, the cell resumes its normal configuration.

As noted earlier, the rate of erythrocyte formation is influenced by the hormone erythropoietin, in a kind of negative feedback mechanism. When the kidney (and to a lesser extent, the liver) experiences a measure of oxygen deprivation, it raises its output of erythropoietin. The hormone is carried by the bloodstream to the bone marrow, where it stimulates increased production of red cells. Chronic anemia, certain lung diseases, and high altitudes cause this sort of boost to occur.

Like all of us, red cells become more fragile with age. They are increasingly susceptible to damage as they squeeze through capillaries or are percolated through the spongy network of the spleen. Damaged erythrocytes are engulfed by cells called macrophages (literally, "big eater"), located primarily in the lining of blood channels in the spleen, liver, and bone marrow. The macrophages break them down, and their parts are later recycled. The rate of erythrocyte destruction is about 1 percent each day. Because they represent more than 40 percent of the blood's total volume, the replacement process must be vigorous and unhalting. François Rabelais, a physician before he became one of the most acclaimed writers of the sixteenth century, said in *Pantagruel*: "Life consisteth in blood; blood is the seat of the soul; therefore the chiefest work of the microcosm is to be making blood continually." The ceaseless activity of the bone marrow ordinarily replaces enough red cells to make up for about a pint of blood each week, but, driven by elevations in the kidney's output of erythropoietin, it is capable of increasing its output to many times that figure when urgently needed to replenish the deficits of significant anemia.

Anemia is defined as a decrease of hemoglobin or of red cells below normal levels. Its causes are either an increase in red cell loss or a deficiency in production, or both. Although there are nearly one hundred different varieties of anemia, the most common cause is either sudden or chronic blood loss. Were it not for the fact that the blood contains factors to promote clotting, or coagulation, we would all bleed to death. The high pressure of arterial circulation pushing the blood out of any small skin wound would virtually guarantee lethal hemorrhage did we not have internal adaptive mechanisms to combat it. Even injured venules and arterioles would continue to bleed were there no such thing as clotting.

One of the several differentiated progeny of the stem cell is a structure called the platelet, which is a major contributor to our ability to clot. Platelets are round or oval disks about one-quarter to one-half the size of erythrocytes. When a vessel is disrupted, connective tissue fibers protrude from its broken end, on which platelets collect and clump together, temporarily plugging the leak if it is not too large. In addition, they release substances that attract more platelets, as well as releasing the hormone serotonin, which contributes to a reflex spasm that a vessel automatically goes into when it is injured. In the case of very small vessels, the platelet plug and the vessel's constriction may be enough to stop the bleeding. Larger vessels, too, may be affected in this way. In fact, a completely transected artery, even one of substantial size, may go into such intense spasm that all hemorrhage is prevented.

Among the many colorful moments of my years as a medical student, there was a particularly instructive one involving an artery's formidable capacity to accomplish such an astonishing feat. One morning during one of the first days of my rotation on the surgical service, a hysterically weeping worker was rushed past the reception desk of the emergency room, having just arrived from a nearby lumberyard only minutes after his right arm had been traumatically amputated a few inches below the elbow by the whirring blade of a poorly shielded buzz saw. When the intern, Charlie Harris, and I gingerly removed the blood-soaked dressing that had been hastily applied by his foreman, we found ourselves staring into the closed, cut-off ends of the two major arteries running down toward the hand, the radial and ulnar. As we looked more carefully, we realized that both of the vessels, each about a fifth of an inch in diameter, were in such intense spasm that they had effectively shut themselves down. A trickle of blood oozed from a few of the larger veins, but the blunt end of the mangled stump was otherwise no more than moist.

In our haste and inexperience, we had forgotten to place a tourniquet on the upper arm, a maneuver that would have been done by an ambulance crew had the patient not run the four blocks directly from the lumberyard to the hospital. No doubt activated by his own adrenaline and serotonin, Charlie wasted no time shaking his head in disbelief. Unlike me, the flabbergasted and mouth-agape medical student at his side, he knew exactly what needed to be done, or at least he thought so. "Let's get a couple of clamps on those suckers before they change their minds and start pissing blood at us" was what he blurted out in his excitement. While he was whipping rubber gloves onto his eager (albeit a bit shaky) hands and shouting for a pack of sterile instruments to be opened, the head nurse very quietly applied a blood pressure cuff well above the patient's elbow and pumped it up high enough to stop all circulation to the arm, thereby preventing the anticipated hemorrhage. The terrified patient was saved the additional trauma of having two clumsy amateurs chew up his wound still further by poking about in the raw unanesthetized stump, and Charlie and I were spared that other kind of traumatic chewing that would certainly have been our lot a few minutes later when the surgical resident appeared and saw what we had wrought.

After the senior surgeons had arrived and the patient was expeditiously packed off to the operating room, I walked into the small alcove where the emergency room bulletin board was and pulled up a chair directly in front of it. I sat staring for a long time at a large rectangle of cardboard that filled most of its corked surface, on which one of the more cerebral of the internal-medicine residents had written in calligraphy a pithy aphorism of Goethe, precisely appropriate to occasions such as this one: "There is nothing more frightening than ignorance in action." As for Charlie Harris—he dropped out of surgical training at the end of his internship year and went into psychiatry. He is now safely employed by a large HMO on the West Coast.

Although marked constriction of vessels, as in the case of my emergency room patient, is commonly seen, platelets and spasm are not ordinarily enough to prevent continuing blood loss, and the body's clotting mechanism must come into play. The process called hemostasis, the stoppage of bleeding, is enormously complex, involving not only platelets and spasm but a multitude of other factors and chemical substances coordinated with the intricacy of a high-precision drill team. Were coagulation to involve less than multiple interlocked steps, it would be difficult for blood within vessels to remain in its liquid state.

The various chemical substances taking part in coagulation are called clotting factors. Factor I and Factor II are plasma proteins called fibrinogen and prothrombin (a thrombus is a clot), respectively. In the presence of calcium and certain other substances released by injured tissues, prothrombin is converted to thrombin, which in turn acts as an enzyme to change fibrinogen into long threads of fibrin. The fibrin threads adhere, as do platelets, to injured parts of blood vessels to form an entangling meshwork in which red cells, white cells, and more platelets are caught. The resultant jellylike mass is the clot. All of the other clotting factors take part in this process as well, to the extent that a proper clot cannot be formed unless the entire panoply of biochemical events takes place.

When a good clot has formed, tiny strandlike processes that have meanwhile extended out from the cell membranes of the platelets attach themselves to the fibrin threads. This begins a series of events called retraction, in which the blood clot becomes smaller and firmer, pulling the edges of the broken vessel together and squeezing out a yellowish fluid called serum, which is plasma without any of its clotting factors.

It doesn't end here. In addition to everything else they do, platelets release yet other substances that promote repair of the injured blood vessel wall. Soon afterward, the clot is invaded by nearby cells called fibroblasts, which start the process of building up scarlike connective tissue to complete the healing.

This is what happens every time we nick our chins or shins while shaving, or even when we sustain an injury as tiny as a pinprick. It is this remarkable sequence that keeps the life from draining out of us when a finger is cut. Although a score or more of diseases are attributable to defects or absence of one or another of the sequence's components, the truly impressive statistic is the rarity of those diseases. Classic hemophilia, which is due to a deficiency in factor VIII, is the most frequent clotting disorder, and yet it occurs in only one of ten thousand of the male population and essentially none of the female.

One of the most humbling experiences of my thirty-year surgical career involved a patient with a clotting defect. Several lessons are to be found in this story, but they are lessons I should not have needed. The fact that I had long ago learned to avoid every hazard presented by situations like this one did not prevent me from being entrapped by it, succumbing to it, and endangering my patient's life.

No matter how I look back on the events of that day, I return to the certainty that the cataclysm was as much my fault as the fault of my unfortunate patient's pathology. It came about because I broke a few inviolable rules that I had set for myself.

During my years of practice, I have performed many sigmoidoscopies. In this screening or diagnostic procedure, a lighted tube called a sigmoidoscope is inserted through the patient's anus and passed upward to allow visual inspection of the lowest segment of the intestinal tract, the rectum and last few inches of the sigmoid colon. During most of my practice years, these tubes were rigid steel cylinders ten inches long. With the advent of fiberoptic technology, they later became flexible snakelike instruments able to reach the eighteen-inch mark or higher. The event to be described occurred in 1980, while the rigid scopes were still being used.

The patient was a healthy unmarried man of fifty, referred to my office because he had been finding streaks of bright red blood on the surface of his stools. His history was otherwise unremarkable and his physical examination was without worrisome findings except for a bit of blood on the tip of my gloved finger when I inserted it high into his rectum. When I told him it would be necessary to use a sigmoidoscope to inspect the area I had touched, he reacted as though I had just announced that I would take out his gallbladder without anesthesia, right there in the office. He became flustered and upset, and he refused to let me go further. "In the first place," he told me, "I have a terrible fear of anyone doing something to me down there, and I just barely got through the rectal exam. And now you want to ram a steel tube up my ass? Not on your life!"

Patients rarely use even minimally off-color language in a doctor's office. This man in particular seemed hardly the type to resort to as much as a relatively inoffensive word like *ass*. He was well educated and quite polished—urbane, in fact—and held a responsible executive job. Until that moment, he had seemed very self-possessed, other than some obvious apprehension when I put my examining finger into his rectum, hardly an unusual response. But now, the thought of further manipulation was agitating him to the point where some of his self-control had been lost. He would not hear of going any further.

In such a situation, it is hardly a surgeon's place to launch into a deep discussion of psychodynamics and the arcana of the hidden subconscious significance of the anus in one's emotional life, but I did feel compelled to tell my reluctant patient, as diplomatically as I could, that he was risking the presence of a cancer if he did not let me complete my evaluation. As it so often does, the mention of the dreaded word stopped his protests. With good reason, its effect is similar to that described by Samuel Johnson with regard to a man about to be hanged: The possibility of cancer "concentrates his mind wonderfully." My patient quickly agreed to the sigmoidoscopy, and he mounted the table, hurling instructions at me over his shoulder that I not forget his fearfulness, his sensitivity, and his downright disgust with what he was about to endure.

Taking care to be as gentle as I could (and quick, too), I lubricated the canal above the anal opening and slipped the scope inside. A small polyp was clearly visible on the front of the circular rectal wall, about five inches up. A polyp looks like a little tree, with a short, narrow trunk and a rounded top. To the naked eye, this one had no characteristics of being cancerous, but it demanded removal nevertheless, not only to stop the bleeding but also because polyps are the progenitors of malignancy. I removed the scope and told my patient what I had seen.

"Why didn't you just go ahead and take it out?" was his annoyed but very appropriate question. I told him that I have always felt strongly that polyps should be removed only under controlled circumstances. Although it is among the most minor of surgical procedures, polyp removal, or polypectomy, is nevertheless an operation done indirectly through a scope and usually anatomically beyond the surgeon's ability to stitch. For this reason, it is subject to unexpected, even if extremely rare, complications.

"Such as?" he inquired. Such as brisk bleeding, I told him, which can be difficult to control. I also pointed out that because polyps like his seem so simple to deal with, the temptation is great to just go ahead and snip them out. But since my office is not a fully equipped gastroenterology suite prepared to handle all eventualities, I resist the temptation by not so much as owning one of the specialized long-handled biting instruments called a biopsy forceps. It is my custom to remove polyps in my hospital's outpatient department, where all facilities exist in the unlikely—for me, unprecedented—event that something should go wrong.

My patient—I'll call him Mr. Trouble—listened carefully to all I had to say, then announced that he had made up his mind. Under no circumstances would he go to the outpatient department, but he was willing to have the polyp removed if I agreed to do it at a later date in the office.

Not only that, but he would not hear of being referred elsewhere, to a doctor whose approach to office polypectomy was less unbending than mine. "I'm too scared at this point to start again with somebody else. I won't do it, and there's no way you can talk me into it." And right at that moment, I made my grievous error in judgment. Knowing full well that he meant what he said, and feeling strongly that I would not be fulfilling my obligation to Mr. Trouble if I allowed the polyp to remain in his rectum, I agreed to do the procedure exactly as he demanded, right there in the office.

But first, I asked Mr. Trouble a few questions. Q: Has anyone in your family ever had a bleeding problem? A: I was adopted. Q: Do you bleed excessively when you cut yourself shaving? A: I use an electric razor. Q: How about cutting your finger? A: I haven't had any trouble. Q: Have you ever had a tonsillectomy or any other operation where you've bled badly? A: I've never had an operation. Q: Do your gums bleed when you brush your teeth? A: No. Q: Do you bruise easily? A: No different from anybody else.

I had seen no bruises on Mr. Trouble's extremities or trunk when examining him. With a negative inspection of his skin and answers like these, the likelihood of a clotting problem seemed remote.

As I left my hospital's operating room late the next morning, the head nurse very kindly lent me a biopsy forceps. I carefully examined its jaws to be sure their biting surfaces were so sharp that they would cut the polyp cleanly off at its base, with minimal injury to surrounding tissues. I have removed great numbers of polyps and anticipated no difficulty, in spite of my hesitation about the wisdom of what I was preparing to do.

The following afternoon, with my patient in perfect position, I inserted the scope up to the five-inch mark and threw a shaft of light directly on the polyp. I slid the forceps into the scope's barrel and positioned the distant jaws at the point where the polyp was attached to the rectal wall. Then, squeezing down on the instrument's handle, I made a clean instantaneous snip that bit off the little structure. With the polyp held in its jaws, I withdrew the forceps from within the barrel. But I kept the scope in place to be sure there was no bleeding.

Not a drop of blood appeared in my field of vision. The inner layer lining the rectum looked as though a tiny circle had been punched out of it, less than a quarter of an inch in diameter. Its circumference was clean and healthy-looking. I kept my eye on it for about a minute and then withdrew the scope.

Looking back on that moment after a span of more than fifteen years, I remember exactly the thoughts passing through my mind: You really are much too compulsive, boy. You see how easy this was. Those precautions you've taken all these years aren't really necessary—the whole complicated business of arranging to do this in the hospital outpatient department; the dislocation you put the patient through, and yourself, too; the days of unnecessary worry that people have, anticipating what they begin to believe is a dangerous procedure. And just think of the added expense. Maybe you should just buy yourself a biopsy forceps and stop worrying about things that never happen.

Mr. Trouble was happy, and so was I. The polyp was obviously benign and almost certainly the entire source of his bleeding. He should have been scheduled for a scoping of his entire large intestine, called a colonoscopy, at some future time, but he had already assured me that he would refuse to have it done. Nor would he consent to an X-ray study, the barium enema, of his bowel. "I had one rape in me, Doctor," he said, "and this was it. Thanks but no thanks."

I felt satisfied with my compromise. Short of a general anesthesia, there was no other way I could have gotten the polyp out, and all had gone well. So well had the affair come off, in fact, that I was still illogically wondering whether I had been overly careful in all the twenty preceding years of refusing to cut corners on what I considered the most meticulous and safest clinical care. Of course, these were foolish thoughts, unbefitting an experienced physician. The reason for my scrupulousness had been to avoid the one in a hundred or a thousand possibilities of complication that might be serious enough to put a patient's life in jeopardy. I had always considered such a course of action to be good management, and it was illogical to think otherwise on the evidence of my single success with Trouble.

As matters turned out, I was in the midst of reenacting an old story: Compromise your principles or cut corners just once, and that is precisely the time some unforeseeable catastrophe occurs. In late afternoon, about three hours after Trouble had left the office in a fine fettle of self-congratulation, my beeper went off. "Call Mr. Trouble right away," said the answering service lady. "He says he's hemorrhaging."

Trouble answered the phone before my first ring had completed itself. His voice was frightened and shaky. "I had the urge to move my bowels and a ton of blood poured out into the toilet. I feel like I have to go again." I tried to reassure my terrified patient that the bleeding was probably less than it seemed—a little blood in the toilet bowl stains the water a deep red and makes it look like a lot more than it is. In any event, it could easily be stopped. But this was Mr. Trouble—nothing would convince him that he would not die in the next ten minutes. I told him to lie down in bed and I would have an ambulance at his home right away. He was to get up and empty his rectum if the urge was uncontrollable, but otherwise his instructions were to lie still.

As soon as I called the ambulance, I quickly finished speaking to the patient I had been seeing on my evening rounds and hurried down to the emergency room to await the arrival of Trouble. When he was brought in on the gurney, he was agitated and almost incoherent. I spoke to him as soothingly as I could, but to no avail. His voice alternated between mumbling and a near-shout, and it was easy to see why. This was no small bleed. Trouble's trousers and underpants were soaked in fresh blood and the backs of both legs were covered with it. Although the intravenous line inserted during the ambulance ride was running at full blast, he complained bitterly of thirst.

Thirst is a common symptom of people who have lost a considerable amount of blood. One of the body's compensating mechanisms in such situations is to shift water from the interstitial fluid into the capillaries in an attempt to maintain volume (see pages 37–8); this dilutes the blood and concentrates the interstitial fluid, thereby raising its osmotic pressure. In the hypothalamus of the brain resides a group of cells called the thirst center, containing sensors for osmotic pressure. When these so-called osmoreceptors detect an increase in osmotic pressure, they cause a feeling of thirst. The thirst is enhanced by a decreased flow of saliva, resulting from the lowered amount of water in the tissues. A feeling of thirst plus a dry mouth adds up to a desperate urge to drink water, a lifesaving adaptation if ever there was one, because it provides more water to fill the depleted circulation.

With Trouble so frantically thirsty, it was no surprise to be told that his blood pressure had begun to drop and his pulse was up to 120, not only owing to the blood loss but also because of the additional factor of his agitation. There was no doubt that the entire gendarmerie of his fight-or-flight responses had been called into action. His sympathetic nervous system was firing away like the United States Marine Corps responding to attack.

A rapid-acting sedative injected into the intravenous line tranquilized Trouble so effectively that he could be turned on his side to allow me to insert a sigmoidoscope into his rectum. As soon as it was in place, a large volume of fecal-smelling blood poured through it onto the floor and my trousers, as well as splattering the shoes of a resident doctor and one of the nurses assisting me. Even though there were large mushy clots in it, I was beginning to be suspicious—suspicious and concerned about the possibility of a coagulation defect. The artery running up the center of a small polyp is tiny, and even had it begun to bleed long after being cut, the normal mechanisms should have closed it. I suspected the existence of the kind of clotting defect that prevents platelets from adhering to the torn edges of small vessels, so that they cannot seal off. If that proved to be the case, the presence of clots in the pool of hemorrhage was easily explainable: Once out of the little artery, the shed blood mixes with the fluids within the intestine, which start off the process of clot formation, but at a point beyond where it is effective in sealing the vessel. In this scenario, the patient continues to bleed even though the blood coagulates once it has left the still-open vessel.

Samples for testing were drawn from a vein in Trouble's arm, but my immediate concern was to get a good look at my biopsy site and do what I could to stop the bleeding. Meanwhile, blood was being cross-matched for transfusion.

Using a long narrow metal aspirator inserted through the scope, I managed to suck away the blood filling the rectal cavity. Within a few minutes, I found myself staring at the punched-out area that had looked so comforting only four hours earlier. In its very center, a tiny spray of bright red arterial blood appeared with every heartbeat, like a diminutive pulsating fountain. Everything around it looked just as clean as it had in the office.

I passed a wad of cotton up the scope, wrapped around the tip of a long plastic stick, and pressed it firmly against the biopsied area. This stopped the bleeding. I kept the pressure up for three minutes by the clock, only to see the pulsing scarlet spray start up again as soon as I let up. I asked that the cotton tip of the next probe be soaked in adrenaline solution, a stratagem intended to make the tiny spouter constrict. This, too, was without effect, even on a second and a third attempt, each time with the wad held in place a bit longer than the last. Trouble's blood pressure had come up ten millimeters of mercury to 90, and his pulse was down to 110, no doubt owing to the large amount of intravenous fluid he had received. At this point, the first transfusion was available from the bank, and it was beginning to drip in.

It was obvious that nothing less than a stitch would stop the bleeding. In order to get enough room to maneuver five inches up into his rectum, Trouble's anal opening would have to be widely stretched to enable me to insert my fingers high enough to reach up with the needle holder that would carry the suture. Only with an anesthetic could this be done. A spinal anesthesia, which paralyzes all sensory and motor nerves below the waist, was the optimal choice.

The operating room had been alerted shortly after Trouble arrived at the hospital. They were now notified that their services would be needed, and the OR nurses replied that they could be ready in twenty minutes. During that time, I sat wearily on the stool at the side of the gurney, resting my burdened head on my patient's uppermost buttock. As those agonizingly long minutes dragged on, it seemed to my dejected mind that no nightmare could possibly have been worse than this foul-smelling, gory reality in which I was entrapped. I was supporting the scope with one hand and with the other pressing an impaled cotton pledget snugly against the murderous little varmint high on the inner wall of my patient's rectum. My trousers soaked in stool-flecked blood, a stinking pool of it on the floor around me, I drearily waited out the twenty minutes until the scenery could be prepared for the next act in this execrable rectal epic. I was hardly the public's image of a confident, heroic surgeon.

The weariness was only partly physical, after a long day that had started eleven hours earlier with two operations. Mostly, it was a weariness of soul. With good reason, I blamed myself for everything that had happened. Had I not, this one time, abandoned my better judgment by giving in to a patient's demands, none of this would have occurred. In retrospect, I blamed myself for not being sufficiently insistent, for not trying harder to sweet-talk Trouble into compliance; I blamed myself for so much as thinking, after the apparently innocuous biopsy, that I had been overly compulsive in the past and could now become less so; I blamed myself for not digging my heels in and refusing to yield to the importunities of Trouble.

And also, I was worried. I didn't yet know the nature of the clotting defect, and it might be hours before I had that information. A transfusion of platelets was being made ready, and plasma was running into a second intravenous line to provide a general storehouse of clotting factors, but such attempts to restore normal coagulation are like laying down an artillery barrage when the remedy really needed is one accurate bullet from a sniper's rifle. Platelets and plasma hardly guaranteed that a stitch placed in the rectal wall would not cause far more bleeding than it might stop. I could visualize the dreadful scenario of getting to the operating room, starting to suture, and then suddenly finding myself in an uncontrollable sea of hemorrhage five inches up the rectum, where I could barely reach it. And all of this without an operative permit from my heavily sedated patient.

I stanched some of these thoughts by trying to believe that such a catastrophe was unlikely. It also helped that I delivered myself an admonition, for the surgeon's egocentric propensity to see a patient's adversity in terms of its effect on himself—on his conscience, his self-image, and perhaps his standing in the eyes of colleagues. *I* was not the victim of this unexpected complication, *Trouble* was. To arrogate to myself another's misfortune, especially one suffered at my own hands, was the height of precisely the self-absorption I had so often decried in others, and now I felt guilty of it. It helped to remember that, and to be ashamed of it. It was my patient who was at risk, not I. Freed, at least partially, of such unseemly ruminations, I could focus more clearly on the job at hand, and far more dispassionately.

The job at hand was not made one whit easier by the anesthesiologist's refusal to administer a spinal. Certainly, her decision was correct, and I had been anticipating it. To stick a needle into the spinal canal of a patient with an undiagnosed bleeding disorder is beyond courting further complication—it is foolhardy, and I could hardly have expected that it would be done.

It would be necessary to put Trouble to sleep. To add to his problems, he had consumed a substantial dinner just before releasing his first hemorrhagic deluge; he was full of partially digested food and the several glasses of wine he had drunk in solitary celebration of the safe removal of his polyp. The powerfully reeking redolence of his stomach contents was competing for prominence with the malodorous rectal blood—the OR nurses were not happy with me for bringing him to their antiseptic domain. It would take considerable skill on the part of the anesthesiologist to get him deeply asleep without a huge upsurge of sickening vomit flooding his windpipe and lungs, not to mention the operating table. The entire OR team was palpably relieved when my anesthesiology colleague did the job perfectly.

Unlike spinal, general anesthesia does not of itself provide paralysis of muscles. A strong dose of relaxant drug had to be administered before the tight circular muscle around Trouble's anus, the anal sphincter (the word originates with the legendary Sphinx—the strangler of Greek mythology—who killed people by squeezing them into asphyxiation should they fail to solve her riddle), could be dilated widely enough for me to get my instruments in and then to advance them so high that I could accomplish my task.

With assistants holding the anus widely open with steel retractors, I was able to observe the bleeding artery directly. It was almost, but not quite, beyond the farthest reach of my needle holder. Straining mightily, I just managed to get a heavy catgut stitch into the bowel wall. It took a colossal stretch of my far-extended fingers to tie it down, but I finally succeeded on the third try. The bleeding stopped.

To my immense satisfaction, there was no ooze from the needle hole or from the tissue crushed by the stitch. Trouble was out of trouble. I stood staring at the dry field for at least five minutes before the metaphoric weight fell from my shoulders.

It took three pints of blood to tank my patient up to an approximation of his normal level. There was no recurrence of his bleeding. When all the data were in and he had been seen by a hematologic consultant, Trouble was found to be the victim of von Willebrand's disease, a rare coagulation disorder due to deficiency of a plasma protein without which platelets do not adhere normally to sites of injury on blood vessels. The protein circulates in the plasma in combination with another clotting factor, Factor VIII, the substance absent in patients with classic hemophilia. They exist together as a complex structure called factor VIII–von Willebrand protein.

Von Willebrand's is a genetic disease, but serious spontaneous bleeding problems are seen only in those patients who have inherited it from both parents. Those who get the gene from only one parent tend to have no difficulty at all. If they do experience a bleeding problem, it is only with injury or surgical procedures. This seems to have been the case for my patient.

In reconstructing the events of the biopsy, it would appear that the trauma of being cut caused the little transected artery to go into immediate intense spasm at first, preventing any blood loss. When the spasm wore off, perhaps ten or fifteen minutes later, the vessel opened up and began to spurt. With the platelets unable to adhere to its injured end, no clot could form there, with the result that the artery was not plugged. It remained wide open and continued to drain the life out of the man it was supposed to protect. The blood collected in the copious reservoir that is the rectum, until its pressure gave Trouble the urge to evacuate.

In the postoperative days and thereafter, Mr. Trouble proved to be an exemplar of a common phenomenon well known to experienced physicians: Sometimes those for whom we accomplish the most paradoxically display no gratitude or appreciation for what we have done for them, even to the point of denigrating what may have been a bit of medical wonder-work; on the other hand, it sometimes happens that those who have somehow survived the jaws of a dilemma we ourselves inflicted on them, whether by our mismanagement, ineptness, or bad luck, look on us with admiration and even awe, as though their rescue proves that we are some rarefied form of demigod marvel-makers. When I had finished telling Mr. Trouble the details of his saga of my poor judgment and his bleeding, he grabbed my hand and kissed it. In spite of everything, he thinks I am his savior.

The real savior of Mr. Trouble was the animal economy of his body. Had he been without responsive stabilizing mechanisms, the rapid loss of so much blood would have resulted in injury or perhaps failure of his heart or brain before I had the chance to stitch up the pesky perpetrator in his rectum. As almost invariably happens in cases like his, the signals were properly transmitted to those command posts equipped to compensate for the assault being faced by his tissues; from there came orders to action. Vessels constricted, flow was diverted, fluid shifted into capillaries, and pumping strengthened and sped up—all coordinated to maintain the efficient transport of oxygen-carrying hemoglobin to vital structures, even in the face of a much-decreased total volume of circulation. In a somewhat lesser way, it was Marge Hansen's story all over again, the story of that wise body taking care to make up for one of its own inadequacies.

Though enormously dependable, the safeguards are not perfect. It is not inevitable that every injury is held within limits or every loss compensated. Sometimes the forces falter in their attempts to rally when fighting off incipient disaster. But these failures are rare, and usually the result of aging or destruction of crucial factors in the body's defenses and response mechanisms. Unless the attack is simply too powerful, as in massive trauma, overwhelming infection, or the depletion of usual resources, the body's economy moves toward gathering itself up and finding within its huge array of self-righting remedies the wherewithal to persevere and conquer. Adjustments may only need to be temporary or they may be permanent, but a return from jeopardy is to be expected almost always.

In a large proportion of such instances, the blood is the vehicle of perseverance and salvation. Whether by the action of one of its own elements or as a thoroughfare for the transportation of signals and correctives from source to site of action, blood is at the center of homeostasis. The blood of mammals is the result of eons of nature's gropings, the special juice that nourishes and interconnects the variegated elements in the enterprise of life.

The specialness of the juice is evident in all manner of its activities. Among those that best illustrate its involvement in righting the wrongs inflicted by a hostile environment is its role in fighting infection, which it fulfills for the most part by means of its white cells, also known as leukocytes, from the Greek *leukos*, meaning "white," but also "light" or "brilliant" or "clear." The Greek name Loukas is the origin of Luke, the patron saint of physicians.

Although they arise in the marrow from stem cells, as do erythrocytes and platelets, leukocytes do the vast majority of their work after they leave the capillaries and enter the tissues. Like so many other of the body's components, they are hard at work doing general housekeeping all the time, but they are capable of putting up a huge show of force when called upon to solve an urgent problem. The urgent problems they commonly address are invasions by outsiders, such as bacteria, viruses, fungi, and other foreign structures. Their show of force is intended to destroy the foreign agents.

There are three kinds of white blood cells, called granulocytes, monocytes, and lymphocytes. The granulocytes, which are about twice the size of red cells, are so called because their cytoplasm contains large numbers of tiny granules. There are three varieties of granulocyte, each being named for the color assumed by their granules when stained with certain laboratory dyes: neutrophils (pinkish or neutral), eosinophils (deep red), and basophils (blue). The most common variety—the neutrophil—makes up 60 percent of all white blood cells, and it will be the one granulocyte in the spotlight on the following pages.

The three kinds of leukocytes, or white blood cells:

GRANULOCYTES—NEUTROPHILS, EOSINOPHILS, AND BASOPHILS
MONOCYTES
LYMPHOCYTES

Although neutrophils are transported through the body by the circulation's flow, they are quite capable of leaving the capillaries and traveling through tissues for short distances on their own, by an ameboid sort of motion. In their movement forward, they extend long feet into which the cytoplasmic granules flow. Neutrophils are attracted by chemical substances produced by injured or infected tissues. Once they have been drawn to their point of action, they engulf the bacteria or other organisms, or whatever foreign bit may be the intruder. Because the granules contain digestive enzymes, these living or dead particles are eaten up and destroyed.

The bone marrow maintains a large reserve of mature granulocytes at all times, amounting to as much as a dozen times the number in the blood. When the need arises, a boost of them is released into the circulation. This is why a blood test done during an active infection shows an elevated count of white cells. Once in the circulation, the granulocytes make their way into the tissues by squeezing out between the cells of the capillary wall. In this way, huge numbers of neutrophils and their two granular cousins, eosinophils and basophils, can accumulate rapidly in areas of infection or injury.

Cells that, like neutrophils, carry out this process of ingestion and digesting are called phagocytes (literally, "cells that eat"); the process itself is called phagocytosis. Eosinophils and basophils, which account for a total of about only 3 percent of the leukocytes, are also capable of phagocytosis, but to a lesser extent. Eosinophils do function with particular efficiency in killing the cells of certain parasites, and they play a role in decreasing the force of allergic reactions. This is virtually all that will be said about them in this chapter.

Monocytes, which comprise 7 percent of the circulating leukocytes, are the largest of the blood's cells, being approximately three times the size of an erythrocyte. They, too, are phagocytic. After leaving their origins in the bone marrow, monocytes ride through the circulation for a few hours and then divide over and over. Their offspring are the macrophages, or "big eaters," that ingest worn red cells, among their other duties in the blood and tissues. Granulocytes have a short life span and disintegrate or leave the area within a few hours, but macrophages can live as long as seventy-five days. This, along with their increased phagocytic powers, makes them ideal members of the cleanup crew. Acting like scavengers, monocytes and their progeny—macrophages—arrive at locations of infection or injury later than the neutrophils, and they sweep the place up.

Monocytes are relatively nonmobile. The macrophages that are the result of their cellular division collect within areas spread diffusely throughout the body such as the spleen, lymph nodes, liver, lungs, digestive tract, and tonsils. The totality of these multiple locations is known as reticuloendothelial tissue, or sometimes as the reticuloendothelial system. A reticulum is a fine meshwork; the reticuloendothelial tissue is a meshwork of delicate fibers and capillarylike vessels, in the webbing of which can be found huge number of the various cells that take part in the body's defenses against infection and invasion. It is the activity of the reticuloendothelial tissue spread throughout the body that removes foreign particles from the lymph. Any particles that escape its cells will be returned to the blood when the lymph reenters the circulation at the subclavian vein. Once in the blood, they will fall victim to the phagocytes in the vessels and in the tissue of the open spongy organs through which the blood passes: the spleen, liver, and bone marrow. If you have ever wondered why you develop lumpy and tender swollen glands in your neck when you have a sore throat, wonder no more. These are your lymph nodes, small nut-shaped packages of reticuloendothelial tissue enlarged with the tremendous swarming activity of phagocytes gobbling up the bacteria entrapped in their filtering network, ever on the alert to protect you against the depredations of alien forces.

One in every hundred of the approximately 75 trillion cells in the human body is a lymphocyte, which gives some idea of the vastness of their numbers and the magnitude of their responsibility. Early in their differentiation from stem cells, immature forms of lymphocytes are released from the bone marrow and travel through the circulating blood to reach reticuloendothelial tissue all over the body. About half of them come to rest in the thymus, a large, flat grayish structure that lies at the front of the chest, between the breastbone and the heart. Because they continue to differentiate and mature in the thymus before returning to the bloodstream, they are called T lymphocytes or T cells (a subset of these, known as T4 or CD4 cells, are those that are destroyed by HIV, the human immunodeficiency virus, in patients with AIDS; the CD4 cell is a major factor in immunity).

The other half of the lymphocytes mature elsewhere and differentiate into what are called B lymphocytes; their source is the bone marrow. By the time the maturing processes are completed, the total circulating population is 20 to 30 percent B cells and 70 to 80 percent T cells. Although they are present in large numbers in circulating blood, the real home of both types of lymphocytes is in the reticuloendothelial tissue.

It is as though nature has distributed reticuloendothelial tissue within the body in two ways: concentrated in the form of a large bunch in distinctive pulpy structures like the spleen, lymph nodes, and tonsils, whose main function is specifically to contain it; and elsewhere diffusely scattered within organs or locations whose main function is something else, such as the liver, lungs, digestive tract, bone marrow, and other, lesser places and patches.

The filtering and infection-fighting system of reticuloendothelial tissue is everywhere, its leukocytes always engulfing and destroying cells and foreign particles that might harm the body. When not overwhelmed or weakened, its ubiquitous network cannot be avoided by the tiny living organisms it seeks out or by the noxious bits of biological flotsam and jetsam carried by blood and lymph. Within the mesh of its webbing are hosts of macrophages and both B and T lymphocytes, dedicated to protecting the body against the unwelcome invaders that would destroy it, including cancer cells recognized as foreign. It should be pointed out that the cells of many malignancies carry antigens that cause them to be perceived as "other," with the result that the body mounts an immunologic reaction to them. Although these mechanisms are still poorly understood, researchers are attempting to develop anticancer therapies based on these aspects of the host's response. See chapter 2 for a more complete description of the body's resistance to cancer cells.

Immunity is the process by which the lymphocytes make their contribution; it is based on their ability to recognize materials that do not belong in the host. By a method not yet fully understood, the cells of the embryo carry out what amounts to an inventory of all the proteins and certain large lipid and carbohydrate molecules in the body and come to "know" that these are the only ones that belong—these are "self," and everything else is "foreign" or "other." One's cells, too, are recognized by one's immune system. This is because each cell carries on its surface a protein distinct to that individual, which functions as a "self" marker. As B and T cells mature, they come to have receptors on their membranes that recognize molecules which are "other." When it encounters them, the lymphocyte generates an immune response. A molecule perceived as foreign, to which a lymphocyte responds in this hostile way, is called an antigen.

The B and T cells respond to the presence of antigens in different ways from one another. The B cell does it by producing immunoglobulins, which are gamma globulins called antibodies—literally "bodies against." The production of antibodies like the synthesis of proteins in general, is controlled by genes activated within the cell, in this case the B cell. An individual B cell can respond to only one antigen; it can produce only one kind of antibody; accordingly, each antibody responds to one specific antigen and no other. When an antigen comes into contact with the proper B cell, the B cell begins to divide into progeny called plasma cells, rapidly producing vast numbers of copies of the particular antibody molecule—about two thousand per second. An antibody molecule locks onto its antigen as though to tag the foreign substance, and macrophages do the rest. Or the antigenic substance becomes coated with a protein called complement, which can either attract phagocytes or promote dissolution of the foreign cell's membrane. The process by which the B cell does its job is called humoral-mediated immunity. It acts primarily against bacteria and also against viruses at the phase when they are not within cells.

T cells, on the other hand, are involved in a process known as cell-mediated immunity, directed primarily against substances, such as viruses, that have penetrated the host's cells and are therefore inaccessible to antibodies.

When a virus enters one of the body's cells, it somehow combines with the "self" marker on the surface of the cell membrane. A type of T cell called cytotoxic or killer T cells have surface receptors that recognize this combination as something foreign, and they bind themselves to it. The killer cell then secretes proteins that punch holes into the foreign cell, destroying it. It is this sequence of events that causes transplanted organs to be rejected, unless measures are taken to mute the immune response.

Other types of T cells act as helpers and suppressors of the various steps in immunity. Helper T cells stimulate the rapid response of B cells described above, which results in a massive increase in specific antibody; suppressor T cells secrete substances that slow the immune response once the foreign attack has come under control. Still other lymphocytes function as what might be called "memory cells." These are B and T cells that are produced during the episode of infection but not used. They circulate in the blood, ready to respond quickly should they perceive the return of the antigen to which they are antagonistic, even decades later. It is for reasons like this that we do not get chicken pox twice. Without memory cells, it would be impossible to immunize against viral diseases.

As with so many of the processes previously described in these pages, immunity involves far more substeps (and sub-substeps, and sub-sub-substeps) and interactions among its various components than are possible to take up in a book of this kind. Like other of the body's intricate balancing acts, the signaling system is the key to everything. All kinds of molecules take part in immunity, whose function it is to coordinate events, to tell cells what to do, and to switch various of the multitudinous parts of the proceedings on and off.

There is no more dramatic example of the body's response to injury than its ability to produce inflammation. Although we moan and groan when a traumatized finger becomes red, swollen, hot, and painful, these annoying symptoms are actually evidence that the nearby tissues are mounting an effective defense to whatever it is that we have done to ourselves. As early as the first century A.D., the Roman medical encyclopedist Celsus already knew enough to catalogue these four findings as the cardinal signs of inflammation. He called them *rubor*, *tumor*, *calor*, and *dolor*, but it took almost two millennia to elucidate their origins and to ascertain whether they were good or bad for the person who has been hurt.

Assume a splinter, or a good hard bang on the thumb, or a burn, or a frostbitten ear. Think of more serious problems, such as pneumonia, or peritonitis, or a bullet wound. By far the most common form of injuries I have witnessed in a long medical career consisted of those I inflicted every day with my scalpel. Although fashioned in the ultimate interest of a patient's health, a surgical incision is nevertheless an example of an injury, albeit a controlled one. The study of the healing of surgical wounds by thousands of research laboratories over the past hundred and more years has been a fruitful source of knowledge about the body's defenses against assault.

The so-called inflammatory response consists of a series of events whose purpose is to treat the local trauma; get rid of the badly injured or dead tissue that is its cause, its result, or both; wall the area off from the rest of the body; and begin the process of healing.

Immediately upon injury, capillaries dilate and become more permeable—so permeable that protein molecules can leak through their walls. This allows protein-rich plasma to enter the tissues in addition to the blood spilled from cut or crushed capillaries and tiny veins. White blood cells migrate through vessel walls that have become more permeable. Antigens are produced. Phagocytes from the vessels and the blood are added to those attracted from local tissues in order to begin engulfing and digesting damaged cells and any bacteria that have entered the site of trauma. Meanwhile, chemical factors in the tissues convert fibrinogen to fibrin, whose strands form a meshwork to close the injured vessels and also wall off the area, as well as to begin the repair process. All of this is abetted by the local release of histamine from basophils, and probably by the presence of other hormones, too. The histamine helps to dilate the capillaries and adds to their permeability. Macrophages appear later to help clean up the debris, and in time there is an influx of fibroblasts, the immature cells from which scar tissue will eventually form. The redness and heat (*rubor* and *calor*) are caused by the dilatation of the capillaries; the swelling (*tumor*) is due to the leakage of fluids into the tissues; and stimulation of local nerve fibers results in pain (*dolor*). Celsus may not have known why, but he certainly did know what.

When this process goes well, the end result is clean healing, with no visible aftereffects if the wound is small, or a scar if it is larger. Except in certain tissues, such as those of the nervous system, surviving healthy cells divide repeatedly to reconstitute the area of injury, which is supplied by new blood vessels that grow into it from nearby tiny arteries and veins. This, with scar tissue to help, completes the sequence of repair.

But such efficient repair is not always what happens. Sometimes there is too much damaged tissue, or the wound is so dirty that there are insurmountable numbers of bacteria; sometimes there is no way for the rich mixture of plasma, blood cells, and debris to be reabsorbed into the local tissues. In such situations, infection worsens and the collection of fluid becomes pus. Pus is a thick mix of putrified local tissue, bacteria, the residues of the plasma, and white cells in various stages of decomposition. If it finds access to the surface, whether by bursting open or being lanced, it drains off and the structures behind it heal. If not, it forms itself into an abscess, which the body attempts to wall off with a thick coat of fibrin and fibrous materials. A frequent outcome of this is that the bacteria within the pus enter the bloodstream and are carried off to all parts of the body. This is the generalized infection called sepsis (or what many call blood poisoning), which is fatal if not vigorously treated with antibiotics and other measures intended to prop up the host's beleaguered defenses.

But always the purpose of treatment is only to restore nature's balance against disease. There is no recovery unless it comes from the force and fiber of one's own tissues. The physician's role is to be the cornerman—stitch up the lacerations, apply the soothing balm, encourage the use of the fighter's specific abilities, say all the right things—to encourage the flagging strength of the real combatant, the pummeled body. As doctors, we do our best when we remove the obstacles to healing and encourage organs and cells to use their own nature-given power to overcome.

We have always known this. Every system of so-called primitive medicine I have ever encountered views disease as the imbalance of certain factors, whose proper interrelationships must be reestablished if recovery is to take place. The ancient heritage of Western scientific medicine is no different. Hippocrates and his followers inherited from earlier healers the belief in the four humors, whose equilibrium maintains health: blood, yellow bile, black bile, and phlegm. Although we have long since abandoned those seemingly fanciful conceits, their symbolism remains, and some of us have begun to wonder whether they will prove, after all these centuries, to be more than symbols. We speak nowadays of such things as hormones, and transmitters, and tissue factors floating around our bodies, and we have even come to introduce terminology that sounds eerily familiar, as though emerged from some cobwebbed cranny in the long-forgotten cellar of our history—such as humoral-mediated immunity.

I have spent the adult years of my life being nature's cornerman. I have provided it with whatever boost was needed, cheered it on, and felt the exhilaration of watching its formidable powers wheel into action once I have helped remove the impediments. An inflamed organ is excised, an obstruction is bypassed, excessive hormone levels are reduced, a cancerous region is swept clean of tumor-bearing tissue—and the wrongs are redressed, thus allowing cells and tissues to take over the process of reconstituting equilibrium. Surgeons are no more than agents of the process by which an offending force may be sufficiently held at bay to aid nature in its inherent tendency to restore health. For me, surgery has been the distilled essence of W. H. Auden's perceptive précis of all medicine: "Healing," said the poet, "is not a science, but the intuitive art of wooing nature." In all of this, the blood *is* the life.

11 | A VOYAGE THROUGH THE GUT

Long before Louis Pasteur uttered his famous phrase in 1854, "Where observation is concerned, chance favors the prepared mind," dozens, if not hundreds, of examples of that memorable maxim had already been recorded in the annals of science. Pasteur was referring to minds so well trained and experienced that they could take advantage of any serendipitous event that might come their way. But he did know of momentous instances when his own aphorism did not hold up. From time to time, the unforeseen opportunity to study some unique property of nature has fortuitously presented itself to someone whose background was without evidence that he was equipped to make good use of it, and yet a great contribution has nevertheless been the result. As a chemist, Pasteur was aware, for example, that virtually everything then known about the stomach's power to digest had been discovered by an ordinary army doctor who, having obtained his training by the apprenticeship system, had never set foot in a laboratory or studied science in any systematic way.

The serendipitous event was a shooting. It took place on the morning of June 6, 1822, at one of the trading posts of John Jacob Astor's American Fur Company, on Lake Michigan's Mackinac Island. Someone inadvertently discharged a musket into the lower chest of a wiry little eighteen-year-old French-Canadian boatman named Alexis St. Martin, who had the bad luck to be facing the gun from a distance of less than three feet. The blast of powder and buckshot blew a wide gash through the front of St. Martin's left fifth and sixth ribs. Continuing on its path, the charge ripped away the lowermost portion of lung and a substantial piece of diaphragm in the course of tearing into the front wall of the boatman's stomach, through which his half-digested breakfast immediately poured forth.

When the thirty-seven-year-old Dr. William Beaumont arrived a few minutes later, he could do not much more than extract whatever pieces of clothing and shot were visible in the depths of the mangled tissues. He cleaned away the thick mess of disintegrating food and the bloody mucus draining from the mutilated lung as best he could, then gingerly lifted off the nearby bits of destroyed muscle and skin. When he was through, he dressed the wound with a poultice of carbonated liquid and kept it wet for the next twenty-four hours by periodically soaking it with an astringent lotion containing ammonium chloride and vinegar. As he left his moribund and barely conscious patient after the first dressing, Beaumont predicted that the young man would be dead within thirty-six hours.

Beaumont's use of a variety of chemical irritants was intended to arouse as great an inflammatory response as possible within the first day. Although nothing was known of neutrophils, fibrin, or monocytes in that era, it was well appreciated that the only hope of survival for a man so gravely wounded was to do everything possible to support and encourage the body's own defenses, namely its ability to mount a fierce counter-attacking inflammation against the extensive damage. In medicine as elsewhere, the best defense is often a strong offense. Although he despaired of success, Beaumont was determined to utilize every means he could muster, to help nature in its attempt to ward off what appeared to be virtually certain mortality.

As the doctor returned again and again to minister to his suffering patient, he became increasingly encouraged by the magnitude of the inflammatory response the young man was able to achieve. St. Martin's course was stormy, the power of his gritty constitution raging against the massive insult to its integrity. Fever persisted for ten days, bits of smashed rib and tissue began to slough out, foul-smelling pus accumulated and spontaneously drained itself into each new dressing—but by the fifth week, the earliest evidence of regenerating tissue was beginning to be recognizable in the depths of the wound. Everything was now proceeding as it should: Macerated bits of dead flesh and rib had been cast off; the macrophages were doing their job; a fibrin barrier had formed to separate the wound from surrounding structures. Healing could now commence.

New tissue gradually appeared as fibroblasts and capillaries grew in. In time, the entire area began to scar down, finally contracting into one large block in which the recovering and adjacent uninjured parts adhered to one another. The process drew the edges of the gaping gash in the stomach snugly up against the healing chest wall. Finally, what remained visible from the outside was an expanse of about twelve inches in diameter of scarred skin over an irregular mass of the thickened fibrous muscle beneath it.

From the center of the entire contracted expanse of deformed but healed chest and upper abdomen protruded a rounded red structure some two and a half inches in diameter, resembling nothing so much as a puffy wide-open rosebud about two finger breadths beneath the distorted left nipple. An observer peering through the pouting lips forming its crimson circumference would realize that he was looking directly into the inside of Alexis St. Martin's stomach.

So effectively had the perforation been caught up in the intense inflammatory process and subsequent contraction of the wound that its spread edges jutted up through a hole in the scarred muscle and skin, kept open by the continuous leakage of the stomach's corrosive contents. The rosebud aperture had to be capped or plugged at all times to prevent chemical digestion of the surrounding surface.

There could not have been a more graphic example of the body's extraordinary capacity to respond to trauma by walling off the injured area and then healing it. The entire process had taken a year. It was a year of slowly subsiding *rubor*, *tumor*, *calor*, and *dolor* for St. Martin, but a year of prodigious performance by his young body's natural processes of recovery and regeneration. For Beaumont, it was a year of marvel and wonder as he watched his patient's tissues gradually pass through the entire sequence he had so often observed in smaller wounds, but which now exceeded anything he had ever encountered or anticipated.

Perhaps it was Beaumont's very awe at the revealed power of nature that led him to his decision to use this unique opportunity to study its mysteries further. Curiosity, wide reading, and infinite patience were the qualities that would have to compensate for his lack of formal training. In the little book he published eight years later, *Experiments and Observations on the Gastric Juice and the Physiology of Digestion*, he never tells his readers precisely what motivated him to begin a series of experiments on his patient in May of 1825. So strong was his persistence during his studies that even his transfer five months after beginning them, to Fort Niagara in northern New York State, did not stop their progress. He simply took St. Martin along with him.

Soon after arriving at the new station, the boatman, neither then nor at any subsequent time a willing subject, absconded to Canada, and four years passed before Beaumont could find him. During that time, he had married and fathered two children. By then, he was known to his cronies as "the man with a lid on his stomach."

The experiments that were resumed in 1829 continued for two years, while St. Martin worked as a servant in Beaumont's house and found enough free time from the testing to sire two more children. Then he left once more, and thereafter it was an on-and-off, catch-as-catch-can relationship, with the doctor intermittently convincing or bribing his recalcitrant subject to remain with him long enough to participate in more studies.

The perennially grumbling Canadian lived to be eighty-three, long outlasting his eager scientific pursuer. When he died in June of 1880, his widow, grimly intent that no doctors would get his wizened little body, kept it in the family home during an ensuing spell of particularly hot weather, until it was badly decomposed and stinking. Even then the determined widow was not satisfied, finally arranging for the rotting remains to be buried eight feet below ground. By then, nobody cared to unearth them. To this day, Alexis St. Martin lies undisturbed—in life a bondsman to science, the price he paid for being a beneficiary of nature's wondrous astonishments.

Fistula is the name given to an abnormal passageway between two internal organs, or between an organ and the surface of the body; it is a Latin word for "pipe" or "tube." *Gastric* is an adjective referring to the stomach, derived from the Greek *gaster*, "the belly." The great fascination of Alexis St. Martin was his gastric fistula, the hole that, when uncovered, provided a clear view of his stomach lining and allowed samples of his gastric juices to be collected. As Beaumont would write in his 1833 publication, "I had opportunities for the examination of the interior of the stomach, and its secretions, which has never before been so fully offered to any one."

Teaching himself as he proceeded, Beaumont conducted hundreds of experiments through St. Martin's gastric fistula. He studied it under conditions of thirst, hunger, and satiety; emotional stress and tranquillity; after the ingestion of stimulants and of alcoholic drinks; empty and with an enormously wide variety of foods within it; in the morning, at night, and in between; under various conditions of St. Martin's sometimes mercurial temperament; in times of fever and in times of normal temperature; in weather rainy and weather clear; in sickness and in health—and he succeeded in the objective so clearly stated in his book: "to be the means, whether directly or indirectly, of subserving the course of truth and ameliorating the condition of suffering humanity." If there is a single definitive theme in the published outcome of Beaumont's wide variety of experiments, it is this: The stomach responds differently each time conditions change, and every response has a purpose.

William Beaumont is popularly called "the Backwoods Physiologist," but he is more commonly known as the father of gastric physiology and the very first American in that proudly long list of those who have made lasting contributions to medical science. His collection of studies became the launching pad from which all further investigations of gastric activity and function would originate. Virtually everything to be found in the following pages of discussion of the stomach is in one way or another a continuation of Beaumont's original work. None of his observations has ever been shown to be erroneous, and a majority of his interpretation and opinions have been substantiated by later research. He was correct in predicting that "their worth will best be determined by the foundation on which they rest—the incontrovertible facts."

Actually, the stomach is best understood by being seen as a large bag near the upper end of what is otherwise a hollow muscular tube some twenty-five feet long from mouth to anus, the central portion of which is coiled up inside the abdomen. Except in those two places, mouth and anus, the wall of the tube is composed of four concentric layers that vary somewhat in structure and function at different areas along its length. From top to bottom, the tube, otherwise known as the gut or alimentary tract, consists of the pharynx, esophagus (or gullet), stomach, small intestine, large intestine (or colon), and rectum.

The tube's four layers, from inside out, are as follows:

1. The mucous membrane, or mucosa, which consists of a layer of lining epithelium at its surface, with some thin underlying connective tissue. Various of the epithelial cells are specialized to secrete mucus (which is why this layer is called the mucous membrane), enzymes, or other chemicals directly into the cavity of the gut. Other cells produce hormones, which enter the bloodstream.

2. The submucosa lies directly under the mucosa, and it contains blood vessels, networks of autonomic and local nerve fibers, and lymph vessels. This is where most of the lacelike reticuloendothelial tissue of the gut is to be found.

3. The muscle layer has two parts, an inner circular component and an outer longitudinal one that runs along the gut's length. Because the stomach has to do a great deal of churning, it alone is additionally provided with an oblique layer to aid in the general mashing of food.

4. The serosa is a very thin, smooth outer layer, so slippery that it allows the various parts of the gut to slide across one another and the inside lining (peritoneum) of the abdominal cavity. Actually, the pharynx, esophagus, and rectum lack a serosa because, lying outside the peritoneum-lined abdominal cavity, they do not need it. Nature, while not exactly abstemious, does not often waste its resources where they would serve no purpose.

The cells in the mucosal, submucosal, and muscular layers of the alimentary tract do their various jobs in response to signals from the autonomic nervous system, from local interconnecting nerve fibers of a sensory and motor nature, and from hormones. In general, these signals are determined by the volume, chemical nature, and other characteristics of the material passing through. Here, as in the heart, automaticity modified by response to changing needs is the rule.

There is so much independence in the gut, in fact, that the term *enteric nervous system* has come into use to describe it. This system consists of the 100 million nerve cells in the alimentary tract (just about the same number as the spinal cord!), their neurotransmitters, and a wide variety of different kinds of protein molecules that carry messages back and forth locally and over long distances. So autonomous is this widespread complex of circuitry and chemicals that it is sometimes called "the brain of the gut." The real brain and the gut's brain are in constant communication with each other through the intermediary of the autonomic nervous system, but also by virtue of the fact that hormones and other signaling molecules produced in various parts of the alimentary tract cause responses in one or another part of the central nervous system.

Long before *Homo sapiens*, the enteric nervous system existed in primitive animals without brains, such as wormlike creatures, to control digestion. As the animal tree was ascended, increasingly complex central nervous systems appeared, and connections between them and the earlier "brain of the gut" were established, leading to the situation in humans where conscious thought and autonomic activity affect the alimentary tract, even as signaling molecules and nerve impulses originating in the alimentary tract affect the brain and cord. All of this contributes to a partly insentient awareness of the inner workings of our bodies. No wonder we sometimes feel as though we respond with our guts to every emotional wind that blows our way.

Some of the signals by which the alimentary tract governs itself are components of reflexes. Most of us, including physicians, tend to use the word *reflex* rather glibly, assuming that it is part of everyone's general vocabulary, even though it is one of those words that we, when pressed, may be hard put to define with precision. As used in this book, it refers to an involuntary action of a highly specific form, automatically resulting from a particular stimulus. Although we are sometimes conscious of the automatic action, it is so programmed that we have no control over it because it goes from point of stimulus to completion of response without passing through the brain's consciousness. Its pathway is called the reflex arc.

Since the knee jerk (or patellar reflex) is so well known, its arc may be used as an example of the genre. As one of a class called stretch reflexes, it depends on the tendon beneath the patella being slightly stretched by a brisk tapping with a rubber hammer. Specialized receptor cells in the tendon sense the slight stretch and send messages through sensory nerves to the spinal cord. The cells of these nerves connect to other neurons in the cord, which send motor messages back down to the thigh muscles, causing them to contract. The result is to jerk the lower leg and foot forward; the participation of the brain is not required for this to take place. The purpose of such tendon reflexes is to resist inappropriate stretch on muscles, which if not controlled would throw off the perfect coordination of mechanical forces necessary to maintain posture. There are untold numbers of reflexes in the body, each of which causes an action that is part of the general economy. The arcs of some pass through the spinal cord or even higher, while some restrict their path to local circuits within the tissues they affect.

Before food enters the esophagus (literally, "the carrier of that which was eaten"), it is bitten, torn, and ground by the incisor, canine, and molar teeth, respectively, each structurally specialized for its own distinct purpose. It is being mixed at the same time with the saliva, which contains mucoid material and an enzyme that breaks down starches and sugars to begin the process of their digestion. The end result of all this activity is that a tasty bit of appetizing food is converted into a slimy mass of pulped mush called a bolus. The mere thought of looking at such a disgusting thing would turn a diner's stomach, and yet there it is at the back of the tongue, ready to be swallowed. A glimpse of the muculent makings of a well-chewed bolus or its stringy wet extensions in an eating companion's open mouth has taken away many an appetite and brought more than a few budding romances to a grinding halt.

Here it may be appropriate to consider the control of salivary secretion, since it will give some idea of the kind of activity that goes on all through the gut, responding to various needs as they make themselves known. Saliva is produced by several sets of glands whose ducts empty into the inside of the mouth at various points. The glands are controlled by both sympathetic and parasympathetic nerve fibers. Sympathetic stimulation results in thick, scanty secretion, as anyone knows who has ever experienced the dry mouth of fear or apprehension. Being attuned to the vegetative functions, on the other hand, parasympathetic stimuli produce large amounts of thin saliva, especially in response to pleasant smells or thoughts of culinary delight. It is the (literally) mouthwatering morsel that brings out the full parasympathetic force.

Swallowing is voluntarily initiated when the tongue forces the bolus back down into a short, muscular tube behind and below it, called the pharynx. Once swallowing starts, the entire passage down as far as the anus is thereafter guided by a combination of sensory receptors in the wall of the digestive tract as they respond to chemical and physical stimuli from the traveling food. When receptor cells in the wall of the pharynx sense the presence of a bolus, its muscular wall constricts and pushes the bolus into the esophagus, which continues the wavelike propulsion until the stomach is entered. The pharynx and esophagus act only to push the food along; they take no part in digestion.

Although the back of the pharynx empties into the esophagus, its forward portion leads into the windpipe, or trachea. Almost always, the bolus unerringly passes into the esophagus, because the mechanism of swallowing involves a series of movements in which the upper end of that organ opens to receive it just as a single-cusped valve called the epiglottis flaps down over the entrance to the trachea. But should the diner start a giggling fit or be suddenly frightened or otherwise distracted, the timing of this reflex may be thrown off just enough to make the bolus enter the wrong orifice. Should this happen, its presence irritates the sensitive lining of the trachea, setting off the coughing reflex, which propels the foreign material upward into the pharynx or even the mouth—or perhaps the shirtfront of the fellow at the next table.

The esophagus is a straight tube that drops directly downward from the pharynx, passing behind the larynx and traveling down the chest, and so lying directly on the front of the vertebral column. When it reaches the level of the aorta, these two, aorta and esophagus, continue alongside each other, finally passing through separate but juxtaposed openings in the diaphragm and then entering the abdomen. In such intimate contact with each other are they that I have seen cancers of the esophagus slowly grow into the aorta and erode their way through its wall until they finally perforate it. At the crucial moment when this process reaches its climactic and unstoppable finale, such patients instantaneously drown in a sudden torrent of their own blood pumping and pouring upward to the pharynx and overflowing into the windpipe.

About an inch below its passage through the diaphragm, the esophagus enters the stomach, but just before it gets there, its muscle fibers thicken to form a sphincter. This as well as the angulation at which the tube enters the gastric pouch serves to keep the bolus and acid gastric juice from regurgitating upward. Should this mechanism not function properly (a hiatus hernia is one possible reason, a condition in which the stomach slides up into the chest and the angulation is lost), heartburn is the result, because the mucosa of the esophagus, unlike that of the stomach, is not resistant to the corrosive effects of the gastric acid.

The rhythmic waves produced by sequentially alternating contraction and relaxation of the muscular layers of the esophagus push the bolus down toward the upper part of the stomach, which lies behind the forward portions of the fifth, sixth, and seventh ribs (this explains how it was that an injury to the front of Alexis St. Martin's lower chest should have resulted in a hole in his stomach). This movement is called peristalsis, from two Greek words meaning "to contract around." Peristalsis is the means by which digesting food is propelled all the way along the gut until its ultimate emergence as feces. Although much of this activity is the result of automatic controls within the gut wall responding to the distension caused by the physical presence of food within, it is affected by sympathetic and parasympathetic messages as well. These messages are sent in response to general conditions throughout the body, including such factors as emotions and state of health.

All along the digestive tract, entering food is being broken down into ever-smaller particles and liquids, then degraded into individual molecules so that it can be absorbed by the cells of the mucosa and pass into the lacteals and tiny blood vessels of the gut wall, which carry it to the bloodstream. A great deal of this process begins in the stomach. Once in that very distensible organ, food is mashed and mixed by powerful contractions of all three muscle layers. Driven primarily by messages from the brain sent via that parasympathetic nerve called the vagus and its neurotransmitter acetylcholine, specialized cells in the mucosa have been stimulated by the thought of food, its smell, or its taste to increase their secretion of acid, mucus, and certain enzymes, thus beginning the digestion of proteins in the entering boluses. The most important of these enzymes is pepsin (from the Greek *pepto*, "I cook, or digest"). We have known that the stomach acid is hydrochloric since William Beaumont sent samples of it for analysis to professors of chemistry at Yale and the University of Virginia.

As Beaumont noted, gastric secretion is stimulated not only by conscious awareness but also by the presence of food stretching the stomach wall and by such chemical influences as caffeine, alcohol, and certain partially digested proteins. The presence of amino acids and other protein products causes certain of the mucosal cells to secrete a hormone called gastrin, which enters the bloodstream and activates the production of more acid. In addition, histamine is secreted by particular cells in the gastric mucosa. The histamine acts together with gastrin and acetylcholine to enhance acid output.

Why the stomach does not digest itself is a puzzlement that has intrigued observers since long before Beaumont began the elucidation of how it digests so much else. A thin protective layer of its own mucus as well as certain alkaline substances secreted by its epithelial cells seem to be the reason. Of course, the defense against injury is not always perfect. If it were, no one would ever have to suffer from peptic ulcer, in which acid erodes a small hole into the lining of the stomach or the first part of the small intestine, the duodenum. Although gastric acid destroys virtually all organisms entering with the food, some people harbor a bacterium called *Helicobacter pylori*, which contributes considerably to the possibility of ulceration. It can be appreciated why such remedies as antibiotics and blockers of histamine action are so effective in combating ulcer.

The mucosal cells of the stomach and duodenum produce not only activating substances but inhibiting ones, too. The total effect is to regulate digestion by providing precisely the proper amount of flow of not only the substances already mentioned but of bile and pancreatic juice, as well. Bile, for example, is manufactured by the liver but concentrated and stored in the gallbladder. The gallbladder is nothing more than a cul-de-sac off the tube carrying bile from liver to duodenum, the common bile duct. Although a constant low flow of the clear yellow fluid enters the duodenum directly via the duct at all times, the gallbladder is activated to cause muscle in its wall to squeeze down and drive additional amounts from its storage reservoir into the duct and thence to the duodenum. It does this in response to the arrival in the duodenum of fatty foodstuffs, which require bile in order to be digested.

A description of the mechanism for contraction of the gallbladder will have a familiar ring, because it is analogous to so many other of the hormonal activities already discussed. The presence of fatty material in the upper intestine causes cells in its mucosa to secrete into the bloodstream a hormone called cholecystokinin (literally, "gallbladder motion-causer"), which not only stimulates contraction of the gallbladder but also results in secretion of enzymes from the pancreas. These enzymes reach the duodenum by a short tube called the pancreatic duct. Because the secretion of cholecystokinin requires food in the intestine to stimulate it, the food's presence in that location means that much of it has already left the stomach, removing the necessity for that organ to continue mashing so hard. Accordingly, one of cholecystokinin's actions is to inhibit gastric motion.

There are also other mechanisms to decrease the stomach's activity, including a sympathetic reflex set off in the upper intestine when acid reaches it; this reflex acts to decrease the amount of gastric juice being produced. By now in this book, these little clevernesses of the body's checks and balances may seem commonplace, but the fact that they are ubiquitous and omnipresent does not make them any less impressive in their implications for the animal economy.

From the foregoing, it is apparent why gastroenterologists are fond of thinking of gastric secretion as taking place in three phases. The first is called the cephalic (or brain) phase, in which the thought of food is the stimulus; the second is the gastric phase, when the presence of food in the stomach sets off the release of gastrin and the other components of gastric juice; the third is the intestinal phase, caused by food reaching the upper intestine, as just described.

The cephalic phase is of particular interest because it provides an example of the way in which the presence of a conscious awareness in a higher brain center causes a signal to be transmitted to a more primitive part of the brain, and finally to nerves that automatically (and with seeming autonomy) carry a message to an internal organ. In other words, the cephalic phase of gastric secretion illustrates a connection between the thinking mind and an unthinking organ. By following the pathway taken by that connection, it can be seen that sometimes what is thought to be thinking (the cerebral cortex) is thinking less than is thought, and what is thought to be unthinking (the stomach lining) seems to be much more under the influence of thoughts than most people might think.

To wit: The consciously aware cerebral cortex (the thinking part of the brain) responds to recognizable stimuli such as smells and the anticipation of a tasty morsel by automatically influencing nerve cells in the medulla (way down in the primitive place called the brain stem) to send messages along the part of the parasympathetic system called the vagus nerve, telling the stomach to secrete. The medulla can do this because it contains the cells in which the vagus nerve originates. Thus, mindless gastric secretion is more mindful of the thinking part of our brain than might be anticipated. As is so often the case, it is the autonomic nervous system that functions as the go-between or middleman. An even more overt example of this kind of thing was described in chapter 7, when the sexual act came under scrutiny.

On every level of our comprehension of the world within us and the world without, influential signals are constantly moving to and fro, bringing into the deepest parts of us messages from outside our bodies, while at the same time allowing even our cellular needs to be known, whether directly or indirectly, by the conscious mind. It is the integration of each separate part into the whole that constitutes the leading characteristic of the wisdom of the body.

Although we are acutely aware of the presence of the smells or thoughts that set off the cephalic phase of gastric secretion, the sequence proceeds without the necessity for us consciously to intervene—we seem to have no overt control over what occurs. And yet it is well known that appetites change with the experiences of the years; a single good or bad episode associated with some particular food may thereafter alter one's perception of whether it is delicious or disgusting—a kind of mental retraining takes place, sometimes planned, but usually arising of itself. This sort of adaptation to experience is characteristic of a whole range of responses throughout the various systems of the body.

The planned variety of retrainings are, in fact, the basis of certain forms of psychotherapy, wherein a new orientation is sought toward a stimulus that is causing difficulties for the patient because it sets off a string of unpleasant reactions, among which some may be as physical as palpitations or a rash. Based on experiences with thousands of these patients (not to mention the confirmable examples of such as the Indian mystics who can change their heart rate or blood pressure by an act of will), it is safe to say that there are circumstances, very few of which have yet been explored, in which it may be possible for deliberate conscious thought to affect autonomic and therefore perhaps even cellular responses. If this is true of *deliberate* thought, how much more must it be true of the constant stream of mental processes that ceaselessly churns through our minds, much of which lies in the realm of what may be called the subconscious? These, by reiteration and increasing reinforcement, become major themes in the inner life of the psyche. It is in this zone of grayness, in which the autonomic nervous system is the intermediary between thought and cells, that the transcendent powers of the human mind seem to provide the potentialities with which we have for at least a hundred millennia connected our deepest tissues with the world that surrounds us.

Incidentally, the medulla also contains the cells in which the nerves to the salivary glands have their origin. This explains why Pavlov's dogs drooled when they heard the dinner bell, even though they were given no food. In their conscious minds, the sound of the bell had taken the place of the other anticipatory stimuli, so off went the signals down to the brain stem and onward to their final destinations. You can be sure that their eager stomachs were secreting just as enthusiastically as their mouths.

But canines are quite limited in their intellectual comprehension of signals and their ability to act on them with the responses of free will. Dogs cannot take a desire for food and make of it something by which they lift themselves beyond being mere servants to the instinctual need to survive. Though they possess a sensation of taste that in many ways exceeds ours, they are incapable of that other, refined sensibility which also goes by the name of taste, so well exploited by such discriminating palates as have been served by a spectrum of gustatory authorities from Anthelme Brillat-Savarin to Julia Child. Starting with the primitive instinct for food, *Homo sapiens* has, using methods both deliberate and unwitting, created an entire culture around the act of eating, much as it has done with so many other of the basic survival drives of multicellular organisms.

Back to fundamentals. The pancreas, which figured so prominently in the saga of Marge Hansen, plays a large role in the various parts of the digestive processes. It is actually two glands in one—it is both exocrine and endocrine, which means that it produces substances that are excreted into a duct that takes them directly to their point of action, and it also produces several hormones which enter the bloodstream to be carried to their point of action at a distance from the source of manufacture.

As an exocrine gland, the pancreas produces enzymes that aid in the digestion of fats, carbohydrates, proteins, and nucleic acids. The cells that make these enzymes form the lining of tiny tubes that are tributaries to increasingly larger passageways which finally merge into the pancreatic duct, leading into the duodenum. The cells are stimulated to secrete by the very first hormone ever to be identified as such, the secretin of Bayliss and Starling (see page 38), which enters the bloodstream from the mucous membrane of the duodenum as soon as the acid quality of chyme is recognized there, after having passed from the stomach through the pylorus. Prior to this, parasympathetic impulses have reached the pancreas during the cephalic and gastric phases of the stomach's activity, inducing it to secrete a fluid high in bicarbonate, thus neutralizing some of the acid lest it erode the intestinal lining.

The endocrine function of the pancreas is due to specific cells clustered in little islets throughout the gland, secreting three hormones, called insulin, glucagon, and somatostatin. To understand the function of these hormones it is necessary to introduce and describe a carbohydrate molecule called glycogen. Glycogen is the most important means by which carbohydrate is stored in the liver and, to a lesser extent, in voluntary muscle.

Carbohydrate molecules range in complexity from simple sugars called monosaccharides to much larger entities made up of a number of sugar units bound together, called polysaccharides. Glucose is a monosaccharide and glycogen is a polysaccharide. In addition to having carbohydrates as components of their structure, all cells use them as a source of immediate energy and as a way to store it. Because the immediate source is glucose, both the maintenance of dependable availability and stable levels of glucose in the blood are therefore absolutely essential for homeostasis and survival.

When the level of glucose in the bloodstream rises after a meal, the pancreas detects the change and responds by increasing its secretion of insulin. Insulin acts to facilitate passage of the glucose into certain cells that may urgently need it as a quick energy source, such as voluntary and cardiac muscle. Also, it stimulates the liver to use it to form glycogen, so that the energy source may be stored for later use. When called upon by glucagon, the liver converts its stored glycogen back into glucose and also manufactures glucose from amino acids. Because this latter process generates new glucose, it is called gluconeogenesis. In the overall picture, then, insulin decreases the amount of glucose circulating in the blood and glucagon increases it, in response to available supply and the moment-to-moment needs of the body. The secretion of the hormones will either increase or decrease as necessary to keep the blood level of glucose at its set point and to accommodate the ever-changing requirements of the tissues.

When glucose is broken down into carbon dioxide and water in the cell, it releases energy that is transferred to ATP so that it can be used in chemical reactions. Since glucose provides the wallop in ATP, it is therefore the body's primary source of cellular energy (see page 124). Accordingly, insulin and glucagon are two very important hormones. Disturbances and deficiencies of normal insulin function are the cause of so-called sugar diabetes (diabetes mellitus), which is usually due to defective activity of the pancreatic islet cells. The third hormone, somatostatin, is a substance with several functions, among them the inhibition of glucagon secretion.

In this and several previous chapters, reference has several times been made to one or another of the several functions of the liver. Already discussed have been the part it plays in glycogen metabolism and in the body's secretion of bile; its detoxification of many of the poisonous byproducts of the body's functioning; and its role in providing housing for some of the filtering and infection-fighting tissue of the reticuloendothelial system. In addition, this large multipurpose mass of various sorts of tissues synthesizes certain proteins, is active in the metabolism of fat, and rids the body of the excess nitrogen resulting from protein metabolism by making urea for excretion by the kidney. It is truly an organ of many parts.

And now on to the intestine. As important as the stomach is in starting the digestive process, its mucosa is not well suited to absorbing nutrients other than small amounts of glucose, a few salts, water, and alcohol. The real job of taking in the molecules resulting from digestion is done in the small intestine, a narrow tube of some twenty feet in length, all but the first ten inches of which lie loosely coiled in the abdominal cavity, suspended from the peritoneum behind by a transparent double fold of filmy tissue called the mesentery, in which run all of its arteries, veins, lymphatic vessels, and autonomic nerves.

Those first ten inches of small intestine comprise the duodenum, which is fixed to the back of the upper abdomen and looks somewhat like a C with a greatly elongated lower sweep. The C begins to the right of the vertebral column, arches downward, makes a left turn, and crosses over the midline before rising up just a bit to form the end of the letter. Here it is continuous with the small intestine's next portion, the jejunem. Snuggled within the C-shaped loop thus formed by the duodenum, the pancreas lies transversely across the abdomen (see page 17).

Before the semidigested food can get into the duodenum, it must pass through a narrowed channel formed by a dense thickening of the circular muscular layer at the outlet of the stomach. This powerful ring of muscle is called the pyloric sphincter, and its purpose is to close the end of the stomach tightly so that the process of mashing is maximally effective, and also to act as a valve preventing backward regurgitation once the digesting material has passed into the duodenum. Onrushing peristaltic waves push one after the other against the liquified pablum (now called chyme, although it is essentially vomit) resulting from the stomach's action. As the chyme accumulates near the sphincter, the muscular ring relaxes somewhat, allowing small squirts to go through. It not only allows their passage but actually helps them on their way by squeezing rhythmically and thus adding a little pumping of its own.

Having brought up *vomit* as a noun, it needs to be described as a verb. This sometimes lifesaving act results from a complex series of reflexes capable of being initiated by all manner of unpleasantnesses, such as the presence of some chemical irritant in the stomach, an overstuffed gut, or by sensory stimuli from the inner ear resulting from rapid positional changes, as in car or seasickness—or even by the sight of a bolus of well-chewed fettuccini Alfredo in a friend's gaping mouth. As every scheming third-grader knows, a little tickle to the back of the pharynx will accomplish the same thing, or at least produce enough retching to attract attention.

The stimuli resulting from such entities go to a vomiting center in the medulla of the brain, which sets off the entire group of reflex responses. The first is a sensation of nausea and a decrease in muscular movements of the stomach. Then the prospective vomitor takes a deep, sighing breath, which fills his lungs and lowers his diaphragm until it is pressing on the stomach. The epiglottis closes off the trachea, the soft palate moves up to obstruct the back entrance to the nasal cavity, and the poor victim of this concatenation of uncontrollable events involuntarily finds himself squeezing down his abdominal muscles. This is accompanied by a powerful wave of reverse peristalsis in the stomach and spasm of the upper intestine. The sudden increase in pressure within the abdominal cavity abets the peristaltic waves, forcefully driving the chyme up into the esophagus, whose sphincter has meantime obliged by relaxing.

It is apparent, then, that more of the gut than just the stomach takes part in the act of vomiting. Depending on the degree of stimulation, this can include the jejunem, which comprises 40 percent of the remainder of the small intestine, the rest being the ileum. The duodenum, jejunem, and ileum are a triumph of architectural design. Obviously, this can be said about every other organ of the body as well, but the humble gut is so rarely acclaimed that it has consistently been undervalued in the pantheon of nature's masterpieces. Being one of its most intimate admirers, I feel called upon to redress this grievous wrong. A consideration of the structure of its wall alone is convincing evidence of the small intestine's extraordinary fitness for its role in the economy of the body.

During those rare occasions on which nonmedical observers are present in the operating room when it is necessary to open the small intestine (they are usually such as cameramen, neophyte nursing students, and an occasional equipment technician), they are likely to express surprise that its inner lining is not smooth. Instead, it is thrown into a succession of thousands upon thousands of delicate and quite beautiful circular folds (called *plicae circulares*, which is simply the Latin for "circular folds"), as though the inside of the gut were lined with a tremendously long series of fine, mucosa-covered wreaths, or—for those of a less aesthetic inclination—perhaps feathery-soft pink rubber washers. The mucosa on the surface of the wreaths has a velvety appearance, the reason for which becomes obvious when the tissue is put under a low-power microscope lens: It is covered with millions of fingerlike projections called villi. Running up the center of each villus is a capillary, a venule, and a lacteal, the tiny lymph channel that absorbs nutrients. Because of the presence of the countless plicae circulares and the astronomical count of villi on each, the surface absorbing area of the intestine is enormously increased over what it would be were the inner lining of the gut monotonously smooth, like the inner lining of the cheek.

Cells in the mucosa of a villus secrete mucus and a variety of enzymes that split protein, fat, and carbohydrate into simpler molecules like glucose and amino acids, which can then be absorbed. The chemical and mechanical stimuli of chyme, as well as the distension of the intestinal wall, activate the local reflexes and the parasympathetic nervous system to regulate secretion and peristalsis.

As noted earlier, carbohydrate digestion begins in the mouth and protein digestion in the stomach, but digestion of fats starts in the duodenum. The breakdown of all three into absorbable molecules is completed in the jejunum and ileum. The molecules are absorbed into the villi along with water and such chemical substances as potassium, sodium, chloride, nitrate, bicarbonate, calcium, magnesium, and sulfate. The first five are absorbed easily, the final three much less so. The products of fat digestion go into the lacteals and those of carbohydrate and protein digestion go directly into the tiny blood vessels in the center of the villi. The intestinal venules empty into a large conduit called the portal vein, which brings the digested materials to the liver, where they undergo a variety of chemical transformations, resulting in products usable by cells all over the body. At the same time, the newly absorbed materials are filtered and sanitized by the liver's reticuloendothelial tissue. After the liver has had its way with them, the blood and its much purified content of changed nutrients leave via vessels called the hepatic veins, which lead into the inferior vena cava to go directly into the heart.

The whole intestinal mishmash of fluid, chyme, enzymes, and mucus is called *succus entericus*, from the Latin, meaning "intestinal juice." Peristalsis takes its time in moving it along, to allow maximum opportunity for digestion to be completed and absorption to take place. After about eight to ten hours, it finally reaches the terminal portion of the ileum, which empties into the right side of the colon at a point located in the lower lateral part of the abdomen. Perhaps it was with thoughts of his colon in mind that Friedrich Nietzsche wrote, "The abdomen is the reason why man does not take himself for a god." With its flatus and feces always at the ready, the colon would seem the basis of what the philosopher believed to be the ultimate anti-superman symbol.

This capacious and odorous cavern of intestine begins at a blind pouch called the cecum, from which the narrow wormlike appendix projects downward. The appendix has no known certain use, although there is some evidence that it may have some minor function in immunity, since it does contain a small amount of reticuloendothelial tissue. From here, the cecum rises up the side of the abdomen as the right colon until it is at the level of the lowermost margin of the liver, where it takes a ninety-degree turn to the left and crosses to the opposite side of the abdomen as the transverse colon. When it reaches a point just below the spleen (at the bottom of the rib cage in the left flank), it abruptly turns downward to become the descending colon. About a foot below this, it takes an S-shaped, or sigmoid, curve into the pelvis, where it becomes the wide, straight tubular pouch six inches long that is the rectum.

The rectum lies immediately in front of the sacrum, which is the lowest part of the vertebral column except for the little tailbone, or coccyx, beneath it. At the bottom end of the rectum, the gut narrows into a tight skin-lined tube about one to two inches long, called the anal canal, at whose very end is the anus. I find myself recalling one of the less estimable of the mnemonics by which medical students have for centuries been memorizing the arcana of anatomy. Although this particular anatomical circumstance hardly qualifies as an abstruseness difficult to remember, it has nevertheless entered the canon: "The *de*scending colon is the *ass*-ending colon."

The lowermost rectum and the anus pass through the center of three powerful sheets of muscle that act as a sphincter. The most internal of these sheets is composed of involuntary muscle, and the other two are voluntary. So acute is the sensation in this particular sphincter that it can tell the difference between feces and flatulence, and it chooses to let the latter pass while retaining the former for a more propitious time and place. The presence of the voluntary component explains why we can usually hold back the disquieting urgencies of both, even though the parasympathetic nerves stimulating the uppermost sheet are importuning us to expel them right there in the crowded elevator. In this effort, of course, we sometimes enlist the aid of the buttock, or gluteal, muscles and hope no one notices the physical evidence of our discomfiture. This is one of those situations where brute force is required to circumvent the body's mindless demands.

By the time the succus entericus reaches the colon, all the nutrients have been absorbed into the bloodstream and lacteals, but a great deal of water is still left. To absorb it is the job of the colon. As the water is removed, the intestinal juice is gradually being dehydrated and formed into stool while moving toward the anus. The presence of stool, or feces, in the rectal pouch reflexly produces the urge to discharge it by defecating. If you have ever wondered why you so often feel the need to have a bowel movement directly after a large meal, wonder no more. A distended stomach is apt to set off the so-called gastrocolic reflex, manifested as a strong urge to excuse oneself during brandy and cigars.

The act of defecating is actually the result of another reflex, but it is one of those comparatively few that we can sometimes set off and sometimes deter with conscious will. When we bear down, a strong peristaltic wave occurs in the rectum, the sphincters relax, and the fecal material, at its best a cylindrical mass of undigested or indigestible food, mucus, bacteria, and water, is forced out. Its brown color is due to pigments in the bile, and its foul stench is a contribution of the breakdown products of bacteria.

Inflammation of the intestine can instigate a reflex similar to the gastrocolic, and in addition it may cause a marked increase in intestinal peristalsis. The overall result may be that the succus entericus is moved along so quickly that diarrhea ensues. Among the most common causes of acute intestinal inflammation is bacterial action. Usually, this requires the presence of organisms that do not ordinarily live in the gut, but sometimes even the regular denizens can be the source of the difficulty. Unlike the stomach and small intestine, the colon harbors many millions of certain varieties of bacteria. These are normal inhabitants in this location, and they aid in the breakdown of certain undigested bits. The price we pay for this is that their action contributes to the colonic gas, which is in part swallowed air and part the result of everything that has been happening along the entire length of gut.

A few years ago, I was involved in the care of a young woman whose normal intestinal bacteria turned on her in such a bizarre and overwhelming way that they all but took her life. Except for the long scar of a surgical incision and a barely perceptible shuffle in her gait, Hope Kuziel has been left with no physical traces to mar the perfect image she presents nowadays of vibrant good health. It comes as no surprise, for example, to learn that she placed third in a county beauty pageant only a year before the near-catastrophe I am about to describe. Even forewarned with the knowledge that since the age of eight she has required twice-daily insulin injections to control her diabetes, it is a great deal easier to think of her as a wholesome, smiling beauty contestant than it is to imagine her mottled and swollen, in a delirium of fever and near death, being rapidly wheeled toward an operating room one spring afternoon six years ago. She had been assessed a Class 5 risk for anesthesia, in the opinion of every physician who saw her. To this day, she's not sure whether to credit her survival to the flabbergasting marvels of modern scientific medicine or the spiritual intervention of her long-dead father, and there are those of her doctors who still sometimes wonder.

The American Society of Anesthesiologists describes a person in Class 5 as "a moribund patient who is not expected to survive without operation." No one, doctor or otherwise, seeing Hope Kuziel just before those preoperative moments had reason to dispute that description, and most observers would have projected her survival period to be hours rather than days. I was her surgeon, and I've now had plenty of time to think about it. I'm absolutely convinced that I have never taken a sicker patient to the operating room, even if I include in my recollections those few who didn't leave it alive.

The reason for Hope's survival is no easier to pin down than is the origin of her sudden catastrophic illness. Although our clinical team was later able to trace the details of the process that made her so sick, we're still puzzled by the why of it. We know the culprit but have no idea how it managed to get as far as it did as fast as it did. Even the instructions we gave Hope after her recovery were based on guesswork: She was told never to eat pork again. The injunction had no scientific basis. In fact, it was nothing more than a kind of clinical rabbit's foot that none of us was willing to throw away, probably because it was the only piece of advice we could think of. Hope never shared our concerns; I recently discovered that she eats pork whenever she can.

Actually, the amount of pig meat consumed by our patient in the days before the onset of her illness was not enough to indict it. She'd had a Chinese dinner about forty hours before her first symptoms, and it included pork fried rice and spareribs. Other than that, she has no recollection of having eaten anything at all different from her usual fare.

Hope's medical saga began in May, on the Monday of final-exam week at the state university where she was completing her sophomore year as an education major. She had just taken the performance exam for a dance course in which she'd been enrolled that semester. It was about three o'clock in the afternoon, and she was walking across the campus, feeling pleased with how well she'd done and thinking about the coming series of finals.

> *All of a sudden, I was on the ground—I couldn't imagine how I got there. I got up quickly because, of course, there were a thousand people around the campus, and I was thinking, Dear God, I hope no one saw me fall. I looked around and there were no stones, no sticks, no cracks in the sidewalk—there was absolutely nothing that I could possibly have tripped on. It was like my legs gave out,*

and I thought it must have been because I'd just danced for an hour. My roommates said later that I was such an idiot I must have tripped over my own two feet.

On the following morning Hope awoke feeling sick.

I was vomiting, I was running to the bathroom with diarrhea, and I was sweating. I thought, Oh great, I caught some kind of a grippe, some kind of a flu. I went back to bed, but I kept going in and out of it. Finally, my roommates began to get worried, because there had been several times in the past when I got dehydrated and had to go to the hospital because my diabetes went out of control. But when I tested my sugar, it wasn't any higher than usual.

Finally, the girls started to get scared. They called my mother at work, and she took me home. I drank lots of diet ginger ale the rest of the day and used suppositories to stop the vomiting. That whole night I was dizzy and throwing up, and drinking water and vomiting again. By early the next morning, my abdomen was aching and I couldn't feel my arms and legs. I tried flapping my arms around and I still couldn't feel them. No matter how weak I'd been in the past, nothing like that had ever happened before. I was in hysterics—it was like a nightmare.

Joan Kuziel, Hope's mother, has been teaching elementary school for more than twenty years. After her husband, Tony, died suddenly of a coronary when her only child was ten, Joan became not only the small family's sole breadwinner but Hope's entire support system. After Hope's juvenile diabetes was diagnosed in 1978, Joan took it upon herself to learn all she could about the disease and to become something of an expert in the various ways it manifested itself in Hope. Her job wasn't always easy.

Like many diabetic kids, Hope had a way of breaking the rules, and it sometimes took all of Joan's accumulated wisdom to extricate her child from the consequences. Occasionally, her efforts failed, and it would then be necessary to rush the dehydrated girl down to Yale–New Haven Hospital's emergency room. Over the years, actual admission had been necessary seven times, always to treat acidosis, the rapid buildup of acid metabolic products in the blood of diabetics, which can lead to air hunger, coma, and finally, if not reversed, death. The last admission had been only six weeks earlier.

Diabetes is a Greek word meaning "siphon." People afflicted with its common form, diabetes mellitus, pass urine in such large quantities that it seems to be siphoning out of them. They do this because of a deficiency of insulin function, which results in a lessening of the passage of glucose into the cells and decreased glycogen formation. These two factors combine to raise the level of the sugar circulating in the blood, a condition that worsens during situations of acute stress, as in infection. When the sugar level becomes sufficiently high, the kidneys begin to excrete it into the urine, thus raising the urine's osmotic pressure. The taste of sugar in the urine was well known long before the physicians of ancient Rome added *mellitus* to the disease's name, Latin for "sweetened with honey." The heightened osmotic pressure pulls more water into the urine, thereby increasing the output of the kidneys. The consequent water loss is the reason for the dehydration.

Knowledgeable as Joan was about the way Hope's diabetes behaved, on that May morning she found herself facing an entirely new symptom. "When she woke me at about five-thirty and said she couldn't feel her arms or legs, I knew I had to get her down to the hospital right away. While I was helping her to the car, she told me she couldn't even feel her feet touching the ground."

There was no prolonged wait in the emergency room when the Kuziels signed in at 6:19. As Hope recently told me, "Generally, you can come in holding your head in your hands and they tell you to wait. But when you're a diabetic, they take you right away."

Blood samples were drawn and intravenous fluids were started without delay. About an hour and a half after her arrival, Hope was told that her test results seemed reasonably satisfactory. But she couldn't be reassured, and she began to feel herself become increasingly panicky. Soon she was shouting.

> *Nothing felt right. At that point, the doctors and nurses were changing shifts, and no one was paying any attention to this screaming person. I was yelling, "Won't someone listen to me? There's something wrong!" My abdomen really, really hurt, like it was a tight, tight muscle spasm and everything was all squeezed*

together. That frightened me, but I tried to blame it on the twenty-four hours of vomiting. But what really scared me was that I had no body perception. I didn't feel like I was there. It was that same spacy feeling I've had when I've had a tooth filled and been given gas. It was like I had no body at all.

My mother kept talking to me all the time, trying to calm me down because I was yelling and thrashing around. And then she asked me if I knew I was going to the bathroom—I didn't. Then I heard her yell, "My God, it's blood!" and she began calling out, "Nurse. Nurse!" The nurse came right away, and after that my only perception was dribs and drabs of the faces of the doctors and nurses around me.

In fact, Hope's blood tests had not been normal at all. The most striking abnormalities were a markedly elevated white blood cell count of 28,500 per cubic millimeter (the normal level is about 5,000 to 10,000) and what is called a shift to the left, which refers to a proliferation of mature and immature granulocytes, which increase in number when an acute infection must be fought off. At 654 milligrams per deciliter, the blood sugar was elevated to some seven times its normal level, and a moderate degree of acidosis was present. To combat infection, the cells need more energy, but they are deprived of it by the deficiency of insulin. Not being able to get energy from glucose, they try to compensate by breaking down fats and proteins. Acids called ketones are a product of fat breakdown, and their presence rapidly leads to acidosis, which is lethal if not expeditiously corrected.

The entire picture was characteristic of the abnormalities that rapidly appear when a diabetic develops serious infection. Once the proper cultures had been taken, the emergency room physicians started Hope on several intravenous antibiotics, having earlier begun treatment with insulin.

The most unusual aspect of the blood studies was the extremely high white cell count, which was approximately twice what might be expected in the ordinary kind of infection. But far more worrisome than the laboratory results was Hope's appearance. She was throwing herself around the gurney and shouting for help—evidently not fully conscious and seemingly unaware that she had just passed half a pint of bloody stool.

The skin of her entire body had become mottled, with great purple-gray blotches appearing everywhere, separated from one another by small patches of stark whiteness. Her body temperature was a full degree below normal and her blood pressure was beginning to fall.

The sequence of events added up to the clinical picture of sepsis, or septic shock, a massive bloodstream infection that leads rapidly to inadequacy of the circulation, often followed by organ failure and then death.

On the presumption that the bloody stool and abdominal pain might be clues to finding the infected site from which bacteria were being hurled into the circulation, the resident physicians sought consultation with Suzanne Lagarde, a gastroenterologist on the hospital staff. When they phoned the patient's physician, Murray Brodoff, to tell him their plan, he said he would also contact a surgeon, on the chance that some remediable intra-abdominal event might be the cause of his patient's strange symptoms. I was draping a middle-aged man for a hernia repair when Brodoff's message reached me. Because I was decked out in sterile regalia, the nurse held the phone to my ear, and Brodoff described what he had been told. I was committed to the operation about to begin, so I had the nurse page the resident on my surgical service, and I delayed the operation just long enough to ask him to go directly to the intensive care unit, to which Hope had by then been transferred.

Sue Lagarde is a thin, bespectacled woman in her late thirties whose stylish good taste in clothes seems pleasantly incongruous with her studious face. She is a skilled clinician, and so enthusiastic about her work that she approaches diagnosis with a certain cheerful ebullience, manifested most directly by a rapid verbal delivery in which the words tumble out so closely on one another that she seems always on the verge of stuttering. When Lagarde examined Hope, she saw a young woman with what she called a hysterical personality throwing herself around the intensive care unit bed, complaining loudly and sometimes incoherently of diffuse body pain. The girl's skin was cold and broken out with the purplish blotchiness doctors call livedo reticularis. Although she shouted bitterly and above all about belly pain, there was no abdominal tenderness and only a mild degree of distension at the time of Lagarde's examination.

Of all the confusing, indeterminate findings, the most disquieting was the acidosis—it was worsening in spite of vigorous treatment. Whatever the obscure nature of the disease process might prove to be, it was obvious to Lagarde that she was dealing with a desperately ill young woman whose condition was deteriorating rapidly. The situation was not only dire; it defied diagnosis. Lagarde recommended what she called a fishing expedition, including a CT scan and a neurological consultation to help point out the proper diagnostic direction. Unless some sense could be made of Hope's bewildering set of symptoms, she would soon reach a point beyond retrieval.

By that time, I was completing the hernia repair. As soon as the dressing was applied, I paged the surgical resident. He responded in less than a minute and assured me that the young woman had "no surgical problem." His examination of her abdomen, he said, was without findings that might suggest the need for an operation. "She's a hysterical kid," he said, "and whatever she's got is medical, not surgical. You don't have to see her." He went on to describe Hope's bizarre behavior, her blotchy skin, and the laboratory findings. It was clear that he was a bit irritated at being asked to consult on a patient who, in his opinion, so obviously didn't need to be seen by a surgeon.

I stopped by the waiting room to have a few words with my patient's wife, then headed up to the medical intensive care unit (or the MICU, as such places are acronymically and universally called by those who work in them). One of the nurses quickly briefed me on Hope's condition; though it had already been very bad when Lagarde examined her a short time earlier, it was now worsening rapidly. Despite the attempted clinical detachment of the nurse's description, it was obvious that she was upset, even distraught. The best intensive care nurses never do become inured to the daily tragedies they witness, and helplessness in the face of imminent catastrophe, especially when it involves a patient not much younger than oneself, is unbearable even for the most stoic of professional personnel.

After our brief discussion, I sat down with Hope's chart to run a quick eye over the lab reports and the previous medical and nursing notes. As I scanned the three pages of neurology consultation, my eye caught the word *hysterical*, this time used to explain the arm and leg symptoms, which fit into no clinical pattern that made sense. The overall impression of the neurologist was that the patient's symptoms were the effects of diabetic acidosis. His final recommendation was: "Suggest close clinical observation. Consider L-S [lumbosacral] CT if symptoms persist."

On the next page, I found the surgical resident's note, describing his findings. At the end, under "Impression and Plan," he had written, "Benign abdomen in setting of acute neurological event and GI bleeding suggest vasculitis [underlining his] though sepsis possible . . . possibly meningococcus. No evidence for acute abdominal process." There was a tone of finality in the note. Even the customary concluding words of the surgical consultation, "Will follow," were absent. In their place was "Will speak with Dr. Nuland."

I stood for a moment at the entrance to the glass-enclosed cubicle where Hope lay attached to electronic monitors, a nasal oxygen line, and a tangle of plastic intravenous tubing. As I observed her from the foot of the bed, I asked the nurse to pull the sheets away so that I might look at the patient's entire body from that perspective. Even to me, a case-hardened veteran of other people's afflictions, the exposed sight was harrowing. The mottled object on the bed looked like a bloated corpse somehow preternaturally animated by the terror of yielding to eternal stillness. Its thrusting chest was straining up and down like a perverse bellows sucking air into itself, while the head and all four extremities were flinging about in a frenzy of attempted escape. In the glare of the brilliant MICU illumination, the skin looked almost eerie. Although I had been told of the livedo reticularis, I was unprepared to see the depth or extent of the large violaceous bursts, especially as they were so harshly revealed by the many foot-candles of piercing light. The pattern of blotch and pallor involved every visible inch of body and was much deeper in its purplishness than I had ever encountered, except on the freshly dead. The thing's legs were quite swollen from the knees downward, and even its face had become puffy. Almost paradoxically, the swelling of the lids made the open, frightened eyes seem to bulge—very likely, the tissues behind them were also swollen.

The abdomen was so distended that it partially obscured my view of the heaving rib cage. In answer to my question, the nurse said that the abdominal girth had reached its grossly protuberant size over the previous two hours. When I stepped to the bedside and tapped on the belly in the diagnostic maneuver called percussion, the amphoric boomlet of resonance that filled the small cubicle had the pitch that might be produced by a felt hammer hitting a kettledrum. With good reason, clinicians call such a note "tympanitic." Hope's tympanitic abdomen told me that her intestines were blown up with gas, and the absence of gurgles when I listened through my stethoscope meant there was no peristalsis, no rhythmic contractions that normally push along the intestine's contents. This is the kind of thing that happens when the gut has been subjected to some severe trauma, whether chemical or physical. The local nerve circuits in the intestinal wall are the primary source of the peristaltic motions, and they shut down under conditions of assault. In addition, the sympathetic system harkens to the alarm by further inhibiting the intestinal muscle, and the usual stimulating effect of the parasympathetic stimulation disappears. The result is a flaccid, motionless gut.

When I pressed down, as gently as I could, into the abdominal surface, the grimace on Hope's puffy face let me know I was hurting her. She had stopped speaking some time earlier, but her bulging, uncomprehending eyes stared fearfully at me.

When an abdomen is expanded by a large volume of intestinal gas in a short period of time, it rises like an overyeasted loaf of bread. Years ago, clinicians used the word *meteorism* to refer to this rapid belly ballooning, which is encountered only in certain unusual circumstances. The word seems to be archaic these days—it is not to be found in my 1974 edition of *Dorland's Medical Dictionary*, although, curiously, it does appear in my much more recent *Webster's Unabridged*. Perhaps it has been preserved for literary use rather than clinical. In any event, I haven't met a medical student in at least two decades who knows what it means.

What it means, almost always, is a belly with dead bowel in it, in which the nonliving part has lost its resilience and the remainder has shut down its muscular activity so completely that it no longer retains any power to resist being blown up like a balloon. In my clinical experience, almost no other acute abdominal disease will raise the white count as high as the intestine does when it is in the process of dying. A drastically risen white count in a patient with a drastically risen belly is a surgical call to action. Unless something is done quickly, the patient will not survive.

The presence of dead bowel furnished a logical explanation for Hope's sepsis and also explained why all the vigorous measures being applied were not resulting in any improvement in her acidosis. As long as a major source of infection remains untreated, there is no way to stop the process of decline. Obviously, Hope needed an operation—right away.

As I was quickly writing my consultation note, Mike Bennick walked hurriedly into the MICU. Mike is an intense, fast-paced young gastroenterologist who had trained with Sue Lagarde and was now her associate in practice. He hadn't seen Hope before, but Lagarde had described her condition to him. He instantly recognized how much had changed since his partner's examination a few hours earlier and concurred that an immediate operation was mandatory. In that tone of euphemistic detachment that even the most caring doctors use in hospital charts, he wrote, "Emergent surgical intervention holds only valid option for this patient, whose risks for demise are severe."

When he had completed his note, Bennick and I went out to speak to our patient's mother. We found Joan Kuziel standing with her two aunts and an uncle just outside the door of the MICU. Bennick had recently treated Joan's mother, and he knew that she would have confidence in his recommendations. He also knew that this very forthright woman, frightened though she was, would be impatient with explanations redolent of chart jargon, or with any hint of evasiveness. As I would increasingly come to appreciate during the next few weeks, she wanted the truth flat out, and Bennick now gave it to her as directly as possible, softening the harshness of his message by the gentle tone in which he delivered it: "Joan, Hope is dying, and we don't know why. We have to look inside—it's her only chance."

During the few seconds it took Bennick to speak and then to introduce me, I looked hard at my new patient's mother, trying to evaluate how she might respond to the details of what I would now have to spell out for her. Even when optimism is impossible, some measure of hope must be found, and it must be transmitted to those who will wait. In desperate circumstances, a surgeon speaking to a family facing the imminence of loss can usually point out that he has seen patients survive even though they were sicker than this one, but I couldn't say that to Joan with any honesty. In the atmosphere of futility that surrounded Hope's rapid decline, and the accumulating evidence of clinical helplessness, what was needed just at that point was some sense of equanimity, and perhaps even of control. I have children Hope's age, and I knew what Joan, without saying it, was expecting of me. If I could do nothing else, I would at least cloak myself in the aura of calm assurance that is the surgeon's armor against impending calamity.

Joan is a large woman, not particularly tall, but roundly and firmly heavy. Even when she is distressed, there is stolidity and determination about her. She listened to me carefully, and her face revealed nothing. She kept her gaze fixed on me, and when she occasionally blinked, it was done very slowly, as though she was momentarily closing her eyes to keep her thoughts from being observed. She seemed by force of will to be separating herself from anxiety in order to focus her mind's entire attention on each successive detail of what she was being told. In a way, each of those long blinks closed a distinct file on a package of newly processed information and sealed it into her hidden mental store. She never looked away, even when I concluded by telling her that the operation would kill her daughter if we were wrong—if no source of sepsis was found in Hope's abdominal cavity. When she had heard me out, she simply nodded, and the slight downward motion of her head punctuated my final word with a full stop. Then she said, "Please operate right away."

I called the OR and asked for the next available room. Within minutes, an anesthesiologist was at Hope's bedside, trying to determine whether she was already too far gone to tolerate his gases and drugs. The last paragraph of his scrawled consultation note summarized the pessimism we all felt: "Class 5. Critically ill, undergoing resuscitation—insulin, fluids, oxygen. Plan rapid sequence intubation. Patient has poor prognosis—heroic measure to attempt to save life."

Fortunately, one of the hospital's eighteen ORs was about to open up, and the nurses quickly got it ready for us. Hope was having few lucid moments by then, but she clearly remembers the brief period when she was in the holding area just before being wheeled in for the surgery. She was still thrashing about and trying to find some comfortable position on the gurney.

> *I wanted to be on my side because I thought that would relieve some of the pain. I was thinking, I'm going to die—I'm twenty years old, and I'm going to die. A priest came and was praying. He was making the sign of the cross, and I thought, Oh my God, this is the last rites—that's a sure sign that you're on your way, you know. I began to say that to my mother, and she was crying—so were my uncle and my aunts, and they were trying to tell me I'd be okay, even though they were crying.*
>
> *I believe in the power of God, and I've always had some kind of relationship with my father, even though he's not here on*

Earth—I feel his presence all the time, and I know when he's there. My uncle Ray passed away when I was five. I've always believed my father and my uncle Ray are in heaven. I've also always believed that someday, when I die, I'll go to them, just like I believe that the smile on my father's face when we found him dead was for my mother—I believe his parents greeted him in heaven. To me, that meant there's something good out there, and that his parents came to him. So I was lying there, and I felt like my whole body was being pulled. You know, when you vacuum a rug and you put your hand over the open hose to be sure the suction's working, and you feel that pull—it felt like my whole body was being pulled forward. I thought, Here I come; this is it—I'm dying.

And then I saw my father and my uncle. They were just standing there, and I was thinking, Okay, God, I'm dead. There weren't lights—I think you have to go all the way to get the lights. Well, I really believe that either they came to me or I came to them, and my mother tells me I was saying, "Tony, Ray!" Of course, I never called my father Tony and I never called my uncle Ray when they were alive, which makes me think things must be different in the afterlife. But I did say, "Tony, Ray," and my father put up his hand in front of him, and he said, "No, not yet." And I sat back. And then I looked up at my mother and I said, "I'm going to live," and those were the last words I said.

True to their plan, the anesthesia team got Hope to sleep very rapidly. With the surgical resident and a medical student assisting me, I made a long up-and-down incision in the midline of Hope's very distended abdomen. As I opened the innermost layer, the peritoneum, a gush of malodorous yellowish fluid poured out onto the drapes. When we had finished sucking it into several large trap bottles, the nurse told us it amounted to some six pints. With Hope positioned on her back, the gas-filled gut had been floating on top of the fluid, explaining the drumlike resonance produced by percussion.

We inspected the small bowel. Although most of it was alive, there was a length of about fifteen inches just beyond the point where the duodenum merges into the jejunem that was either dead or barely viable. It was suffused with a dusky bluish hue and was completely without peristalsis, even when I tried to stimulate it into some kind of action. The discoloration gradually faded out at the upper and lower margins of the involved segment, so that there was no definite line of demarcation between healthy and sick tissue. The vessels entering the darkened piece looked normal, and the arteries pulsated vibrantly. When the electronic listening device called the Doppler was applied, we heard the healthy whooshing sound of good circulation. And yet the bowel looked asphyxiated.

I explored every portion of the abdominal cavity, seeking an instigating factor for the imminent intestinal gangrene, but when I had concluded my probing and peering, I knew no more than I had at the outset. No obvious cause revealed itself that might explain the rapid death of an otherwise-normal-appearing length of intestine in a youthful, pristine-looking abdominal cavity. The gut's blood supply appeared perfect, there were no adhesions or similar fibrous bands that might have pinched off the involved segment, and the bowel wall seemed free of inherent pathology—nevertheless, it was near death. My puzzlement is summarized in a sentence of the operative note I dictated shortly after the conclusion of the surgery: "It is very difficult to know the cause of this ischemic [lack of blood] pattern, which is a form that no member of the operating team has seen before."

There is only so much time to cogitate when the belly of a failing patient is wide open and begging that some action be taken. I fired a surgical stapler across the intestine an inch above and then an inch below the dying segment, divided its blood supply, removed the specimen, and handed it off to the pathology resident, whom I had summoned to the OR on the slim chance that he could add something of value. He looked at the piece of gut, made a few cuts into it, and pronounced himself as stymied as we were.

Nothing was left but to reestablish the continuity of Hope's digestive tract. In a brief series of steps, the surgical resident and I reconstructed the gut, again with staples. The entire sequence of removing and restoring took less than fifteen minutes. Before stapling came into common use about a dozen years ago, this part of the procedure had to be done by hand, and it would have taken at least three times as long. I still prefer old-fashioned manual cutting and stitching, not only because I relied on it for two decades before the current mechanical era but for the simple reason that I love the way delicate steel instruments feel between my fingers. I find the technical sequences of cutting, suturing, and tying to be such aesthetically pleasing exercises that I've been loath to abandon them. Nonetheless, stapling provides the same result and is much faster. Hope's precarious condition required speed as much as it did technical precision, and in such a situation, aesthetics must yield to expeditiousness.

As soon as I was satisfied that I had made a good reconstruction of the gut and its blood supply, I poured at least ten quarts of warmed antibiotic-laced saline into my patient's gaping abdomen in order to rinse out as much bacterial and other debris as possible. We sucked it clean and ascertained that we had stopped any oozing of blood. Then we removed all sponges and instruments and began to close. I passed a heavy polypropylene stitch through all layers under the skin of the topmost part of the incision, then whipped it quickly all the way down the length of the wound until I reached the bottom. I stepped back from the table, and the resident placed a row of some thirty staples into the skin.

I went out to the waiting room to tell Joan that the operation had gone well, at least from a technical point of view. Hope was still septic and not much further from death than she had been when we wheeled her into the OR suite, but at least the source of her infection was removed, and she now stood some chance of recovery. When I was through speaking, Joan asked me the obvious question, and I had no answer for it. "No," I said, "we have no idea why this happened to her intestine. Maybe the pathologist will be able to tell us after he puts it under the microscope."

The specimen of jejunem I had handed to the pathology resident had the appearance of a segment of organ that had lost its blood supply, and yet I knew that the flow into it was normal right up to the very wall of the gut. I expected the explanation to be found in the microscopic vessels that traverse the bowel wall. For reasons yet obscure (but in some way related to Hope's diabetes), the tiny arteries, I supposed, must have become acutely occluded by an inflammation called arteritis, or vasculitis. The diagnosis provided a neat, all-inclusive package, because it would also explain the livedo reticularis and neurological symptoms. If the vessels to the skin and the central nervous system were similarly involved, everything we were seeing would make sense. If our patient could be proved to have some form of vasculitis, the surgical resident might still save some face, even though he had completely missed the diagnosis of dying bowel. Over the next twenty-four hours, vasculitis or one of its close nosological relatives became the fashionable diagnosis agreed upon by nearly every one of the doctors hovering around Hope's bedside in the MICU.

A consultation was obtained with the chairman of the dermatology department the next morning, because the livedo reticularis had not lessened as much as we might have liked. His list of possible diagnoses reads like a tabulation of esoterica, a group of diseases I've almost never encountered in any patient during my entire clinical career: livedo vasculitis; polyarteritis nodosa; Wegener's; cryoglobulinemia. He added a much more familiar entity at the end—collagen vascular disease—but obscured it beyond my recognition by parenthetically adding, "including Sneddon's syndrome," as though he expected anyone other than his own staff to know what he meant. Like the rest of us, the professor was looking for a rare disease to explain Hope's rare symptoms.

When it was finally revealed, the true diagnosis proved far more esoteric than even the dermatologist anticipated. While on rounds that afternoon, I received a phone call from Brian West, the pathologist in our hospital whose specialty is disease of the gastrointestinal tract. West had been at our institution for less than four years, yet in that relatively brief period he had quite transformed his department's capabilities in the area of his expertise. With his arrival from the medical school of Dublin's Trinity College, Yale GI pathology had become what West's colleagues call world-class.

I find the lilting ups and downs of West's soft County Cork brogue to be one of the most reassuring sounds I ever perceive in the jumbled cacophony that is the background noise of a modern university medical center. If hearing could somehow be transformed to vision, it might be said that West speaks with a gentle smile. The first time I saw his voice, it was over the telephone, and I distinctly remember visualizing not only the smile but also his blue eyes and the reddish beard that makes his still-young face appear craggy and wise. Soft though it may be, West's smiling voice speaks with persuasiveness and the authority that comes from an impressive ability to interpret the arcane microscopic clues left in the gut by obscure diseases.

"Do you have a minute to come over to the lab?" he asked, and there was a hint of expectant promise in his rising rhythm that told me it would be well worth my while to get there right away—the words weren't said with any sense of urgency, but more in the tone he might have used to invite me in for a pint of some long-awaited brew just arrived. "I want to show you what I've found in the specimen you sent me yesterday." Anticipating his discovery, I burst in before he could continue: "What do the microscopic vessels look like?" His answer surprised me. "The vessels are fine. What I think she has is enteritis necroticans—the thing they call pigbel."

It was an embarrassing moment, and I was grateful to be on the end of a phone in a far-distant part of our sprawling medical campus. I'm sure there must have been a confused look on my face while I paused for just a second, uncertain of how to reply. But speaking to West, I knew I'd stumble over my tongue if I feigned familiarity with these abstruse terms, so I confessed my ignorance. "Okay, Brian, what's that?" He gave me a brief explanation, but it wasn't until I had made my way over to the lab about ten minutes later that I really began to understand what he was talking about.

Peering down the twin barrels of West's microscope, I could see that almost the entire mucosa of the specimen was dead, although most of the main layers of encircling and longitudinal muscle were still within the definition of being viable. The most striking structures on the slide were the thousands upon thousands of rod-shaped bacteria forming a lengthy rank along the surface of the mucosa, palisaded like an irregular picket line of soldiers standing at attention. Their appearance and later lab tests showed that they were a genus of bacillus called *Clostridium*, closely related to the organisms that cause tetanus and gas gangrene. In fact, microscopic gas-filled spaces were visible within the layers of the bowel wall. The toxins produced by these particular microbes are capable of causing inflammation and necrosis (death and decay) of the intestinal wall—hence, the process is called enteritis necroticans. Hope's sepsis was caused by the clostridia, and all the bowel, neurological, and skin symptoms were the result of the bacterium and its toxins.

By this time, I knew that Hope had begun to exhibit various signs indicating destruction of the cells of some of her voluntary muscle tissue, a process called rhabdomyolysis. This, too, was attributable to the toxins. The combination of massive clostridial growth in her intestine, sepsis, rhabdomyolysis, and the resultant diabetic chaos was the explanation for the entire spectrum of destructive events that our patient had been experiencing. We could only hope that the removal of the nonviable bowel and the consequent diminution in the volume of bacterial load would enable our antibiotic and other treatments to reverse the process.

Clostridia in moderate numbers are normal inhabitants of the gut. Ordinarily, they live in harmony with other bowel organisms and with the various physiological substances with which they come into contact. But when some event occurs to disrupt the balance among the gut's organisms and chemicals, it is possible—though unusual—for the clostridia to become sufficiently numerous to be a source of danger. For those of us involved in Hope's care, the clinical challenge was to pull her through; but the intellectual challenge was now to figure out what had so upset the intestinal homeostasis that a massive overgrowth of clostridia occurred. For this, Brian West didn't have a definitive answer, but he had identified a disease model that so closely resembled Hope's that I was persuaded they were one and the same. Within a few days, and especially after West's diagnosis was confirmed by an expert in Southampton, England, the conclusion had become inescapable.

During our discussion that afternoon, West answered an important question without my having to ask it: How did he know that this huge increase in the population of clostridia had not occurred between the time I excised the gut during the previous afternoon and the time it was put into the germ-killing preservative, which was perhaps not until the next morning? This is, after all, what happens to a fresh corpse, in which these and other organisms rapidly increase in number because there is no longer an equilibrium with the life-sustaining mechanisms of the body. The result is that the corpse festers and rots, and in the process becomes bloated with gas, which is precisely what Hope's jejunem had been in the process of doing when it was discovered and removed.

"The pathology resident was in a hurry. He had promised to take his fiancée to dinner and the theater that evening, and he had to finish the day's work quickly in order to pick her up on time. As soon as he got back from the OR, he dropped the gut into formalin. If he hadn't done that, it might have putrefied overnight, and our finding all these clostridia would be meaningless. But this way, we can be sure that what we see here was the specimen's exact condition when you cut it out of the patient. This gut really does have all the earmarks of enteritis necroticans."

It's not easy to tell the mother of an attractive young woman that her daughter has a disease whose name is pidgin English for "pig belly," but no more likely diagnosis has appeared in the years since Hope's narrow escape. Except for the complications added by diabetes, the clinical course of Hope's disease and the microscopic appearance of the excised tissue are exactly the same as they are in the thousands of New Guinea tribespeople who have died of the same process. Acute pigbel is a major cause of premature death in the highlands of Papua New Guinea, with a mortality rate among those contracting the disease of almost 85 percent. Second only to respiratory disease, it is a leading killer of children in the area. Its prevalence is highest at times of the year when ceremonial pig feasting takes place, and the disease has been so carefully studied that it is possible to describe its evolution with considerable certainty.

The pig feast is an integral part of many of the ceremonials attached to various kinds of highland celebrations and sacrifices. The meal is always prepared in a traditional manner. After the animals are clubbed to death, their intestines are removed, washed, and wrapped in leaves. Alternating layers of filleted carcass, guts, fern fronds, banana leaves, and breadfruit are placed into earth pits along with sweet potatoes or bananas, chopped greens, and stones that have been preheated. Tier by tier, a mound of the ingredients is fashioned, with insulation provided by a final packing of pigs' quarters and flanks. After a large quantity of water is poured into the vapory, structured mass, more leaves and an outer layer of earth are added as a covering. In this way, a large steam oven is created whose internal mean temperature, when visiting health officers have tested it, has been 172°.

Not only does such a heating system result in inadequate cooking of the meat; it also provides plenty of opportunity for bacterial contamination. After all the festive cooking is completed, the banquet takes place under conditions that would throw a sanitation inspector into fits of apoplectic convulsion. Those conditions are ideal for the proliferation of dangerous organisms, particularly clostridia.

Ordinarily much of the clostridial toxin would be destroyed in the intestine by one of the protein-splitting pancreatic enzymes called trypsin, to which it is very sensitive. Unfortunately, sweet potatoes contain a chemical that inhibits the action of trypsin, and sweet potatoes are not only a major constituent of the pig feast but also a staple of the highland diet. Thus, the body's own attempt to defend itself is frustrated. The ingestion of large amounts of clostridia-rich meat accompanied by plentiful doses of trypsin inhibitor provides the perfect concoction to induce fulminating outbreaks of pigbel. The situation is made even more egregious by the common presence in local children of the intestinal roundworm *Ascaris lumbricoides*, a parasite that secretes its own brand of trypsin inhibitor, adding to what is already in the poisoned food.

As for the clinical aspects of the disease, they are precisely those that were exhibited some ten thousand miles away in New Haven, Connecticut, by Hope Kuziel, absent of course the components attributable to diabetes.

If massive overgrowth of clostridia is the cause of enteritis necroticans, the disease might be expected to occur in places other than Papua New Guinea, and without the necessity for ingesting a witches' brew quite so potent as the one cooked up during pig feasting. This is in fact the case. An epidemic disease of identical nature made its appearance in northern Germany shortly after World War II. It was called Darmbrand, or "fire bowels," and the doctors who studied it concluded that it was caused by unaccustomed intake of excessive amounts of protein-rich food by a malnourished population. Outbreaks of the same thing have been reported sporadically in several African countries, China, Bangladesh, the Solomon Islands, and at an evacuation site for Khmer children in Thailand.

A contributing factor in such areas is that chronically undernourished people do not ingest enough protein to have the amino acid building blocks that would enable them to make sufficient quantities of trypsin. When access to meat is suddenly provided, the meal may for one reason or another be contaminated, and then the levels of clostridial toxin become very high in the bodies of people with not enough trypsin to counteract it. This is consistent with an observation made by several of the first investigators of Darmbrand, which was that it seemed to have made its appearance when the diet was suddenly changed. The German patients were indeed chronically malnourished during the terminal phases of the war and for an extended period afterward. Those who become sick had very likely overeaten on occasions when meat, perhaps contaminated, was made available to them.

Although there have been scattered reports of individual patients dying of enteritis necroticans in prosperous Western countries, no real epidemics have occurred among populations living in areas where sanitation levels are high. But a few of the single cases are instructive because they illustrate some of the most dramatic aspects of the disease. In 1983, for example, a surgeon and a pathologist at England's Royal Liverpool Hospital described in the journal *Gut* (the British tend to be quite direct in their medical terminology, and this is the name of their most highly regarded gastroenterology journal) the case of a twenty-three-year-old photographer's model who walked into their hospital's emergency room at eight o'clock one Sunday morning complaining of abdominal pain and bloating. She told the staff doctor that she ordinarily tried to remain very thin but periodically went on an eating binge. Between midnight and four that morning, she had eaten the following: two pounds of kidney, one and a half pounds of poorly cooked liver, a half a pound of steak, two eggs, a half a pound of cheese, two large slices of bread, one whole cauliflower, one pound of mushrooms, two pounds of carrots, ten peaches, four pears, two apples, four bananas, two pounds of plums, and two pounds of grapes. She had then gone to sleep for a few hours and been awakened by the abdominal pain.

As the medical staff tried unsuccessfully to empty the young woman's stomach with a wide-bore tube, her condition rapidly deteriorated, and she had to be rushed to the operating room. When her abdomen was opened, it was seen that a section of the upper small bowel appeared to have lost its blood supply. As the surgical team watched, doubtless horrified, the area of ischemia gradually extended until it involved most of the length of the gut. Soon small gas bubbles became visible in the intestinal wall. Their patient died shortly afterward. Autopsy revealed massive clostridial overgrowth in the esophagus, stomach, and upper portion of the small intestine.

The microscopic appearance of the young model's digestive tract fit exactly the description of enteritis necroticans. As the authors of the report stated in their discussion, "The features of this case are strikingly similar to pigbel." They considered their patient to have been bulimic, and her chronic undernutrition to have been the cause of a presumed inadequate level of the trypsin that might have counteracted her sudden huge intake of protein, at least some of which was undercooked and possibly contaminated.

The unanswered question about Hope Kuziel is not whether she was the victim of enteritis necroticans—it seems almost certain that she was. What is not known is the underlying reason for the unchecked growth of clostridia in her intestine. The amount of pork she had eaten before her earliest symptoms was not excessive; she was not malnourished; she had not ingested any significant volume of food containing a trypsin inhibitor. The only possible clue is her diabetes. The disease is well known to be capable of causing a degree of immunodeficiency, which is one of the reasons diabetics are more infection-prone than the rest of us. But any indictment of a diabetic immunodeficiency is weakened by the absence of previous or subsequent evidence that Hope is particularly susceptible to abscesses, inflammations, or other manifestations of decreased resistance to bacteria. In searching for some underlying cause, we were left with the succinct summarizing comment entered in Hope's chart by Ann Camp, one of the interns who took such good care of her in the MICU. She called her patient's disease "interesting and mysterious."

Another of the few individual case reports of pigbel in the medical literature describes a young diabetic nurse in the Netherlands who, in 1984, died twenty-four hours after being admitted with characteristic symptoms. He had eaten an unspecified quantity of pork at a party the day before becoming sick, but no other guests were affected. As the paper's authors noted, "It is well known that diabetic patients have a lowered resistance to infections. It is therefore tempting to speculate that this may have been a contributing factor." Neither the Dutch doctors nor those of us who treated Hope can go any further than that.

Hope improved only transiently in the hours following her operation. At first, her acidosis responded to treatment and her blood pressure stabilized. She developed a sepsis-related condition of inadequate blood clotting, called disseminated intravascular coagulation, but it wasn't severe enough to cause serious trouble. Her subnormal temperature rose to 103, indicating a more appropriate response to infection. On the morning after surgery, we were cautiously hopeful, even though the mottling had decreased only slightly, and blood tests continued to show evidence of rhabdomyolysis. But it became increasingly difficult to maintain the balance of minerals and fluids in her body, and the generalized swelling of her tissues progressed as her kidneys began to fail. Dialysis was begun late that day, shortly after Brian West called me with the diagnosis.

The number of consultants was multiplying. By evening, Hope had been seen by specialists in infectious disease, dermatology, neurology, kidney disease, gastroenterology, surgery, and anesthesia, and every one of us continued to monitor her condition closely. Besides the minerals added to her intravenous solutions, she was receiving five medications, three of which were antibiotics. The intern's summary note takes up seven pages of closely written script, in which fourteen distinct problem areas are identified: sepsis, recent necrotic bowel, blood pressure, kidney failure, difficulty ventilating lungs, rhabdomyolysis, low calcium, low magnesium, the effect of shock on the liver, disseminated intravascular coagulation, diabetes, pain control, skin mottling, and nutrition. The white blood count, which had dropped to 16,000 in the immediate post-operative period, was beginning to rise again and had reached 21,000. By the next morning, the evidence of worsening sepsis was mounting. Almost certainly, the process in Hope's bowel was extending to the area that had appeared uninvolved two days earlier. When I made the decision to re-explore her, there was universal agreement. Her belly had begun to distend again.

By then, the kidney failure was rapidly worsening, complicating the low albumin levels in her blood caused by inadequate nutrition in spite of all attempts to maintain it. The combination of kidney failure and low osmotic pressure in the capillaries meant that Hope was becoming water-logged. Her tissues had retained so much fluid that her presickness weight of 125 had risen to 185—her entire body was bloated and swollen. It was decided to give her another dialysis treatment and then go directly to the operating room.

Again, I went out to speak to my patient's mother, and again I described the situation to her exactly as I saw it. Joan had not left the hospital since Hope's admission, sleeping in the MICU waiting room and eating in the cafeteria. When it was permitted, she would stand at her daughter's bedside, holding her hand and stroking her face—speaking quiet words of encouragement, even though Hope didn't know she was there. In her thoughtful, analytic way, she listened to every consultant and always came to the right conclusion. Joan had added everything up, and before I said a word, she knew what I had come to tell her. Our conversation was a reprise of the one we had had two days earlier, but the outlook was even worse. I had thought it impossible for Hope to have been any sicker than she was before the first operation, and yet the impossible had happened. Joan signed the consent form and took my hands in hers, just for a moment. This time, nothing needed to be said.

When the abdomen had been sterilized and draped, we reassembled on each side of Hope just as we had forty-eight hours earlier, but now there was a larger group around the head of the table. When a patient is very sick, anesthesiologists cluster about, trying to help one another as much as possible. During thirty years of a surgical career, it has been my not-quite-tongue-in-cheek observation that a patient's chance of survival is inversely proportional to the number of anesthesiologists required to get the operation under way; a figure of six or higher is a virtual guarantee of death. As I looked up at the assembled group, I counted six. I made a wry comment that they did not seem to appreciate, then went right to work.

Hope's abdomen was bulging so tightly that it strained against the stitches holding it together. As soon as they were removed, the contained fluid and gut exploded out onto the drapes. Quickly, the surgical resident and I put everything in some approximation of order and assessed the findings. Starting just at the point where we had placed the staples to restore continuity, a bit beyond the anatomic point called the duodenal-jejunal junction, the next eighteen inches of intestine looked exactly like the segment removed two days before. The preoperative impression was correct—the process of necrosis and clostridial overgrowth had extended and would require further excision. This time, Brian West had come to the OR himself. When I completed the removal of the specimen, I handed it directly to him. He scrutinized it silently for a few minutes, and then we spoke briefly about its appearance before he took it off to his lab for further testing.

I carried out the operation much as I had done before, except that this time I closed the wound with a series of individual stitches of heavy nylon, placed in such a way that they exerted a pulley effect—abdominal distension would bring the wound edges closer together. It's a time-consuming and not very pretty closure, but the strongest I know of, and I wasn't taking any chances with the possibility of a burst incision.

Afterward, Hope's improvement was more sustained. Within twenty-four hours, the rhabdomyolysis had decreased, and her kidneys began to function better—she went from nearly zero urinary production to the beginnings of what would soon be a reasonable output. Moreover, her clotting mechanism was satisfactory, the white count had dropped to

15,000, and the acidity of her blood was within the normal range. The evidence of sepsis was much less. The livedo reticularis had begun to recede, and within another day, it would be gone. Twenty-four hours after the surgery, Mike Bennick wrote in his note, "Improvement on all fronts." For the first time, the campaign was beginning to look winnable.

There was to be one more scare, a few days later. Hope's fever began to rise in a sequence of ascending spikes, and her white blood count went up to 33,000 by the fourth postoperative day. I thought the problem was an infection in one of her many intravenous lines, but I couldn't find any proof of it. I then began to worry that leaking intestinal contents might be contaminating the surgical wound, but I couldn't find any evidence of that, either. The most frightening concern was the possibility of yet another extension of the clostridial infestation, now into the remaining length of bowel. To evaluate this, a radioisotope scan was done, of a type designed to light up areas of infection or necrosis. I looked forward to having my fears laid to rest by the absence of any troubling findings, but when I reviewed the study with the superspecialist who had done it, I felt my knees weaken. The entire length of remaining small bowel showed an irregular pattern of involvement with the process I had seen in two successive specimens of excised bowel. My mind's eye could visualize the now-familiar carpet of clostridia lining Hope's gut.

Yet the study, scientifically precise as it was, seemed strangely inconsistent with what I kept finding each time I returned to examine Hope's abdomen, something I did over and over again. Despite the radiographic appearance of necrosis, her belly was flat and she didn't grimace or in any other way display evidence of pain when I pressed deeply inward. Through my stethoscope, I could hear the unmistakable and very comforting sounds of peristalsis. Most important, although she was still quite sick, Hope's general appearance was improving each day. My patient looked hardly at all like the deathly ill girl I had twice rushed to an operating room.

There was a great deal of pressure on me to open Hope's abdomen again. High-tech gadgetry is very impressive to young doctors, and has long since, in the hearts of many, usurped the revered place once reserved for the clinical skills of history-taking and physical examination. Except for the senior infectious-disease consultant, I was about twenty years older than any of Hope's panoply of caregivers, and I decided it was time to pull rank. I went upstairs to the MICU and wrote a long note in Hope's chart, the gist of which was expressed in two sentences: "Her abdomen is simply not the abdomen of a person with necrotic bowel. I do not think she should be operated on." Then I got up to tell Joan. She was standing at the entrance to Hope's cubicle, deep in conversation with Mike Bennick.

Joan recalls that morning's events very well. She had followed every step of the previous days' evaluations and knew that all the doctors were talking about another operation. She also knew that the operator (she now tells me this is what she and her family called me during the first hectic day in the hospital) seemed reluctant. The operator was now leading her and Bennick into the only empty cubicle in the MICU.

I have no recollection of what I said, but Joan remembers the exact words. Thinking back on them now, they sound unnecessarily magisterial, but perhaps that was what was needed at the time. Joan tells me I looked directly at her and said, "I'm going to make a command decision," and then to Bennick, I said, "Mike, come with me." What I do remember is that Bennick and I went to Hope's bedside and carefully reviewed the physical exam of her abdomen. Joan tells me we took a good long time to do it, but when we emerged from the cubicle, we were of one mind. Bennick had agreed that we should sit tight.

During my training years, I worked with a surgical resident who had been a star athlete at a large southern university, and he seemed to have a down-home Carolina bon mot for every clinical occasion. He would have said that this kind of decision-making was "playing guts football"—we absolutely had to be right. Actually, Bennick and I were taking less of a chance with Hope's life than some might have thought. She almost certainly would not have survived a third operation, in which I might have been forced to remove all of her remaining small bowel. I was betting that only the surface of the mucosa of her intestine was involved in the process of necrosis and that her present benign physical exam meant that she had already marshaled the forces necessary to overcome the infection and heal that layer.

Fortunately for all of us, that thesis proved to be correct. By the next day, Hope had improved sufficiently for her breathing tube to be disconnected from the respirator. Twenty-four hours later, in her first fully alert moment since admission, she opened her eyes. Within minutes, she saw her mother looking down at her, holding a large card printed with the alphabet, which Joan had made ready for just such use. Hope gestured for the card and, pointing very slowly to each letter, she spelled out "I have a history exam on Friday." Eleven days had passed since her first operation, and she had lost every moment of them.

> *My mother said, "No, honey, that was two weeks ago." I felt like, Oh my gosh—you know, total amazement. Then she asked, "How did you make it? You weren't supposed to." And I spelled out, "I got my strength from my daddy."*

The improvement continued, although very slowly. It took almost three more weeks in the MICU before Hope was ready to be transferred to an acute-care floor. She stayed there an additional two months and then moved to the hospital's rehabilitation unit. She had lost a great deal of weight and considerable muscle mass in her legs, but she knew that everything was recoverable with hard work. She was finally ready for discharge from the hospital eighteen weeks after she had entered it.

It would be another four months before Hope regained enough strength to return to college. Her mother considers her graduation two years later to have been the final step in a triumph not only of perseverance and luck but of Tony's protecting spirit, too. A few hours after Hope's return to wakefulness on that joyful morning six years ago, a rainbow appeared in the sky, even though there had been no rain. Joan remembers looking at it and being sure it was a good omen.

12 | MINING THE MIND: THE BRAIN AND HUMAN NATURE

I am not the first clinical physician who has undertaken to explain the workings of the body to the general reader. The most successful and effective of previous attempts was that by Dr. Logan Clendening, an articulate and self-assured professor of medicine at the University of Kansas, who in 1927 wrote a beautifully clear exposition of anatomy and physiology, *The Human Body*. But even then, when there was much (much, much, much) less to write about, and what was known was much (much, much, much) less complex, he figuratively blanched as he approached the challenge of explicating the details of the central nervous system. He put if off for 223 pages, and then he began his description by giving words to precisely my thoughts as I begin this intimidating assignment: "It is a hardy spirit which undertakes to give an account of the central nervous system in a book of this size."

Since those words were written, knowledge of all biological systems, and perhaps the nervous system most of all, has been exponentially advanced and complexly compounded. Today, the spirit who sallies forth to elucidate it must be even hardier than ever before. But no matter the complexity and an ever-expanding dependence upon intimate knowledge of other sciences, a single statement of Clendening's still holds, and always will. In two pithy sentences, he managed to summarize everything that matters: "The central nervous system is essentially a number of masses of nerve cells connected to each other by a complex set of fibres. The function of a nerve cell is to interpret the impulse brought by the nerve fibre, and to initiate new impulses to be sent out over other nerve fibres." There it is, in the proverbial nutshell.

Up to this point, much of what has been discussed might just as accurately be used in a description of the physiology of all other mammals. This can with equal validity be said about the forthcoming material concerning the structure and function of the human brain. But it is also true that much else of our brain's structure and function is unique to our species, and is the source of our humanness. The brain is therefore the ultimate key to that innermost sanctum of understanding, wherein we may find the secret of the human spirit.

It has already become clear that the central nervous system consists of a hierarchical and interconnected gradation from the lowermost centers in the spinal cord to those that are most sophisticated and most human, in the highest parts of the brain. This is true not only of our sensory and motor abilities but particularly of the so-called association areas, where messages are interpreted, integrated, and coordinated to allow the various communicating parts of the system to contribute to the outcome. To recapitulate some highlights of earlier pages:

All of the nervous system's work is carried out by neurons of many kinds, via signals transmitted through axons and dendrites and across synapses. Because of the extensive arborization of these two types of elongated processes, any given nerve cell can connect with many others. The synaptic connections of any single neuron in the brain, for example, range in number from a few hundred to ten thousand, averaging about a thousand. Because axons travel together in distinct groups of commonality that after a bit may divide up to go off in separate directions, these messages may be carried across prodigious distances and to several centers virtually at once. The result is that a single external stimulus may move via as primitive a pathway as a spinal reflex arc even while it is simultaneously being projected not only to several centers of automatic control but also up to those areas of the conscious brain wherein reside reason, judgment, and memory—that place that Hamlet called "the book and volume of my brain."

The book and volume, plus all other parts of brain, occupy the space of a quart container and weigh a total of only about three pounds. Though the three pounds represent a mere 2 percent of the body weight of a 150-pound person, the quartful of brain is so metabolically active that it uses 20 percent of the oxygen we take in through our lungs. To supply this much oxygen requires a very high flow of blood. Fully 15 percent of the blood propelled into the aorta with each contraction of the left ventricle is transported directly up to the brain. Not only does the brain demand a large proportion of the body's oxygen and blood but it also begins its life requiring an equivalent share, or even more, of its genes. Of the total of about 50,000 to 100,000 genes in *Homo sapiens*, some 30,000 code for one or another aspect of the brain. Clearly, a huge amount of genetic information is required to create the human brain; the 1 percent of DNA in which we differ from the chimpanzee represents an enormity of genetic potential.

The brain evolved to its present state in humans because its powers of thought equip it in unique ways that allow us to cope with the ever-present dangers of the environment. From nerve centers in primitive animals that were able to respond to stimuli only by performing reflex acts, the brain has sequentially developed higher and higher centers. What had been the rudimentary brain of lower forms eventually developed into a structure having an outermost covering, called the cerebral cortex, in which are located the powers of conscious integration, learning, and memory. It is the human cerebral cortex (*cortex* is Latin for "bark" or "shell")—specifically the highly skilled part called the neocortex (new cortex)—and its connections that distinguish humankind from beast.

Cerebrum is the Latin word for "brain," but in actual usage it refers to the highest of what may somewhat arbitrarily be considered, from the viewpoint of function, as the brain's three parts. The other two parts are the brain stem and a few other structures; and the limbic system.

The cerebrum is the center for skilled motor activities and the higher mental capacities that are associated with thought. It makes up 85 percent of the human brain's weight and has the consistency of tapioca. The lowermost of the three parts of the brain (the brain stem and a few other structures) controls such ongoing automatic bodily activities as circulation, respiration, and digestion, while the third, intermediate, area of function deals with emotions and instincts.

The cerebrum consists of two large masses called the right and left cerebral hemispheres, separated by a deep groove or fissure. In the bottom of the fissure the hemispheres are connected to each other across their lowermost portion by a bridge of nerve fibers called the corpus callosum. Each cerebral hemisphere is composed of a layer of cells, the cortex, covering a large collection of countless numbers of nerve fibers. Because tracts of nerve fibers cross over in the medulla and to a lesser extent in the spinal cord, a hemisphere deals with activities on the opposite side of the body, but the signals are coordinated by messages traveling back and forth through the corpus callosum. In more than 90 percent of people, the cerebral areas responsible for spoken and written language and intellectual functions are dominant on the left side; the left is the linguistic hemisphere. Those that govern nonverbal abilities such as abstract thought and visuospatial (and therefore artistic) skills are dominant on the right. The domination is far from complete, neither side being uninvolved in those abilities that predominate on the other.

The cerebral cortex, the outer rind of the brain, is composed of a layer of nerve cells about one-fifth of an inch thick, covering the large mass of underlying fibers. Because nerve cells in large concentration have a grayish color and fibers appear white, these two areas are commonly called the gray and white matter, respectively. The gray matter has much more volume than the bony confines of the skull would seem to permit, so it is folded and convoluted into multiple curving ridges called gyri. The furrow between one gyrus and the next is called a sulcus. So effective is the folding that the human skull contains an area of cortex nearly two and a half feet square, more than half of which is hidden within the sulci and fissures. Within this two and a half square feet are 10 billion neurons and 60 trillion synapses.

Each hemisphere is subdivided into four lobes, named according to their position: frontal (at the front), occipital (at the back), parietal (top and side, between frontal and occipital) and temporal (lower part of the side). For easy reference, it seems most efficient to list what are in general the functions governed by the cortical cells of each lobe.

FRONTAL: learning; higher psychological processes; motor function of voluntary muscles; coordination of mouth, tongue and larynx, allowing speech; voluntary movements of head, eyes, lids, hands, and fingers

OCCIPITAL: sight

PARIETAL: skin senses, such as touch, pain, temperature; certain cognitive and intellectual processes

TEMPORAL: hearing; some speech function; memory; smell.

In all lobes, the locations known as association areas analyze, interpret, and coordinate sensory experiences. By tracts of fibers in the white matter, connecting the various parts of a cerebral hemisphere, they allow the cortex to act as an integrated whole. From these integrated functions come memory, reasoning, judgment, verbal expression, and certain aspects of the emotional sense. Those association areas that are in the frontal lobes deal with higher intellectual processes. It is here that we are made aware of the possible consequences of our actions.

Association areas in the parietal lobes help us to understand sensory information, including the speech we hear and utter. It is my parietal cortex that allows me to choose precisely the correct words to write this description of the brain, or at least what I hope are precisely the correct words.

Association areas in the temporal lobes help to interpret complex sets of signals such as those needed to understand speech and written words. It is your temporal lobe that enables you to take what I have written and make sense of it. Here, too, is where visual and auditory memories are processed. Memory is, after all, what Lady Macbeth called "the warder of the brain," and its site of custodial authority is in the temporal cortex.

Association areas in the occipital lobes, being close to the visual centers connected to the optic nerve of the eye, analyze what we see and aid in its interpretation.

It should be apparent, therefore, that extensive areas of the cerebral cortex are neither sensory nor motor. Most of the human cortex is, in fact, involved in high-order associative, interpretive, and integrative activities. Equivalents of significant parts of these areas are not found in any other animal, including our closest primate relatives. "Man carries the world in his head," wrote Ralph Waldo Emerson, "the whole astronomy and chemistry suspended in a thought. Because the history of nature is charactered in his brain, therefore is he the prophet and discoverer of her secrets."

If the contents of the cerebral hemispheres incorporate the brain's highest capabilities, the cerebellum and brain stem incorporate its most primitive. The cerebellum, which contains reflex centers that coordinate the movements of voluntary muscles, lies below the rearmost part of the cerebrum. It integrates messages from the eyes, muscles, and skin, enabling it to orchestrate movements of our extremities and trunk. While doing that, it is sending messages to other parts of the brain, informing them of what is going on. Clearly, the cerebellum integrates its activities with the spinal messages concerning proprioception. The great English neurologist Sir Charles Sherrington described it as "the head ganglion of the proprioceptive system."

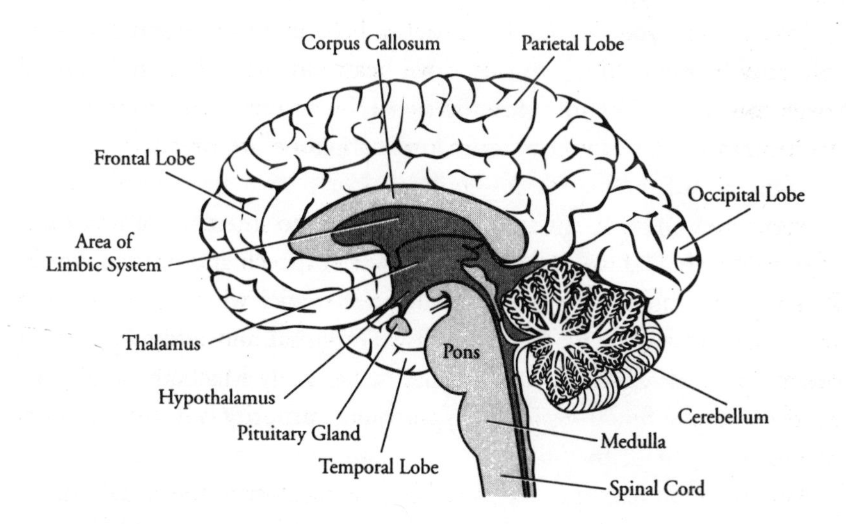

A section cut through the center of the long axis of the brain, showing the locations of the principal structures.

The brain stem connects the cerebrum to the uppermost portion of the spinal cord. It contains many nerve tracts and a number of the clumps of neurons called nuclei. Each nucleus superintends a distinct neurological function. Three of the parts of the brain stem—the thalamus, hypothalamus, and medulla—have come up several times in the course of this book.

In appearance, the medulla looks like a thickened continuation of the top of the spinal cord; all ascending and descending nerve fibers between the brain and cord pass through it. Its most important nuclei are the cardiac center, which can quicken or slow the heart rate; the vasomotor center, which can control the constriction of blood vessels; and the respiratory center, which regulates breathing. Other of the medulla's nuclei serve as centers for swallowing, vomiting, coughing, and sneezing.

Another significant part of the brain stem is a long, thin network of fibers and small foci of gray matter called the reticular formation, which provides connections between the ascending and descending tracts and the hypothalamus, cerebellum, and cerebrum. Activity in the reticular formation not only reaches the cerebral cortex but actually stimulates it into increased wakefulness. When this activity decreases, sleep is induced. It is in this manner that raising or lowering the level of external stimuli determines how good a night's rest we get. Not only does the reticular formation send messages up to the cortex but it also functions as a kind of triage officer to allow only essential signals to go through, those which it is important for the upper brain to know about.

At the top of the brain stem (and considered by many neurologists to be part of it) and under the cerebral hemispheres are the thalamus and hypothalamus. The thalamus lies at the base of the cerebrum and serves as a relay station connecting the cerebral cortex to various lower parts of the brain, particularly the medulla. Recent research indicates that the circuits connecting thalamus to cortex may play a crucial role in consciousness.

Several of the functions of the hypothalamus are by now familiar, in its role as regulator of the autonomic nervous system and the most direct supervisor of homeostasis. It plays a key part in heart rate, blood pressure, water and chemical balance, glandular secretion, hunger, peristalsis, sleep, and in the production of hormones that affect the pituitary gland (colloquially called the master gland), which lies conveniently attached immediately below it. In addition, it acts like a thermostat to maintain the steadiness of body temperature, which it does partly by its effect on the caliber of blood vessels in the skin and its control of sweating (the human skin contains at least 2.5 million sweat glands), and partly by influencing the secretion of thyroid hormone and therefore the rate of metabolism.

By now, it must be evident that there exists in the human brain an enormous amount of interconnection, overlapping, duplication, and complementary function—any given activity is apt to be multipartnered. In an integrated fashion, signals go every which way, bringing automatic together with deliberate; voluntary with involuntary; essential with nonessential; higher with lower; emotional with physical—sometimes allowing choices and sometimes not. Cortex here speaks to cortex there, and calls downward to places below even as messages are ascending. No part of the brain keeps secrets from any other part. Every conscious thought produced by the cells and carried by the fibers of the cerebrum is influenced by countless factors rising up out of the prehistoric depths of mankind's origins and modified by the events of his long journey to the present moment. Tennyson spoke for the brain when he wrote, "I the heir of all the ages, in the foremost files of time."

Somewhat differently than do the lowermost areas of the brain, the intermediate part has its own way of representing the prehistoric depths of humankind's inheritance from the ages. This small group of structures, called the limbic system because it is in a sense in limbo between heaven and the nether regions (the cerebrum and the brain stem), deals primarily with emotions and instincts. In the evolution of vertebrates, many instinctive behaviors have been noted to be initiated by the sense of smell. Close connections between the central areas of smell and the limbic system allow such guidance to be so direct that anatomists call the limbic system of lower vertebrates the rhinencephalon, or "smell brain."

As animals evolved into higher forms, the limbic system's functions and connections became much more sophisticated, to the point where it is in us an intermediary between the brain stem and the cerebrum, involved in central control over instincts, motivations, and feelings. But it has never lost the ancient recollection of its special relationship to smell and the responses that smell can evoke. Marcel Proust understood well that an entire world of childhood memories can rise up out of the unconscious mind under the influence of the chance smell or taste of some long-forgotten essence. As readers, we are drawn early into his *Remembrance of*

Things Past by our own innate perceptions and even our experience that such things, albeit on a lesser scale, have happened to all of us. My mother used face powder only on those very rare evenings when she and my father went out to some social function. But always, she would lean over my bed and kiss me just before leaving the house. Although I have encountered it only two or three times in the more than half a century since her death, the distinctive delicacy of her barely perceptible odor has remained sweetly stored in the recollections of those blessedly protected days. On each occasion, the smell memory has flooded my mind with the love of her, and the remembrance of a safeness never again to be felt once that cocooned time had come to its end.

Proust's experience and mine are characteristic of the way the brain retrieves memories. They are stored as pieces of a large pattern, the parts of which were originally perceived simultaneously. A bit of recall, whether by smell, sight, or some other stimulus, can bring back the entire pattern or large parts of it, as well as linkages to other memories whose relationships to the first may or may not be immediately evident. This explains how psychoanalysts are able to use the technique of free association to bring back seemingly long-forgotten experiences. All of this is dependent on sequences of molecular changes at specific sites in neuronal systems.

The various anatomic elements of the limbic system lie within the deepest parts of the cerebral hemispheres. Named from Greek words for their shapes, they are the amygdala ("almond"), hippocampus ("seahorse"), and several communicating structures. Because of their many connections to both the older and lower parts of the brain, which are anatomically below, and the newer and higher parts, which are anatomically above, the limbic system plays a vital role in learning, memory, and the emotional life. It connects areas of the brain that are instinctive with those that consciously respond to incoming stimuli. The limbic system receives sensory input from voluntary and involuntary centers and sends motor output back to areas where it can be acted upon. Connections between the hippocampus and cortex, particularly, are vital to memory.

One can think of the limbic system, lying deep within the cerebral hemispheres, as being both anatomically and functionally in the center of the brain, as befits its role as an intermediary between higher and lower. A diagrammatic representation of the limbic system's connections and functioning would be replete with arrows leading in and out of it, either directly or indirectly to and from all other parts of the brain. In both the anatomic and functional sense, it is so close to autonomic activities that some neurologists consider the hypothalamus to be included in it, or at least to be its gatekeeper. The hypothalamus can be thought of as the highest part of the brain stem or the lowest part of the limbic system.

The hypothalamic nuclei concerned with emotion are known to have a very intimate connection with certain primitive (that is, prior to the evolution of the neocortex) cortical cells that are contained in the part of the limbic system which borders the brain stem. This is thought to explain why some people can train themselves to trigger these limbic cortical cells and thereby stimulate association neurons to the hypothalamic nuclei that change the heart rate or affect some other function not ordinarily under conscious control. In chapter 4, I referred to such people as "masters of meditation." By studying their skills, it might be possible to find the linkages by which conscious thought can be employed in the service of emotional tranquillity and perhaps even physical health.

Numerous documented examples can be found of swamis and others trained in the appropriate techniques who have been able to slow their heart rates, lower blood pressure, or stop peristalsis by various forms of meditation. A reliable source has even told me of a mystic who could give himself an enema by drawing water into his lower bowel while sitting in a bathtub, probably by reversing the usual peristaltic direction. But the most dramatic example I have ever heard described of the apparent conscious control of internal or vegetative function was seen repeatedly in the late 1950s by young doctors training at a famous university hospital in a large city in the northeastern United States. Told to me by a colleague who was in an internal-medicine residency there at the time, the story involves a man in his forties who made a habit of periodically turning up in the emergency room with one or another complaint. After checking in at the front desk, he would seat himself in a very conspicuous part of the waiting area and promptly proceed to fall unconscious onto the floor. When the doctors rushed to his side, they would invariably find him to be without a pulse and not breathing—in complete cardiac arrest and turning bluer by the second. Because he always responded in less than a minute to resuscitation efforts, it never became necessary to jolt his heart with electricity or to go very far with that era's predecessor to CPR; although he had multiple surgical scars on various parts of his trunk, they were not of the sort through which the chest might have been opened for cardiac massage.

On not one of those galloping stampedes to assist the stricken man did any of the young doctors ever notice that none of the participating nurses seemed particularly alarmed by the life-threatening drama taking place before their eyes. If anything, in fact, they usually appeared amused by the patient's precipitous journey toward the eternal beyond and back, and by the frenzy of the doctors working feverishly over his inert, pulseless body. The reason for the nurses' levity would soon become apparent, however: as the still heart began to beat again and a blood pressure once more became obtainable, they would often break into laughter. They were in on a scam they had seen many times before.

The patient seemed always to know when a new group of novice doctors was scheduled to begin the standard four- to six-week training rotation through the emergency room, and he would time his stunt for one of the first days after they arrived. Though he never revealed the secret of his startling ability to make his heart stop, it has been presumed, in all likelihood correctly, that he did it by an act of will, gradually slowing the frequency until it quit completely or at least until the interval between beats became so long that his brain was deprived of sufficient oxygen. In spite of plenty of opportunity to look for it, no evidence was ever found that might indicate his having taken some sort of medication to achieve his purpose; moreover, had he done it that way, he would not always have recovered so rapidly, nor is it likely he could have completed his performance many times without at some point killing himself accidentally. He was never observed by the vigilant nurses to be carrying out any conceivably precipitating physical act while sitting in the waiting room, such as holding his breath and bearing down. Of course, it is possible that he used some never-discovered stratagem that he somehow managed to keep secret, but no one could imagine anything it might have been. Almost certainly, he had in some way acquired training in the esoterica of willed control over autonomic function and was using it as part of a curious pattern of behavior that has been called the Munchausen syndrome since it was first described in 1951 by Dr. Richard Alter in the English medical journal *The Lancet.*

Alter's choice of name for the condition derives from Baron Karl Friedrich Hieronymus von Münchausen, a probably fictional character whose colorful tales of thrilling but preposterous adventures were the subject of a book published in English in 1785 by the German writer Rudolf Eric Raspe. Defined by Alter and confirmed by the experience of thousands, if not tens of thousands, of doctors worldwide, the syndrome consists of repetitive episodes characterized by the dramatic onset of an alarming symptom or symptoms; a multiplicity of scars on the skin and other evidences of frequent past medical care of a major nature; and a remarkable tolerance for what Alter called "the more brutish hospital measures" for diagnosis and treatment. Typical presentations are sudden excruciating abdominal pain, bleeding from the lungs, stomach, or urinary tract, loss of consciousness, or convulsions of some peculiar sort. Our man appears to have been unique in his ability to induce cardiac arrest, all other such patients having faked their symptoms.

Unlike the malingerer, who has some material objective to gain from his charade, the Munchausen patient has no obvious end in mind, except to deceive and be cared for. No psychiatrist has definitively tracked down what Alter calls "the psychological kink" that motivates such people, but it is common to observe that they appear to be immature personalities who are relentless in their craving for attention: "Their effrontery is sometimes formidable, and they may appear many times at the same hospital, hoping to meet a new doctor upon whom to practice their deception."

But Munchausen syndrome is hardly an example of a way in which linkages between conscious thought and automatic internal functioning might promote emotional tranquillity or physical health. At least as far back as Hippocrates, physicians have known that psychological factors can under certain, usually unpredictable, circumstances affect the course of a disease. Recent elucidation of the interrelationships between various aspects of human physiology has given considerable impetus to efforts by researchers in a variety of disciplines to put the study of such phenomena on a scientifically rigorous basis. The result has been an enlarging body of evidence that at least *some* individuals, under *some* circumstances, may be able to modify the course of *some* diseases in *some* way that is beneficial. Verifiable findings have been reported from the laboratories of meticulous researchers, indicating that measurable changes in components of immunity, for example, can occur in response to changes in such qualities as a patient's attitude or surroundings. What undefinable effect might attitude have had, it must be considered, in the outcome of the disease process of Sharon Fisher or Kip Penn? And how much of Anthony Cretella's long-term success was attributable to his determination to make himself "a step above what I am"? Could the cheerful optimism and devout religious faith of Marge Hansen have been instrumental in her body's astonishingly resistant response to the massive assault inflicted on it, even though she was anesthetized during much of the time? And finally, what really stopped the progression of clostridial overgrowth in the intestine of an unconscious and barely responsive Hope Kuziel? These are unanswerable questions, but with the revelations of the newer experimental findings, it is ever more appropriate to ask them. The implications of this kind of research are enormous, dealing as they do with factors that would seem at first glance hugely disparate from one another, such as immunity on the one hand and the human spirit on the other.

For some three decades, a field of scientific investigation has existed, called psychoneuroimmunology, that explores the relationships among mind, nervous system, hormones, and response to disease. Much specific and much suggestive data has emerged from the laboratories and surveys of researchers in this multidisciplinary arena of study, and more is revealed with each passing year by the work of those engaged in it: neuroscientists, endocrinologists, pathologists, psychologists, social scientists, immunologists, cell biologists, pharmacologists, physiologists. In the first (1981) edition of a large volume devoted to their findings, the distinguished Rockefeller University immunologist Robert A. Good encapsulated the work of these researchers in three sentences, with which anyone who has followed it must surely agree: "I am absolutely convinced that the interaction of mind, endocrines, and immunity is real. Of this there can be no doubt. . . . The question that remains is *how* these three major networks—the nervous system, the endocrine system, and the immunologic system—interact and, *how*, by understanding these interactions in precise quantitative terms, we can learn to predict and control them."

I have several times glibly stated that the limbic system is involved in our experience of emotions and instinctual drives, as though anyone knows how it actually does this. Other than the foregoing, neither does anyone know in any detail how it integrates its own activities with those of, say, the hypothalamus or the cerebral cortex. Although much is beginning to be revealed about such things, present knowledge is a small fraction of what must be discovered before a clear understanding is achieved.

Study of the limbic system is, in fact, one of the most active areas of brain research. It has fascinated the entire gamut of investigators, from molecular biologists to neuropharmacologists, and even psychoanalytic theorists, for the obvious reason that such studies may one day shed light not only on the causes of mental illness but also on the very notion of consciousness and on the eternal conundrums about the relationship between mind and body.

Because the connections to and from the limbic system are capable of so directly initiating activity in other parts of the brain, any pathology within it may be reflected in what would at first glance appear to be manifestations unrelated to limbic function. Among them are certain forms of epilepsy.

By definition, an epileptic is subject to paroxysmal disturbances of the electrical activity of the brain, resulting in transient abnormalities of motor activity, altered consciousness, or derangements that may be psychic or autonomic. One of its common forms, once called psychomotor or temporal lobe epilepsy but now known as complex partial seizures, is most often caused by some lesion, such as a tiny patch of scar, within the limbic system. Characteristically, patients with this problem exhibit mental disturbances during an attack, with automatic, purposeless movements. They may have auditory, visual, or smell hallucinations, and sometimes they experience sudden recall of past events. Though they usually remain conscious, they have no later recollection of the attack.

The ancient Greeks called epilepsy "the sacred disease," and they tried to cure it with an array of mystical medicines that brings to mind Macbeth's witches: blood of tortoise, hair of camel, testicles of hippopotamus—the list goes on, its length testifying to the inadequacy of every treatment. Celsus, the renowned Roman medical encyclopedist of the first century A.D., wrote of occasional cures achieved by drinking the blood of wounded gladiators, although he never witnessed such a cure himself.

Even the name of the disease, derived from the Greek *epilepsis*, or "a taking hold," signifies that a victim is carried off by some mysterious supernatural power. Twenty-five hundred years ago, Hippocrates wrote that the otherworldly image of epilepsy made it appear "not at all like . . . other diseases." Not surprisingly, the great majority of people still regard its seizures as uniquely terrifying. Not only the condition itself but also those afflicted with it are all too often feared and misunderstood. At worst, the victims of epilepsy are shunned as though they harbor within themselves a contagious touch of devil-dealt danger; at best, they suffer embarrassment and prejudice in daily life.

Gloria Mangual was fired from three jobs because of her epilepsy, in spite of the antidiscrimination laws of her home state of Connecticut. She was no stranger to ridicule. As a child, for example, she suffered the humiliation of being abandoned by classmates and a teacher who fled the room during one of her seizures. Time and again, she was subjected to the hurtfulness of well-meaning but ignorant people, such as the nun at her school who offered the misinformed opinion, "It's unfortunate about the epilepsy, because you'll probably be feebleminded someday."

Gloria, then in the eighth grade, had reason to believe the nun's prediction. The medications she took in an unsuccessful attempt to control her seizures often left her in a state of torpor, unable to concentrate. As she told me one day about six years ago in a conversation at Hartford's St. Francis Hospital, where she was working as a public relations officer, "There are years I don't remember. I don't recall anything at all about two years during my teens when I was taking a medication called Mysoline. Having to cope with the medications is one of the hardest things about epilepsy." In spite of the disruptions to her life, however, she had gone on to college to earn a journalism degree, graduating with honors.

As I sat talking with this confident, attractive thirty-six-year-old woman that day, it was difficult to imagine the years she had spent perceiving herself as "not just different but somehow dirty, like I wasn't quite as good as everyone else." I was unable to find any trace of the teenager whose emotional neediness had drawn her into unhealthy, dependent relationships. Neither could I recognize the daughter who until the age of eighteen had lived with her parents' overprotectiveness and dictum that she must, so far as possible, keep secret the nature of her very visible affliction.

Since undergoing surgery at Yale–New Haven Hospital in 1983, Gloria has not had another seizure. She tried not to be overdramatic when I asked her what it is like to be free after so many years of living in constant fear of the capricious whims of her disease. But there was no mistaking the wonder in her voice. "I feel like a walking miracle," she replied softly.

The disease probably had its origin in a brief period of very high fever when Gloria was five months old. Most children who have fever-induced convulsions never become susceptible to epilepsy, but in an unlucky few, the childhood episode may leave a lasting trouble spot in the form of a microscopic bit of scar, or gliosis, in the brain. When that spot becomes stimulated again, it may trigger the sudden burst of abnormal electrical activity that causes an epileptic seizure.

Gloria was five years old when she had the first of these frightening seizures, whose pattern soon became all too familiar to her and her parents. Each of the convulsions was preceded by a ten- to fifteen-second aura during which she felt what she describes as "a ticklish feeling in my stomach that started rising," accompanied by an overwhelming wave of terror. "I'm not sure if my fear was caused by something physical or the knowledge of what was coming. Just knowing that in a few moments I would lose complete control was terrifying." Once the aura began, she would only have those ten to fifteen seconds to get into a safe situation, preferably a quiet place where she could sit or lie down on her own.

Gloria was having complex partial seizures. She did not fall to the floor and thrash her arms and legs, as do people suffering from the more familiar grand mal seizures. "I would twist my right arm, and my eyes would roll a little bit. If I was speaking with someone, I would keep speaking, but I sounded as though I was mildly retarded—my speech would slow down and become mixed up. I wasn't aware of any of this, but I know it from descriptions people gave me." The actual seizure lasted one to two minutes, during which she lost all normal consciousness. It was followed by a period of confusion and lethargy that might last for a half hour. Sometimes, on regaining awareness, she would find herself in an emergency room. Once, at a mall, she awoke to find an oxygen mask on her face. "It wasn't the seizures that were so bad," she told me. "What was so bad was the attitude of people around me when I came out of it."

The attacks came in clusters, with as many as a dozen or more occurring within a week and then disappearing for a few weeks. By age twenty-eight, it was clear that the epilepsy was intractable. Unable to hold a job and living on disability checks, Gloria came to an important decision. She had been under the care of neurologists at Yale–New Haven Hospital since she was a child, so she was well aware of research being done by a young neurosurgeon, Dr. Dennis Spencer, on the operative treatment of epilepsy. She made up her mind to be evaluated as a possible candidate for surgery.

Surgical approaches to the cure of epilepsy are not new. Since the ancient Egyptians drilled holes in patients' skulls to let out the bad spirits, surgeons have tried to come up with a method of localizing the source of seizures within the brain's substance. Until relatively recently, however, identifying the abnormality that triggers a seizure was difficult, and complete cures were few. In the past two decades, a revolution in biotechnology has enabled neurologists to identify definitively a seizure's trigger point.

The majority of those who have poorly controlled complex partial seizures are found to have a triggering focus in the limbic system deep within either the right or left temporal lobe. This may be a tumor, a tiny abnormality of blood vessels, or, most commonly, a scar of the sort that can be the residual of a febrile seizure in infancy or early childhood. As a result of some stimulus that is usually difficult to ascertain (blinking lights and certain music are among those that are sometimes indicted), there is a change in the environment of the neurons near the trigger point, making them more excitable. Excessive neuronal activity is set off and then is transmitted along the normal pathways that lead into other parts of the brain, such as the sensory and motor cortex and the structures of the brain stem.

If the instigating focus of complex partial seizures or the other forms of epilepsy lies in an area of the brain that can be removed without damaging critical functions, neurosurgery can have an amazingly high success rate. Spencer, a leader in developing these procedures, has been able to cure more than 80 percent of carefully chosen patients who fall into that category.

Before considering Gloria for surgery, the doctors gave her a full-scale diagnostic workup. A battery of tests, including CT scans and dye studies of the blood vessels in her brain, was done to observe structure and function. (Today, some of these tests might be replaced by the more specific and high-resolution pictures produced by PET scans and MRIs.) To further characterize her seizures, she was put under audiovisual electroencephalographic (EEG) surveillance: an EEG machine monitored the electrical activity in her brain, while her outward behavior during a seizure was recorded on a closed-circuit television camera. Finally, to map the exact location of her triggering spot, she underwent an invasive study that probed deeply into her brain. Specially constructed monitoring electrodes were placed into her temporal lobes through small holes drilled in the skull. Each time she had a seizure, one of the electrodes would record rapid-fire activity from the same specific area. So precise is this technique that the origin of the seizures can be localized to within a half-inch block of tissue.

The source of Gloria's problem was found in the part of the gracefully curving hippocampus that is deep within the left temporal lobe. It was obvious that the seizures were beginning there and spreading out to other areas of the lobe. One evening, when all the studies had been done, Dennis Spencer went to Gloria's hospital room and sat down on the side of her bed. He had told her that she was a good candidate for the operation she had been thinking about. He drew a sketch to explain the proposed surgery to her, answered questions with gentle patience, and then, holding her hand in his, shared his optimism about the outcome.

Until this point, Gloria's parents had been very supportive. But now, with the imminence of the operation looming before them, they began to have second thoughts: Any surgery in the brain entails a risk of impairing a patient's cognitive or motor skills. Gloria's mother, apprehensive about a possible poor result, went to the hospital to attempt to dissuade her daughter from going through with the surgery. Gloria wavered a bit, but in the end she remained steadfast: "When I thought about how life had been for the past few years, I said to myself, Forget it, I'm having the surgery."

The operation began early on the morning of July 15, 1983. Once Gloria was fully anesthetized, the left side of her head was shaved and sterilized with iodine antiseptic. Considering the mystery, and even mythology, in which this disease has been shrouded over the past two and a half millennia, it is appropriate that Spencer's skin incision was shaped in the form of a question mark. At its base, it rose straight up from a point in front of her ear; then, after curving gradually toward the back of Gloria's head, it followed the midline of her scalp forward, ending just inside the hairline. Spencer's team then peeled back a flap of skin and muscle to reveal the part of the skull lying over the temporal lobe. A rectangle was marked out by drilling six evenly spaced holes through the thick bone; when the holes were joined by cutting from one to the next with a fine-toothed saw, the rectangle of skull was lifted out, exposing the dura mater, the tough fibrous membrane ensheathing the brain.

The dura mater (so called because it surrounds the brain like a strong, protective mother) was then opened to reveal the forward portion of Gloria's temporal lobe. Working slowly, with the aid of dissecting microscopes, Spencer outlined a small block of expendable tissue that he would excise from the tip of the lobe to expose the underlying hippocampus. The brain has such a profuse blood supply that it is impossible to cut into its substance with a blade, as surgeons might do anywhere else in the body. Instead, an instrument is used that combines electric cautery, ultrasound, and controlled suction to delicately burn through and remove minute bits of tissue. The outlined area was dissected with such faultless accuracy that to a bystander like me—a surgeon who has spent most of his operating time in the spongy red wetness of the abdomen—Spencer's technique looked like the eighth wonder of the world.

With the block of brain finally out of the way, the hippocampus came into view. Now three hours into the operation, Spencer and his colleagues worked toward their quarry as though time and fatigue did not exist. Spencer is a tall, bearded Iowan, and his voice has the softness of a late-summer prairie breeze. With his feet in large wooden clogs and his head enveloped in a surgical hood tucked into the top of a long sterile gown, he stood motionless as a draped monument, peering down the barrels of his microscope. Only the small movements of his fingers and the quiet directions he gave his team provided evidence that his towering green-swathed form was engaged in one of the most challenging operations any surgeon can undertake.

At last, having avoided injury to any of the crucial neurological structures that lay within a figurative hairbreadth of their dissection, the surgeons lifted out the offending piece of hippocampus. Their objective achieved, they stitched the dura mater closed and wired the plate of bone into the place from which it had been lifted five hours earlier. To complete the closure, they folded Gloria's skin back over her skull and secured it with metallic staples. Not long afterward, she awakened to the beginning of her new life.

The block of hippocampus that Spencer removed from his patient was barely a cubic inch in size. It was enough—the instigating focus of her epilepsy had been cleanly excised. Still, Gloria did not feel completely certain of the surgery's outcome until some six years had passed. It was only then that her cautious doctors finally allowed her to stop her medications. She has remained free of seizures and life has been good.

As is true for all cured epilepsy patients, Gloria's successful treatment affected more lives than just her own. But the list of those benefited is not restricted to individuals who knew her before the operation or even to those who were alive at that time. About four years ago, her gynecologist told her that she appeared to be undergoing the changes characteristic of early menopause. Thirty-six years old at the time and unmarried, Gloria determined not to miss out on her lifelong dream of being a mother. With an excellent job as manager of public relations for Connecticut's Hartford Visiting Nurse Association, and eleven years of healthy, seizure-free life behind her, she had no reason to deny her instincts any longer—she took herself off to the sperm bank, looked over the qualities of prospective fathers, and made her choice. The artificial insemination took, and Samantha Mangual was born in 1994, to the delight of her grandparents and her beaming mom. If ever a medical case history had the makings of a happily-ever-after scenario, the story of Gloria and her Samantha certainly seems to be it.

Gloria Mangual's complex partial seizures originated in a tiny area of pathology in her brain, which utilized perfectly normal pathways to transmit signals to several other areas. Thus, a system of communication meant to help her became instead the source of her distress. The basis of the brain's marvelous ability to coordinate widespread and disparate phenomena proved also to be the basis of a disease that threatened the fabric of her life. In its own way, Gloria's is one of those examples, like the aneurysm in Marge Hansen's splenic artery, of a disability caused by the very mechanisms that have evolved to enable us to cope with the dangers or requirements of everyday life. Without the brain's richness of connections that bring various of its parts into communication with one another, temporal lobe epilepsy would not exist. In these ways do our bodies sometimes betray us; in these ways are we miracles with built-in flaws.

But the magnitude of that richness and the variety of possibilities it allows are the wherewithal that makes us human, even though the magnitude brings an occasional problem with it. The brain of *Homo sapiens* is like an offering that nature has brought to our species, encouraging us to make the most of the biological equipment that is so splendidly ours. What has been offered is exuberant with opportunities, overflowing with possibilities: immense numbers of cells and pathways, and immense numbers of junctions. All kinds of electrical journeys may be taken. Sometimes the junctions are convergence points for signals arriving from several different directions and multiple sources; sometimes the junctions are convergence points at which a single message is distributed in several different directions and to multiple destinations. Signals originating at a variety of sources meet and then proceed together; signals originating at one distinct source separate and then proceed to a variety of areas. Signals loop back and return to some site they have just quitted; signals reverberate through the same circuitry again and again, and augment their effect. The brain is a huge convergence field that processes information from everywhere and issues the ultimate orders of our existence.

From all of this emerges the brain's overarching responsibility—it is the chief means by which the body's activities are coordinated and governed. Without it, many of nature's animals would be no more than a collection of cells and tissues, unable to survive for long without the overseeing mastery which only that unique organ can provide. It is the integrated action of all parts of the body under the supervision of the brain that makes higher animals what they are. Without the distinctive qualities of his brain, *Homo sapiens* would be merely another of those higher animals.

But complete integration involves far more than the brain can provide on its own. Messages transmitted along nerves are not the only signals to which cells respond. A great deal of the body's integration is carried out through the agency of signaling molecules. A signaling molecule is a hormone or other chemical substance for which some particular kind of cell has receptors in its membrane. The molecule binds to the receptors, causing the cell to react in some specific way.

A neurotransmitter such as acetylcholine or noradrenaline is such a chemical substance. Another type is the so-called *local* signaling molecule, which certain cells secrete in response to some change in their immediate environment, in order to produce a necessary effect in nearby tissue. The various prostaglandins, of which there are almost twenty, fall into this category, affecting such functions as local blood flow to a small area and changes in the diameter of the minute air passages leading to the alveoli in some bit of lung. Other examples of the vast assortment of the body's local signaling molecules are those that regulate the rate at which nearby cells divide, thereby influencing local growth of tissue.

In a category by themselves are the signaling molecules produced by a special class of neurons found in the brain and spinal cord, called neurosecretory cells. In addition to functioning like ordinary neurons, they act like endocrine cells in the sense that they produce protein or polypeptide substances called neurohormones which enter tiny blood vessels, allowing their dispersal to distant structures. Some sixty neuropeptides have been identified in the human cerebral cortex, and the action of some of them is well known. Among them are the endorphins, morphinelike compounds that decrease pain and may induce states of calm or even euphoria. (Incidentally, the signaling function of peptides is not restricted to the neurohormones. Active peptides are also secreted in other parts of the body, such as the intestinal tract—as noted in chapter 11.)

Not uncommonly, a neurohormone's action is to make an endocrine gland secrete. Accordingly, the neurosecretory cell allows direct communication between the endocrine and nervous systems. A stimulus reaching a neuron can thus be converted into an action caused by a hormone. By now in this narrative, such interdependent linkages between systems have become familiar, as characteristic of the ways in which the body gets all of its jobs done.

The relationship between the hypothalamus and the pituitary gland is particularly important in this regard, especially because the hypothalamus is involved with autonomic regulation and the states of the emotions, and it does contain certain neurons that secrete neurohormones. The interaction of the two structures is facilitated by their anatomic one-on-top-of-the-other juxtaposition at the base of the brain. So closely integrated is their functioning that they have together been called the neuroendocrine control center.

The pituitary has two parts. Its front, or anterior, section is an endocrine gland. Neurohormones from the hypothalamus cause it to secrete FSH and LH (see chapter 7), thyroid-stimulating hormone, growth hormone, and a hormone called adrenocorticotropic hormone, or ACTH, which stimulates the outer shell of the adrenal gland (the adrenal cortex). The relationship among the hypothalamus, pituitary, and adrenal has been formalized in the writings of medical researchers by calling it the hypothalamic-pituitary-adrenal axis, or HPA. Reciprocal relationships among these three structures have been shown to influence the body's response to stress and the secretion of certain neurohormones that affect the immune system. This is one of the fields of research under the general heading of psychoneuroimmunology, and it has shed considerable light on possible ways in which the brain and even conscious thought may play a role in immune responses, including those involved in a variety of diseases in which the immune system does or may have a part, including cancer.

The back, or posterior, part of the pituitary is actually composed not of endocrine cells, but largely of glial tissue and nerve fibers coming down from neurosecretory cells in the hypothalamus. The axons of these cells extend down into the posterior pituitary, carrying the neurohormones oxytocin (see chapter 8) and vasopressin (chapter 2).

As sketchy and incomplete as the foregoing description of signaling molecules perforce must be, even the inclusion of much more detail would still add up to a single statement: A great deal of the behavior and response of the body's 75 trillion cells is determined by signaling molecules, whether carried afar by the bloodstream or active locally. In complex and complementary ways, such chemical substances work in coordination with the wondrous array of responses within the nervous system, all with the aim of maintaining that dynamism of constancy, that exuberance of seeking, which is the substance of human life.

The result of the interaction of electrical and chemical messages is that every part of the body is able to know what every other part needs and is doing. There is within us a sense of what can perhaps best be called awareness of our inner selves, which would seem to ascend and descend to multiple levels at once. From an awareness of our bodies—which is as conscious as the attention we pay to a growly stomach, all the way to an individual cell's adaptations to what are ultimately the needs of the entire organism, there are subtle interconnections that shade one level into the next. Far beyond the capacity of any other animal, this internal sensibility determines more of our overt and unconsciously based behavior alike than we may know.

We respond not only biologically but emotionally, intellectually, and culturally to the rhythms and requirements in the deepest tissues of our bodies. What *Homo sapiens* has created in the world around him and in that other world, the world of the mind, which faces both inward and outward, is in harmonious response to the churnings of his cells' ceaseless quest for homeostasis and ongoing life. The equilibrium and harmony of all the struggle for constancy within our biology provide the rising and falling strains of the ever-adapting melodies heard by the physicochemical agglomeration of sensation, processing, and action that is the human brain. With the vast array of potentialities made available by its cellular and molecular activities, the brain responds. One of the multiple ways in which it expresses its response is to create patterns we have chosen to call the mind.

The mind is a man-made concept, a way to categorize and contemplate the manifestations of certain physical and chemical actions that occur chiefly in the brain. It is a product of anatomic development and physiologic functioning. What we call the mind is an activity, made up of a totality of the innumerable constituent activities of which it is composed, brought to awareness by the brain. The brain is the chief organ of the mind, but not its only one. In a sense, every cell and molecule in the body is part of the mind, and every organ contributes to it. The living body and its mind are one—the mind is a property of the body.

If the mind is a constant within which to understand certain manifestations of the biological activities of the brain and our cellular selves; if the brain is the overseer of the body's neurological responsiveness in the interest of maintaining homeostasis; if the body's responsiveness is dependent on the physics and chemistry of molecules—then the mind is, no matter how remotely, the ultimate product of molecular interactions. But what a product! It is a product of unimaginable glory, which has built upon itself millennium by millennium to create a psychic structure that has, in a sense, taken on a life of its own even while remaining subject to all of the processes that sustain its organic effects. It is in this sense that I claim for it the quality of being greater than the sum of its parts.

Will we ever fully understand the mind? Will we ever agree about its origins, its meaning, its intent? Will we ever agree, in fact, whether it even *has* meaning or intent? I am hopeful, but I must counter my optimism by quoting the cynical response of Ambrose Bierce, defining the word *mind* in his 1911 catalog of causticity, *The Devil's Dictionary*: "Mind: n. A mysterious form of matter secreted by the brain. Its chief activity consists in the endeavor to ascertain its own nature, the futility of the attempt being due to the fact that it has nothing but itself to know itself with."

Notwithstanding Bierce's pessimism, nowadays we do know some things that seemed beyond understanding when he composed his dismal prospectus. We begin with what was known in his day: to develop the quality we call the human mind required the evolution of a new part of the primate brain, capable of forethought, analysis, abstract reasoning, and language. *Homo sapiens* emerged in a relatively short evolutionary period, endowed with a hugely enlarged neocortex and the synaptic potentialities to fulfill the requirements of mind. Only in recent decades have the combined studies of an entire spectrum of research scientists begun to elucidate just how it is that the brain and mind do their work.

The findings of recent studies in neuroscience have necessitated considerable re-evaluation of previous long-standing theories about brain function. Unsurprisingly, there remains controversy about details, but by using a clinician's criteria, it is possible to construct a synthesis that is useful in practice and internally coherent. Readers familiar with the writings of such investigators as Donald Alkon, Francis Crick, Antonio Damasio, Gerald Edelman, Walter Freeman, Donald Hebb, Ronald Kalil, Paul MacLean, Israel Rosenfield, and Gordon Shepherd will detect their influence in what follows. The work of these researchers and others concerned with consciousness and human thought is representative of the finest of that combination of verifiable laboratory investigation and informed speculation that has always been the acme of the scientist's contribution to our understanding of nature.

Almost certainly, the circuitry in much of the brain retains a great deal of plasticity throughout life. Though lessening with advancing age, there is constant reorganization of function, serving to strengthen some kinds of synaptic connections and weaken others. How much of the basic structure laid down is genetically determined seems to depend on anatomic level. The lower parts of the brain, regulating basic homeostatic mechanisms necessary for survival—namely, the brain stem and hypothalamus—appear to be "hardwired," without significant likelihood of meaningful variation among individuals. They are probably as stable and innate as the length of your nose, being determined by the genome and present before birth. The other parts of the brain retain in various degrees the capacity for change, ceaselessly adapting to the signals reaching them. This is more true of the cerebral hemispheres than of the limbic system.

It would appear, then, that the higher the part of brain, the more adaptable it is to influences entering it. The adaptations occur by means of new synapse formation, strengthening of existing synapses, and changes in the effectiveness of neurotransmitter interactions. The more a set of connections is used, the stronger it becomes. This means that pathways conducting large numbers of signals more readily transmit subsequent signals. For similar reasons, those synapses that are unused cease to function, or disappear.

Those parts of the brain that are evolutionarily new are thus seen to be much less under the influence of genetic predetermination than the more primitive lower sections. At any given moment, what determines the functioning of the higher brain is therefore largely the result of experience. The newest parts of the neocortex are readapting for each moment's stimuli and simultaneously readapting for the long haul as well. We do this in response not only to messages from the outside world but also to messages from within the depths of the body. Such messages are transmitted via nerve impulses, neuropeptides, and other signaling molecules.

Perhaps the most influential of the stimuli entering from the outside world are those that come in the form of heard language. The earliest *Homo sapiens* were the first of our evolutionary ancestors equipped with a voice box, or larynx, capable of producing the sounds of articulate speech. Over the millennia, tiny incremental refinements took place that finally enabled fully modern man to achieve sophisticated communication. As words and then languages appeared, the brain absorbed all the input and responded by the formation of new synaptic connections, which in turn led to ever-more-complex speech and the development of the type of conceptual thought characteristic of humans. The increasing complexity of heard input and the increasing complexity of synaptic connections had a reciprocally constructive effect on each other. The result was an ever-enlarging brain. Between 200,000 and 20,000 years ago, the modern vertical forehead appeared, indicating a cerebral capacity capable of making use of all that was about to be attained. By about 20,000 to 30,000 years ago, we had the brain required for all of today's activities.

Not only sounds, images, and other such input influence the development and changes in circuitry. Emotions also play a significant role. The higher brain is perpetually shuffling and reshuffling its impressions, perceptions, memories, and responses. Influenced by all of these factors, it shuffles and reshuffles the workings of its circuitry. The total process adds up to a perpetual machinery of learning, in which new synaptic connections are always being assembled and others are strengthened, while those not often used are weakened or eliminated. Altering and realtering neural circuits never stops, even as we age. As so well put by Oliver Wendell Holmes, "Men do not quit playing because they grow old; they grow old because they quit playing." We learn by living.

All of this is consistent with everyone's common daily experience—it explains how we learn and how we forget. But there is much more to the mechanism than that part of which we become conscious. The signals reaching the cortex may emanate from anywhere, including such disparate structures as our muscles, our lungs, and our viscera—from wherever cells live, messages of a chemical or electrical nature are being exchanged with the cerebral center of coordination and control. When Sir Charles Sherrington coined the term he would later use for the title of a book published in 1906, *The Integrative Action of the Nervous System*, he could not have predicted how far-reaching would be its implications.

We are dealing here with awareness. As I have said in previous chapters, I believe that subtly graduated levels of awareness ascending through the medium of chemical substances and nerve impulses connect the cells of the cerebral cortex with the cells of every other part of the body. Whether directly or indirectly, mutual interactions are always occurring between the many divisions of the far-flung empire governed by the brain.

The aggregate is input for the cortex, just as the cortex concomitantly provides input to the distant groups of cells. The body is a unity, in which every constituent portion is in uninterrupted, even if indirect, communication with every other. Seen this way, the mind emerges as the product of all these interactions, coordinated and interpreted through the sophisticated modality of the cortex. The cortex may be said to have "a sense" of itself and of the total body of which it is a part. The number of interactions is so immense as to be incalculable.

Just as the behavior of distant tissues is regulated from afar by the brain, the brain's function and its very ultramicroscopic architecture are influenced by what it is told by even its most remote cellular components. It ignores no message—it acts on everything that comes to it.

By the time input attains the level of the cortex, it has already passed through the modifying centers of lower parts of the brain, which have made their contributions to the overall response. Once in the cortex, individual messages are handled by distinctive collectives of neurons, of which there may be as many as a hundred million, each composed of between fifty and ten thousand cells. Such neural groups all over the brain send messages, responses, new messages, and reresponses, reverberating back and forth, entering and reentering, reinforcing and weakening one another, up and down and every which way, like so many chattering multitudes, combining and recombining in various forms to produce images, perceptions, and thought. The endless number of possibilities is the result of the vastness of each neuron's multiplicity of dendritic and axonal connections to the cells near it, and through them to so many other collectives, networks, and regions. The simplest visual image, for example, is composed of a large complex of bits and pieces. They are received, processed, and transmitted along a multitude of axon groups, all being ultimately integrated as though instantaneously into a single coherent scene that enters consciousness.

Among the signals that go into the aggregate forming a perception are those that arise in the emotions, in memory, and in the background or context from which the original stimulus comes. Different kinds of messages appear to be valued differently by the various neuronal groups, and this, too, influences the strength of transmission and affects the outcome of the final conscious recognition.

All of this activity shapes the pathways along which it is taking place. The plasticity of our brain, based as it is not only on receptor-generated stimuli but also on memory, experience, context, emotional content, and value, is what makes us uniquely human. This is the physiological basis upon which we have created the human spirit.

To me, just as the mind is a quality of the physicochemical substance of the entire body, the human spirit is a quality of the mind. The human spirit is the moral force of the mind, a product of its multiple levels of awareness, striving always toward an intellectual accord with the search for stability and harmony characteristic of its physicochemical origins. Everything we are is the result, I would suggest, of that vast maelstrom of instability within us, ceaselessly adapting to every challenge to ongoing life. What makes us different from the beasts is the might of the human brain and its comprehension of the signals reaching it, bringing unique responsiveness as well as a unique awareness of the rising and falling rhythms of what is within.

The Devil's Dictionary notwithstanding, the endeavor to ascertain the mind's nature has not been an exercise in futility. The cumulative power of human thought is overcoming the theoretical handicap that "it has nothing but itself to know itself with." When turned inward toward itself, in fact, the magnificence of the mind has exhibited authority over ignorance; there is that in it which accepts the challenge with firm faith in an ability to conjure with its own essence. Human thought is quite enough to solve the mysteries of human thought. It brings its own unique excellence to the solution of the problem. Long before the advent of modern neuroscience, Aristotle already knew of such things. In his *Metaphysics*, he wrote, "It is of itself that the divine thought thinks (since it is the most excellent of things), and its thinking is a thinking on thinking." It is a thinking that thinks successfully. Even a curmudgeon like Ambrose Bierce would have to admit that we have come a long way toward understanding the mind.

To recapture now and then childhood's wonder, is to secure a driving force for occasional grown-up thoughts. Among the workings of this planet, there is a tour de force, *if such term befits the workings of a planet. Wonder is the mood in which I would ask to approach it for the moment.*

Sir Charles Sherrington
"The Wisdom of the Body," 1937

EPILOGUE

The tour de force of which Sherrington spoke was the human body—wonder is the only proper attitude with which to approach it. Knowing its underlying physicochemistry only increases our wonder. In seeming paradox, the stability of this self-regulating organization that is us achieves its stability through the unique nature of its very *in*stability. Its instantaneous readiness to react and return to constancy's baseline makes possible every restorative response toward maintaining the delicate balance of homeostasis that is the foundation of life.

Always on the alert for the omnipresent dangers without or within, ceaselessly sending mutually recognizable signals throughout its immensity of tissues, fluids, and cells, the animal body is a dynamism of responsible consistency. By untold trillions of energy-driven agencies of correctiveness, inappropriate alterations are balanced and changes are either accommodated or set right—all in the interest of that equilibrating steadiness that is the necessary condition of the order and harmony of complex living organisms.

Its capacity to communicate within itself and with its external environment is the basis of an animal's viability in the face of the many unremitting forces that never cease to threaten its existence. The transmission and reception of messages and the response to their urgencies constitute the fundamental qualities that sustain life. As the scale of animals ascends from the simplest to primates, receiving structures and signaling molecules of ever-more-versatile capabilities appear, to function in coordination with increasingly sophisticated powers of the evolving central nervous system. In its own mindlessly blind way, nature has for eons been treading the path toward one day molding a creature so faultlessly responsive that it is equipped to deal with every peril that might threaten its cells and itself, whether from the world by which its skin is surrounded or the world which its skin surrounds.

But responsiveness is not enough. Full independence requires the capacity to *foresee* harm and prevent it, the capability of transforming the relationship with environment from a reactive to a creative one. Were an animal to be completely freed from the perils and necessities of its surroundings, it would be liberated, its inherent drive to seek gratification and pleasure limited only by its own inventiveness. Were the gift to include emotional sensibility, a voice box able to produce the distinctive variety of sounds required for communication by spoken language, and an instinct for community, an entire vista of opportunity might appear for the taking. We—the human species—are the closest approach to such an animal yet seen on our planet.

By our technology and our transmission of an ever-expanding body of knowledge to those who come after, we seem to have rendered ourselves so independent of our surroundings that there is little stimulus to further significant evolution, except in response to relatively minor changes in the environment. We are builders and protectors. We have created a tradition strengthened in every generation and passed on to the next. With our engineering, architecture, medicine, and science, we have with considerable, if still-imperfect, success protected ourselves from the ravages of nature; with our social structures, legal systems, and religious institutions, we have with considerable, if still-imperfect, success protected ourselves from the ravages of one another. Evolutionary needs seem in large part to have been replaced by the evolving of culture, heredity by education.

We have gone beyond mere survival and the essentials of protecting ourselves from the harshness of the surrounding world; we have far exceeded the requirements of satisfying the elemental urges that impel us to reproduce our kind. An entire edifice has been created from the rudiments of instinct. The simple need to stay alive and pass along DNA has become clothed in the enriching robes of culture, beauty, and a spiritual sense.

Consider shelter. A structure meant only to keep the wind and rain away becomes in our hands a triumph of architecture—engineered, designed, adorned, furnished, and burnished with the symbols and values that reflect our civilization and our individual predilections. What began as mere shelter has been transformed into the expression of an evolved society: a sanctum in which to commune with our fellows and ourselves; a forum where the human spirit may soar to the sound of spoken words or a musical étude, or to the splendor of a painting.

Consider sexual reproduction. A means of passing along DNA becomes in our hands a multifarious manifestation of the deepest emotional capacities of our inner selves. What began as mere reproduction has been transformed into the expression of an evolved sexuality: an accompaniment to love; the underpinning of family and, in turn, of society itself.

I believe that much of the pattern of what we have become results from an awareness of our biological selves, from the response to the organization and integrated behavior of the cellular and organic processes within us. We are aware on multiple levels, from the signals sent from molecules to molecules on the one hand to the consciousness that makes us want to solve the puzzles of the universe on the other. I propose that there are smooth and subtle gradations in our awareness, so that there is no boundary between what is conscious and the level immediately below it, the level immediately below that one, and so forth, all the way down to the insides of our cells. It is by responding to the messages from within that we have gradually learned to deal as only humans can with the world without. Lacking our powerful human brain, we could not have attained the final level of awareness that is continuous with every level below it.

In my view, the nearness of destructive chaos against which we can never cease struggling creates in us the instinctive need for order, and every fiber of our being is somehow aware of this. The rhythm and tempo of our bodies tells it to us. The cells of our innermost structure speak in not-to-be-denied voices to the depths of our consciousness, bringing inexpressible knowledge that we can survive only by making a unity of the multitrillion processes of metabolic activity. To do otherwise is lethal. The innate sense of the precariousness of existence—the way our physiology turns its own instability to the service of sustaining us—these are phenomena of which I believe we are aware on a gradation of levels. We know things we do not know that we know. "The heart has its reasons," wrote Blaise Pascal more than three centuries ago, "of which reason has no knowledge."

Think of a behavior as constant and voluntary as regular breathing, to which we pay no heed unless our attention is called to it; think of the uncontemplated images that we unseeingly absorb from the furthest peripheries of our field of vision; think of the sounds that we hear without listening, because they shake our auditory ossicles at frequencies to which our minds are not thinkingly attuned.

The avant-garde composer John Cage once described his visit to an anechoic chamber at Harvard, a room constructed of materials providing an atmosphere in which absolutely no sound can enter. And yet he did hear two, one high and one low. When he asked the engineer in charge about them, he was told that "The high one was my nervous system, the low one my blood in circulation." I have never met anyone who has had such an experience consciously, and I am inclined to doubt its interpretation, but the underlying theme is clear. We sense that events are taking place within us, and we seek overt evidence of their existence, as though to confirm what our much deeper awareness already detects.

There are subliminal messages that reach our brain and make an impression on it which training can sometimes bring to consciousness. But there are lesser signals too, I would argue, originating from sources within our organs and tissues and emanating to them from our cells. Our brains receive information so much beyond conscious perception as to be absorbed and processed in ways not accessible to overt thought. But because of the multitudinous connective circuitry of the human brain, something in us knows.

We are aware, knowingly or not, of the perpetual closeness of death and therefore of the instinctive need never-endingly to resist by undoing the consequences of every aberration in order that constancy be maintained amidst the teeming changes of cellular life. Constancy demands that a harmony of all the infinitely variegated processes of metabolism must be sustained. Our survival depends on harmony and order, and the human neocortex responds to this most fundamental of needs by placing harmony and order in the forefront of its priorities. Harmony and order provide not only the elements of survival but the predictable sources in which gratification and pleasure may be sought and found. They are the basis of our search for beauty and the dependability of human relationships. They are the basis of the depth and richness within love—that distinctive love of which only humankind is capable.

We have sent exploring couriers in the form of electrical impulses up and down the cerebral highways and into remote junctions, paths, and way stations. They have returned after finding or creating linkages and routes that we then employed to provide us with the uniquely human form of memory, foresight, and emotion. With memory, foresight, and emotion, we were able to build a logical pattern of evaluating the evidence of our senses and employing it in decision-making. That power might not necessarily have arisen, though the strictly physiological equipment for it was in place. By a process of chance, by trial and error, by firing and misfiring among the approximately 10 billion cerebral neurons and their 60 trillion connections, we have made unanticipated, albeit not always consciously recognized, discoveries, thereby adapting what nature originally gave us with the intention only of helping us to survive and reproduce.

In coming upon, forming, and utilizing a myriad of local electrical circuits organized into interconnecting regions and systems in our brains, and by turning our hormonal capability and reproductive drive to uses beyond the basic needs of passing on DNA, we have created not only the concept of social relationships and community but what is more basic—hope, faith, altruism, obligation, charity, morality, and even those aspects of love that are selfless and nonprocreative. We have gone so far as to create enriching qualities that would seem totally useless for survival, most particularly our esthetic sensibility, manifest in our appreciation of beauty and our need for order. Certainly, the *capacities* for developing those characteristics we recognize as uniquely human came originally with the molecular equipment assigned to us by natural selection, but it is through

Homo sapiens's gradual cerebral explorations and discoveries that the capacity became reality. This, then, is the ultimate process by which the human spirit has come into being. It is in the way we have made use of our innate physiology and anatomy that we ourselves, the members of our own species, are the real creators.

I refer here to actual organic events, to messages that move along nerve fibers and cells that respond to the stimuli thus transmitted. Attempts to link thought and emotion to cellular and molecular phenomena are still in their infancy, and although rapid progress has been made, the yield has thus far been of necessity too fragmented to provide anything beyond the beginnings of an overarching theory. But what is already known proves the promise in this kind of research.

In searching out the biological basis of the mind, I do not dismiss the notion of certain bulwarks of psychological or even psychoanalytic thought. Rather, I seek to find a way in which they can be explicated on an organic basis. After all, the young Sigmund Freud began his career in the laboratory as a researcher in the physiology of the nervous system. In later years, he never gave up on the cherished hope that his speculations on the dynamics of the mind would one day be seen as only a temporary formulation awaiting refinement and proof by the rigorous methods of pure science. I am encouraged to believe that his hope will prove not to be in vain.

I suspect that the elemental brain of *Homo sapiens* was (and is) the repository of vast possibilities of adaptive circuitry and cellular structure, just as the rest of the body contains vast quantities of excess hormonal and other capacity. Nature everywhere provides its creatures with plenty of reserves of cells, tissues, and even organs—we really do not need two kidneys or such a huge liver, or our more than twenty feet of small intestine. The basis upon which natural selection provides a surplus is self-evident. An injured creature is more likely to survive and reproduce if it has a surplus to fall back on. Perhaps the very bilaterality of animals is the result of such a selectivity, providing paired organs and even considerable extras in so many tissues. Adaptive use of the considerable extra capacities of the endocrine and central nervous systems, I believe, has been a significant mechanism contributing to our creation of the human spirit. Responding to the vast variety of stimuli that are constantly reaching every level of our organic structure, we have never-endingly enlarged the magnitude of our capabilities, as required by the necessity of ever-changing events. It is our extras, to a large extent, that have permitted us to do this.

How else but by newly discovered or newly formed circuits can certain of our recently acquired capabilities be explained, those that would have had no survival value twenty thousand to forty thousand years ago, after *Homo sapiens* seems to have made his last significant genetic advance? How, for example, have some of us mastered the ability to drive a modern automobile safely at seventy miles an hour on a crowded, poorly lit highway? Or to understand the complexities of a computer, or even to write a sentence with a number two pencil? Or to do the intricate fingerwork required to play a violin sonata or transplant a heart? When already endowed with what is essentially our present genome and living in caves, staying alive did not require that we have these skills. Just think of the wide spectrum of cerebral activities in which we engage, using what are in essence the same genes we had when our species underwent its last meaningful mutation.

Such accomplishments of *Homo sapiens* are easily explainable on the basis of the plasticity of the brain's neural circuits, and that is certainly sufficient. But I find myself wondering whether there may be an additional factor, as well. Perhaps, just as we have far more liver, kidney, and gut than we need, we have also been genetically provided with an excess of neurons and synaptic connections. Given what we know about the brain's ability to develop new or stronger pathways of transmission, it is not necessary to postulate such a thing, but it does seem probable nevertheless. Why, in fact, should we suppose the brain to be alone among the body's organs in not having plenty of excess tissue? Between its known plasticity and this hypothetical redundancy, there are certainly immense synaptic possibilities. Voltaire might as well have been thinking of the cerebral hemispheres when he wrote of *"le superflu, chose très necessaire"*— "the superfluous, that most necessary stuff."

The most recent addition to the animal brain is the neocortex. It has enlarged in size and complexity with every evolutionary advance, reaching a sumptuous abundance of cells and connections in *Homo sapiens.* Immediately preceding *Homo sapiens* in the line of evolution was *Homo erectus*, whose skull had a capacity of sixty cubic inches. Some idea of the huge change in brain size that came in a single series of evolutionary steps may be gained by considering that our own skull has a capacity of eighty-two cubic inches. It is to the richness and organization of the highly developed neocortex of humans that we owe our ability to adapt in the intellectual and emotional (among other) senses in a manner and to a degree that is capacious and almost certainly underutilized even now.

Responding to sensory input from the body and its surroundings, delivered over incoming fibers and via chemical messengers, the human brain has, I believe, engaged itself in the instinctual battle between stability and chaos, echoing up from its deepest cellular self. That battle is expressed in the psychological conflict between Eros and Thanatos—the forces of love (and therefore life) against the forces of the death instinct. Because the two are irreconcilable, the central nervous system of man has had, since the time it originally came into existence with the birth of the first *Homo sapiens*, to conjure with itself—to try various combinations of circuitry and chemistry, and to turn to its excess reserve capacity in exploratory ways—until it became what it is today, a vast machine works of intellect, spirituality, and even neurosis.

It might be pointed out, and properly so, that all of the foregoing presupposes a state of constant improvement, and therefore presents Pollyanna's view of the mind and its potentialities. But my definition of the human spirit is not restricted to the sublime qualities developed within our species. It includes, as well, those other characteristics of which we are far less proud, the baser qualities in all that is subsumed under the rubric of humanness. If there is an antonym for everything we customarily associate with *spiritual*, it must surely be *mean-spirited*. The same adaptive use of circuitry and molecular interactions that allows humankind to perform the mental gymnastics leading to our finest accomplishments is also in thrall to our baser instincts. Like all adaptations, some are *mal*adaptive. The maladaptations, the conflict between order and chaos, as well as the imperatives of living in societies in which individualistic drives must be restrained in the interest of community—these are the stuff of antisocial behavior and neurosis. This, too, is part of humanity.

The very instability of the multitudinous mechanisms that maintain our homeostasis is reflected in the instability and ambivalence with which we view our fellows and the universe, but especially ourselves. Echoing his inner physiology, man is engaged in a constant struggle to maintain the equilibrium that permits daily living. The conflict between constancy and consistency on the one hand and chaos and destruction on the other is mirrored in the mind's equally persistent struggle between the goodness that is in us and the dark drives of anarchic catastrophe. That luminous quality of reason that we value so highly is precariously perched on the unsettled knife edge between good and evil. The human mind being some 200 million years younger than the mammalian body to which it can trace its origin, the quality we might call mental homeostasis is not yet as effective as its physical counterpart. We function not only physically but mentally too, in a crucible of conflicting forces; we continue in stable emotional life only because a degree of balance is achieved by the internalized morality that is sustained by our individual and societal equivalents of enzymes and other regulatory mechanisms. Sometimes we lose the uneasy equilibrium we have attained with so much effort. The result is mental illness, injustice, and the maleficence to which we give daily witness.

My rabbinic teachers first made me aware of the Talmudic teaching that man lives in eternal conflict between the *yetzer hatov* and the *yetzer hara,* his good and his evil inclination. Civilization began and persists because the maintenance of what might be called social homeostasis, and therefore a civilized society, demands that the forces of equilibrium—the forces of the good—win out. But the history of the twentieth century and the events of which we read in our daily newspapers tell us that this is an ideal too often unattained. Society's struggle, like ours, never ends.

It is my spiritual sense that makes me human. It enables me to reason, to sublimate my instinctual drives, to be of use to society, and to love in the way that only members of my species can love; it enables me to do harm, to scheme against the interests of others, and to misinterpret the reverberating subconscious and distortedly recollected traumas of my childhood to such an extent that I become depressed, anxious, or a danger to society. The human spirit can be the high road to the fulfillment of my greatest hopes; it can be the grim pathway to my self-destruction.

The process is in part without deliberate thought: chance stimuli, firings, misfirings, electrical journeys made, electrical journeys aborted or rerouted anywhere along the way, and gradually the establishment of message routes that seem most suitable to the needs of the therefore increasingly intellectually capable human being. But there is also in it a large element of will; it is in conscious will that such qualities as responsibility and a developing moral sense can enlarge the vistas of productive thought.

As the newly discovered pathways are more frequently traveled, the passage of messages along them becomes easier and easier, until it is at last virtually automatic, while the resulting thought and behavioral patterns become the accepted characteristics of the person. The ever-enlarging set of responses is so internalized after a while that offspring can learn it from their parents and surroundings during the long period of human childhood.

I do not believe any of this formulation to be fanciful or even metaphoric. I am convinced that it is real. Though it be for the most part unconscious and inaccessible to deliberate reasoning, the very awareness of our body's rhythmic and dependable physiological processes instills a certain rhythmicity of thought, a need for symmetry and order to overcome the constant threat of ruinous disorder and death. Whether in response to such an obvious sequence as the heartbeat or respiration, or to a far more subtle one like the diurnal or circadian rhythms of metabolism, we march in step, needing the predictability of a certain regularity of organization. The trillion trillions of cellular reactions that seem at first glance uncontrolled and without governance are actually all integrated in the interests of the harmonious functioning of the entire organism, the smooth running of the engine as it were. The maelstrom of molecular dynamism is integrated on the level of tissues into a smoothness of systematized function that ticks with the syncopated rhythms of the heartbeat, uncoils with the coordinated undulations of the intestine, or makes itself known with similar predictability in the numerous other manifestations of order within organs. No wonder that *organ*, *organism*, and *organization* are words so similar to one another and derived from the same source. Its Indo-European root is *uerg*, meaning "work," indicating that what is meant is function rather than mere structure.

The organization and harmony begin at the very beginning—in fact, even before the beginning. Prior to fertilization of the ovum, the oocyte from which it arises has already begun the process of distributing its cytoplasm, regulating the orderly process by which its different kinds of molecules are directed to one or another of the four quadrants of the cell.

Commencing with fertilization, an assortment of orchestrated chemical interactions determine the specific location in the embryo to be occupied by each new cell produced during the ovum's repeated multiplications. Everything proceeds in its proper sequence, following the coordinated rhythms of the processes that biologists call by such names as induction and pattern formation. And all the time, little errors are occurring that must be corrected by the tissue's survival necessity that it right the wrongs perpetrated by its surroundings and itself. The result is a fetus, and the glorious moments when it enters the world as an infant. For the rest of that newborn human's existence, the changing constancy that leads to the overall harmony of its biological processes will function like some ceaselessly dynamic gyroscope that maintains life. Until the responsiveness of the system is broken down by aging, severe injury, or some devastating disease, the gyroscope equilibrates every imbalance.

The method by which an embryo's developing axons find the way to their final destinations is exemplary of the entire coordinated process of differentiation of cells and the way in which responsiveness and adaptation are integrated into the changing needs of tissues. In essence, embryonic axons grow along a trail of chemicals laid down in a genetically determined sequence to guide their progress. This means that the general anatomic connections are genetically controlled, but it says very little about the way they are used or how the branchings and connections are perpetually being fine-tuned throughout life. Although genes determine how the basic circuits of the brain are wired together, it is clear that variable degrees of change are possible and that changes in many of the circuits can continue to appear indefinitely. Some of the central nervous system must of necessity become "hardwired" early in life in order to guarantee consistency of response. But neuronal connections in other kinds of circuits retain their plasticity—they can be altered and realtered as long as we live.

Ongoing life requires that instability be in the service of stability, and change serve the needs of constancy. Harmony is the essence of our esthetic sense, and it rises up out of the ultimate harmony and integration of bodily processes. It is reflected not only in our music and in our poetry but in our visual appreciation, as well. No matter the disparate qualities that may enter into its composition, what is esthetically pleasing is what conveys to our sensibilities the same ordered regularity of outcome demanded by the deepest levels of our cellular selves. Our lives march to the molecular beat of our tissues. Our spirits sing to the music of our biology.

Perhaps the greatest feat of the humanizing process is the recognition of beauty, both the beauty we find around us and the beauty we can create. Beauty in and of itself would seem to be of no direct consequence to the DNA's survival needs (nature has provided other ways of attracting members of the opposite sex), and that alone makes its recognition one of the supreme accomplishments of the human mind: Beauty of image, of sound, and of thought give us the sense of enrichment, even of spirituality, that goes well beyond our constant seeking of mere survival and the most elementary forms of gratification and pleasure. The human spirit and its perpetual search for beauty are the defining characteristics of our humanity at its best.

Think of poetry. Its most fundamental characteristic is the line, however constructed; the line is the tissue of poetry. Like a tissue, it exhibits repetition and variation. Although any of its words, like any cell, is insignificant when taken by itself, its presence in the line is essential to the cadence and meaning of the whole; it therefore demands attention in its own right.

Every word is the precise word—every pause is the precise pause, whether indicated by the voice or in the punctuation. Each depends for its significance on the entire poem, and at the same time each gives its own significance to the entire poem. The whole gives meaning to each constituent part and to the specific location of that part within it. Is this not true of every part of the body, perhaps even of every cell? It is precisely the right kind of cell, but standing alone by itself without context, its work has no meaning. The various elements of a poem combine—are organized, are integrated, are unified—into the complex organism we behold. The poetic organism lives because each of its words and pauses and punctuations live. True of a poem, true of a man. We create a poem in our own image.

A poem is a composition of particular resonance: the resonance of poetry in general and the distinctive resonance of the individual poem we are hearing. Its resonance and its meaning arise from emotion and in turn evoke emotion, as the body's resonance and the significance of its activities achieve their highest fulfillment in our emotional life.

In addition to the sounds of its distinctive words and lines, each poem has a total sound of its own, never to be duplicated by another, just as each of us has our own total "sound," our total image in the society we inhabit. The sound of a man is a reflection of what he hears from the physiology and mentation of his inner self.

Scholars who study the history of verbal communication point out that even the earliest forms of culture seem to have had a class of language usage that differentiated it from formal speech—as cultures develop, poetry arises from such uses; the earliest literature is in the form of poetry: Every religion, no matter how "primitive," expresses itself through one or another form of poetry. Rather than believing such observations to be explained by some genetic or instinctive quality of our species, I would propose that the source of such universals is to be discovered in that profound awareness of our inner selves to which all humankind responds with the symmetry and order characteristic of our physiological processes. We live in rhythms, because rhythms live in us.

Considerations such as these form the basis of my conviction that the human spirit arises from the physiology of the human body, just as does the mind of which it is a product. It is my thesis that responsiveness to our internal and external environments and adaptation of our preexisting biological equipment are what make us what we are. The human spirit is, I believe, the generated product of our innate biology, encompassing the molecular behavior of our cellular structure. Nothing more need be sought. There is no need to invoke either a higher power or magic. We need only invoke what is in our human cells—the highest power and the greatest magic that has ever awed a wonder-struck observer of its magnificence.

All of this may seem agnostic, but it is hardly the philosophy of an atheist. In my view, to espouse atheism is to be unscientific. To presume to know defies all logic and flies in the face of reason. Elsewhere, I have written that the proper state of mind of a skeptic is uncertainty—to believe, as I do, that even while questioning everything around us, we must also be prepared to accept the premise that *anything* is possible. Lack of evidence is not convincing proof against a proposition—it is simply lack of evidence and not the presence of powerfully *opposing* evidence, which is quite a different thing. We have no evidence that there is no God, nor is it imaginable that we will ever discover any.

In fact, nothing in my hypothesis about the human spirit necessarily rules out the existence of God. The mere fact that I propose an explanation that does not require His intervention should not be construed to mean that I am at all certain that at some future date evidence will not appear to convince skeptics and questioners of God's existence. Nevertheless, at this moment in history, late 1996, the evidence that would move an objective mind has not yet appeared. Some will continue to hope as others continue to pray. Still others prefer things just as they are.

For the faithful, a hypothesis like mine might be untenable. Or it might be seen simply as changing God's role. He becomes, in such a scheme, the shepherd not of creatures of His own direct making, but of the self-created, free-willed result of the molecular forces He may or may not have unleashed billions of years ago. His job is more demanding of all the wisdom and certainty we expect of a God. It may not be doable, even by Him.

For those whose world is more agreeable without such belief, the potentiality of our species may seem all the greater because there is no end to the series of adaptations along whose continuum we now find ourselves. Without the existence of a God who may decide to stop the process that others believe He set in motion, and barring world or species-destroying catastrophe, the cerebral circuits and the neuroendocrine interactions will never stop discovering new ways to deal with the stimuli presented to them, and they will accordingly never stop enhancing our capacity to solve the puzzles inherent in the physical and mental challenges that constantly surround us.

It has been my observation that people rarely acquire or lose faith on the basis of reason, or what most of us would consider verifiable evidence. Were it otherwise, we would not use the word *faith* to describe the belief system of the religious. Faith or lack of it is the outcome of one's personal needs, nurture, training, psychic constitution—call it what you will, it amounts to the same thing: the predisposition brought to the issue. I do not believe that any of the material in this book will convince a single reader to change one iota of viewpoint about the role of a Supreme Being. Those who already believe will only find strengthening of their affirmation of His majesty; those who do not will choose to see the elements of regulation and automicity as supporting their contention that there is no need to invoke God. Of course, a wide difference exists between "no need to invoke God" and "no God."

Whether one agrees with Wittgenstein that "the human body is the best picture of the human soul" or is convinced that the philosopher had it backwards, there can be no doubt that there is a divinity in it—"a divinity that shapes our ends," as Shakespeare had Hamlet say. Maybe it is the ur-divinity that our Indo-European ancestors discovered in nature;

maybe it is the divinity that is transmitted by the touch of God's hand. Either way, there *is* a divinity from which the human spirit is shaped. Like the wisdom of the body, the real and potential magnificence of the human spirit must be approached not only with wonder but perhaps also with the awestruck attitude of Wordsworth's nun on a beauteous evening: "Breathless with adoration."

GLOSSARY

Acetylcholine: One of the chemical substances that transmit a nerve impulse across a synapse; a neurotransmitter primarily for the parasympathetic nervous system.

acidosis: A pathological state in which there is an overabundance of acid in the blood and tissues.

adrenaline: Also called *epinephrine*; a hormone produced by the adrenal gland; *see* noradrenaline.

albumin: The predominant protein in blood plasma.

alveolus: A thin-walled air sac in the lung, one of many such, where exchange of gases takes place between the blood and the inhaled air.

amino acid: A small molecule that is one of the building blocks from which polypeptides and then proteins are made.

aneurysm: An outpouching in a weakened segment of blood vessel.

antibody: A globulin molecule, produced by B cells in the immune system, that binds to a specific antigen.

antigen: A substance perceived by the immune system as foreign, thus triggering a protective response in the host.

aorta: The main artery to the body, receiving its blood directly from the heart's left ventricle.

arteriole: A very small blood vessel with a strong muscular coat. Arterioles act as gatekeepers between the arterial system and the capillaries.

atom: The smallest unit of any element (an element is the most basic chemical substance—e.g., sodium, carbon, potassium, hydrogen, and oxygen are elements in living things).

ATP: Adenosine triphosphate. A molecule whose function it is to transport energy from one site to another within a cell, releasing it as needed to enable chemical reactions to take place.

atrium: One of the two upper chambers of the heart. The right atrium receives blood from the superior and inferior vena cava and the left from the pulmonary veins.

autonomic nervous system: The part of the peripheral nervous system which controls internal events that take place autonomously, that is, independently of conscious thought, such as the activities of involuntary muscle, heart muscle, and glands. The autonomic nervous system is composed of two parts, the sympathetic and parasympathetic nervous system.

axon: An elongated branched extension of a nerve cell, whose primary purpose is to carry impulses away from the cell.

Baroreceptor: A sensory structure that monitors changes in pressure within a blood vessel.

Capillary: A narrow vessel whose thin wall allows the diffusion of gases, nutrients, and waste products between the interstitial fluid and the blood.

catalyst: A substance that increases the speed of a chemical reaction but is not changed by it.

cell: The microscopic mass of protoplasm of which all plant and animal tissues are composed; the smallest unit of a living thing that can, when the proper environment is provided, live independently. A cell consists of a nucleus and its surrounding cytoplasm, enclosed within a highly specialized envelope called a cell membrane.

central nervous system: The brain and spinal cord.

cerebrum: The uppermost part of the brain, involved with skilled motor activities and the higher mental faculties.

chromatin: Long, threadlike fibers of DNA attached to an equivalent amount of protein scaffolding.

chromosome: Tightly coiled chromatin; a chromosome is a folded package of DNA, consisting of a single DNA molecule attached to an equivalent mass of protein that acts as scaffolding. The forty-six chromosomes of human cells contain all of our genes.

corpus luteum: A yellowish glandular structure on the surface of the ovary, developing from the ruptured follicle left behind after ovulation. The corpus luteum secretes progesterone and estrogen.

cortex: A general term for the outer shell of an organ or other anatomical structure. The cerebral cortex is a thin layer covering the cerebral hemispheres, receiving input which it integrates and coordinates.

cytoplasm: All of the contents of a cell other than its nucleus.

Dendrite: A short branched extension of a nerve cell, whose primary purpose is to carry impulses toward the cell.

diastole: The phase of the cardiac cycle during which the heart muscle is relaxed.

differentiation: The process of development of a cell during which it becomes uniquely specialized to carry out some specific function.

diploid: Containing two complete sets of chromosomes, one contributed by each parent. All normal human cells other than the sperm and unfertilized ovum are diploid.

DNA: Deoxyribonucleic acid; one of a class of chemical compounds called nucleic acids; a molecule consisting of an enormously long chain of structural units called nucleotides, in the form of a double helix. Along the chain, lengthy sequences of nucleotide pairs form genes, which are interspersed with lengths of DNA whose function, if any, is unknown.

Embryo: The early stages of an animal's development. In man, the term is understood to mean the period from approximately the second to the eighth week following fertilization of the ovum.

embryonic induction: A chemical or physical interaction in which groups of cells influence other groups of cells to some action, such as a change in location or the synthesis of a new protein.

endocrine glands: Glands producing substances, such as hormones, that enter the bloodstream to act at some distance from their origin; sometimes called *ductless glands*. Examples are the pituitary, thyroid, and adrenal glands.

endometrium: The thick, gland-filled membrane lining the inside of the uterus.

enzyme: A protein substance that acts as a catalyst, to speed up a chemical reaction by lowering the amount of energy required for the reaction to take place. Accordingly, an enzyme will sometimes initiate a process that might not have occurred in its absence.

epithelium: The covering and lining tissues of the body.

erythrocyte: Red blood cell.

erythropoietin: A hormone produced by the kidneys, and to a lesser extent by the liver, which stimulates the bone marrow to manufacture red blood cells.

estrogen: Female sex hormone; in addition to stimulating the development of secondary sexual characteristics, estrogen is responsible for creating a physical environment in which fertilization and uterine implantation of the ovum can occur, as well as facilitating nutrition during the embryonic period.

exocrine gland: A gland producing substances that are secreted into a duct, through which they pass directly to their point of action.

extracellular fluid: All the body's fluids that are not contained within cells. This includes the blood plasma and the interstitial fluid in which the cells are bathed.

Fibrinogen: A plasma protein that is an essential participant in the process of blood-clotting; also known as Factor I.

fibroblast: A cell that can differentiate into more mature forms which become part of the fibrous and supporting tissues of the body, including the constituents of a healed scar.

Gastric: Of, or relating to, the stomach.

gene: A unit of heredity; a localized region on a DNA molecule, consisting of a long sequence of nucleotide pairs, whose chemical structure carries information that can be transmitted from generation to generation. The great majority of genes have as their function to direct the synthesis of proteins.

genetic code: The relationship between genes and the traits they influence; specifically, the relationship between genes and the proteins or other substances whose synthesis they determine.

genome: The complete set of hereditary factors of a given species.

genotype: The genetic makeup of an individual. See also *phenotype*.

globulin: One of the types of protein found in blood plasma. Some globulins are important components in the immune process.

glucagon: A hormone secreted by the pancreas, stimulating the liver to convert glycogen to glucose, and also stimulating the production of glucose from amino acids.

glucose: A simple sugar molecule which breaks down into carbon dioxide and water, releasing energy that is then taken up by ATP. Glucose is the body's primary source of energy.

glycogen: A carbohydrate stored in the liver and to a lesser extent in voluntary muscle, serving as a source of glucose.

gonad: An ovary or testis.

gonadotropin: A hormone that stimulates the activity of the ovaries or testes.

granulocyte: A type of white blood cell. There are three types of granulocytes: neutrophils, eosinophils, and basophils.

Haploid: Having only one set of chromosomes. The sperm and ovum are haploid cells.

hemoglobin: A large iron-containing protein molecule in the red blood cell, which carries oxygen to the tissues of the body.

hemostasis: The stoppage of bleeding.

homeostasis: A term introduced by Walter B. Cannon to describe a state of internal equilibrium within the body, necessary for ongoing life.

hormone: A type of signaling substance, whose function is to regulate the activity of a target organ or group of cells.

Hox genes: Master regulatory genes that lay out the general geographic plan of the body.

hypophysis: The pituitary gland.

hypothalamus: A brain center that monitors and regulates certain autonomic activities, especially those of the sympathetic nervous system.

Instinct: An inborn tendency to act in some distinctive way, specific to a particular species.

insulin: A hormone secreted by the pancreas, which facilitates passage of glucose into cells and stimulates the liver to form glycogen.

interneuron: A nerve cell located anywhere between the original sensory neuron and the final motor neuron.

interstitial fluid: The fluid surrounding the cells of the body; sometimes called *intercellular fluid.*

Lacteal: A tiny lymph channel that absorbs nutrients from the small intestine.

leucocyte: A white blood cell, of which there are three kinds: granulocytes, monocytes, and lymphocytes.

limbic system: A group of structures deep within the brain, acting as an intermediary between the cerebrum and the brain stem. Along with the cerebral cortex, the limbic system influences the emotions.

lipid: A class of fatty or oily substances with various functions in the body.

lymph: A liquid derived from the tissue fluids, which travels in the lymphatic channels.

lymphocyte: A variety of white blood cells that takes part in effecting and maintaining immunity.

Macrophage: A large phagocytic cell.

medulla: A general term for the inner part of an organ or other anatomical structure. The medulla of the brain monitors and regulates certain autonomic activities, especially those of the parasympathetic nervous system.

meiosis: A modification of mitosis by which a diploid cell produces haploid progeny; specifically, the final process of cell division giving rise to the ovum or sperm.

metabolism: The totality of chemical and physical interactions within an organism.

metastasis: The spread of disease from one part of the body to another. The term is usually used to refer to the behavior of malignant tumors.

milieu intérieur: a term introduced by the nineteenth-century physiologist Claude Bernard, to describe the internal environment of fluid in which the cells of the body are bathed.

mitosis: The sequence of steps by which a cell divides in such a way that two identical copies of itself are produced.

molecule: An aggregate of atoms bonded together in chemical combination to form the smallest unit of a specific substance.

mucosa: The layer of tissue that lines the inside of tubular organ systems, such as the respiratory, digestive, and urinary tracts; also called the *mucous membrane.*

myocardium: The heart muscle.

Neocortex: The newest portion of the cerebral cortex to have evolved, engaged in the most sophisticated of mental functions.

neuron: Nerve cell.

neurotransmitter: A chemical substance released from the end of an axon into the junction between it and the cell to which the nerve impulse is traveling.

noradrenaline: Also called *norepinephrine*; a variant of adrenaline; a hormone produced by the adrenal gland and axons of the sympathetic nervous system. Noradrenaline functions as a sympathetic neurotransmitter, playing a critical role in the "fight or flight" response.

nucleotide: The chemical building block of which nucleic acids, such as DNA, are constructed.

nucleus: As used in reference to an individual cell, the nucleus is the membrane-enclosed structure that is the center of the cell's activities. As used in reference to areas of control in the central nervous system, a nucleus is a collection of neurons sharing a common function and grouped in one location within the brain or spinal cord.

Organelles: Specialized components within cells, each having its own function, such as digestion of nutrients or extraction of energy.

organism: An individual living thing, whose various components function together to maintain its integrity.

osmosis: The passage of water through a semipermeable membrane in order to balance the concentration of molecules of dissolved substances on the membrane's two sides.

osmotic pressure: The force that inhibits osmosis; specifically, the tendency preventing water from leaking out into the tissues through capillary walls.

ovum: A mature female reproductive cell; often used synonymously with "egg."

oxytocin: Hormone produced in the pituitary gland, which stimulates contraction of involuntary muscle in the uterus and breast.

Parasympathetic nervous system: The part of the autonomic nervous system that controls the involuntary regular "housekeeping" functions of ongoing life, such as intestinal movements and the rate of the heartbeat. In general, the parasympathetic nervous system is governed principally by centers in the medulla.

pattern formation: The mechanisms by which specialization and positioning of tissues are governed in the developing embryo.

peptide: A molecule constituted of two or more amino acids; a building block for proteins.

peripheral nervous system: All structures of the nervous system other than the brain and spinal cord.

phagocyte: A cell that ingests and usually digests unneeded or dangerous substances in the blood or tissue.

phenotype: The traits that an individual actually possesses, which are the result of interactions between genes and between genotype and the environment.

pituitary gland: Sometimes called "the master gland," because its secretions affect the activities of numerous other endocrine organs; also known as the *hypophysis*. The pituitary gland is located at the base of the brain.

plasma: The protein-rich fluid portion of the blood, in which platelets and red and white cells are suspended.

platelet: A disc-shaped structure suspended in the blood plasma, integral to the clotting mechanism.

polypeptide: A peptide in which three or more amino acids are bonded together.

progesterone: A female hormone whose principal function is to prepare the uterus for implantation and development of the fertilized ovum.

prolactin: A hormone produced by the pituitary gland that stimulates and sustains the production of milk by the breasts.

protein: A complex compound constructed of amino acids linked into peptides. Proteins are the principal constituents of the protoplasm of cells.

prothrombin: A plasma protein that is an essential participant in the process of clotting; also known as Factor II.

protoplasm: A general designation for the essential matter that is the stuff of all cells. Its major constituents are proteins, lipids, carbohydrates, salts, and nucleic acids.

pulmonary artery: The large vessel carrying blood from the right ventricle of the heart to the lungs for oxygenation.

pylorus: The outlet of the stomach, containing a ring of muscle that acts as a sphincter.

Receptors: (a) Highly specialized protein molecules in a cell membrane, which combine with transmitters or signaling molecules of various sorts to excite or inhibit some action of the cell. (b) Cells or groups of cells that pick up signals from inside or outside the body and transmit them to nerves, in the form of impulses.

reflex: An involuntary stereotyped action automatically resulting from a particular stimulus.

reticuloendothelial system: A group of tissues scattered throughout the body, containing phagocytic cells and functioning as a defense system against infection, debris, and foreign materials.

RNA: Ribonucleic acid—a linear molecule constituted of a single chain of nucleotides for whose synthesis DNA is the template. By this means, the information in the DNA is transcribed into RNA. The RNA, in turn, becomes a template to translate its information into the synthesis of proteins.

Secretin: The first hormone to be identified, in 1902. Secretin stimulates the pancreas to secrete certain digestive enzymes.

sinoatrial node: The internal pacemaker of the heart.

sphincter: A ring of muscle with the capability of closing or opening a tubelike structure of the body.

sympathetic nervous system: The part of the autonomic nervous system that controls involuntary functions of a more or less urgent nature, such as the "fight or flight" response. In general, the sympathetic nervous system is governed principally by centers in the hypothalamus.

synapse: The junction across which an impulse is transmitted from one nerve cell to another.

systole: The phase of the cardiac cycle during which the heart muscle contracts.

Thalamus: A part of the brain lying at the base of the cerebrum, serving as a relay station connecting the cerebral cortex with various lower parts of the brain, particularly the medulla.

transcription: The construction of a strand of RNA using DNA as a template, so that the genetic information in DNA is transferred to the RNA.

translation: The construction of a sequence of amino acids on a strand of RNA, to form a polypeptide chain.

Vagus nerve: A long nerve originating in the medulla of the brain and passing down into the chest and abdomen, carrying parasympathetic fibers to the heart, lungs, and digestive tract.

vena cava: One of the two large veins (superior and inferior) returning blood from the body to the heart.

ventricle: One of the two large pumping chambers (right and left) of the heart.

Zygote: The fertilized ovum.

INDEX

Italicized page numbers indicate illustrations.